Várzea

Várzea
Diversity, Development, and Conservation of Amazonia's Whitewater Floodplains

edited by
Christine Padoch
José Márcio Ayres
Miguel Pinedo-Vasquez
Andrew Henderson

Advances in Economic Botany, Volume 13
Series Editor: Charles M. Peters

THE NEW YORK BOTANICAL GARDEN PRESS

Published by
The New York Botanical Garden Press
Bronx, New York

Composition by Binghamton Valley Composition
Cover photographs: Front, Mario Hiraoka; back: Mario Pinedo Panduro
Cover/interior design by Joy E. Runyon
Manufacturing by BookCrafters, Inc.

The paper used in this publication meets the requirements of the American
National Standard for Information Sciences—Permanence of Paper for
Publications and Documents in Libraries and Archives
(ANSI/NISO Z39.48–1992).

Printed in the United States of America using soy-based ink on recycled paper.

Metropolitan Life Foundation is a leadership funder of
The New York Botanical Garden's Scientific Publications Program.

Library of Congress Cataloging-in-Publication Data

Várzea : diversity, development, and conservation of Amazonia's
whitewater floodplains / edited by Christine Padoch ... [et al.].
     p.   cm. — (Advances in economic botany : v. 13)
   Includes bibliographical references.
   ISBN 0-89327-419-4
   1. Floodplain ecology—Amazon River Valley. 2. Sustainable
development—Amazon River Valley. I. Padoch, Christine.
II. Series.
QH112.V37 1999   98-55089
577'.0981'1—dc21

Contents

Section 3: Conservation

Section 4: Soils and River Dynamics

Section 5: Land Resource Management

Section 6: The Case of the Vanishing Stingless Bee

Introduction

The floodplains of the Amazon River and its major whitewater tributaries, the *várzea*, occupy a mere 2% of the total area of the Amazon basin, but play a role far out of proportion to their relatively small size. The floodplains differ in many ways from other areas of the basin. Generally high in biological productivity, the Amazon várzea includes a great diversity of environments and ecosystems often compressed within a very small area. Many areas of the floodplains are crucial habitats for the reproduction of numerous species and others are home to populations of important and severely threatened endemics.

The várzea is, and has been for centuries, an important area of resource extraction, agriculture, and settlement. The diversity of environments has led to the development of a great variety of agricultural and forest management patterns. The alluvial soils, replenished every year by floods, are highly fertile. However, the same annual floods that make these lands agriculturally desirable also make them difficult to exploit by modern agricultural methods. The accessibility and relatively high human population densities greatly increase the risk of overexploitation of biological and physical resources. The history, diversity, and functional complexity of ecosystems make it imperative that any plans for improved exploitation or conservation of várzea resources take an approach informed and enriched by the divergent contributions and perspectives of many disciplines.

Despite its various singular characteristics, the várzea was until recently the subject of very little scientific research. Several groups of researchers from Brazilian, Peruvian, North American, and European institutions are now working on, or have recently completed, important projects in the várzea. The purpose of the symposium from which the following papers were drawn was to bring together researchers and policy experts from these different groups in order to share ideas and information on the sustainable development and conservation of the resources of Amazon floodplain lands and waters. Held 12–14 December 1994 in Macapá, Amapá, Brazil, this was the first international conference to be devoted entirely to the topic of várzea.

Most of the papers that make up this book are based on papers presented at the Macapá conference. Most were translated by the authors from Portuguese to English, all have been reviewed and revised, and several have been changed so extensively as to be essentially new papers.

The papers are grouped thematically. Each of the five principal sections comprises a diverse selection of articles, reflecting the complex task given our authors of considering both needs for conservation of the várzea's valuable species, habitats, processes, and landscapes, and the imperative of enhancing the livelihoods of floodplain farmers, fishers, and forest managers. The variety of each section mirrors not only a heterogeneity of research concerns among várzea scientists but also the great variety of approaches to conservation and development that are being proposed and implemented in Amazonia.

The first section, devoted to the discussion of várzea fish and fisheries, provides a good example of the interdisciplinary nature of this book and of the Macapá conference, as well as of the diversity of interests and opinions of the authors. The articles making up this section discuss not only particular fish species and their habits, but also various types of human exploitation, their effects on várzea fish populations, their outlook for the future, and the uncertainties, difficulties, and successes of an assortment of conservation efforts. Each subsequent section—on Forests and Forestry; Conservation; Soils and River Dynamics; and Land Resource Management—also includes a mix of approaches, emphases, viewpoints, and conclusions. The diversity of views and disparity of opinions found throughout each section and the volume as a whole represent, we believe, the strength of current research and conservation efforts in the Amazon floodplain. We are witnessing and participating in a time of great research and conservation activity and resultant new insights into várzea realities and futures.

We were disappointed not to have had the opportunity to publish several very interesting conference papers; their authors either declined to submit written versions or found that they could not revise them in time for publication. We also regret not being able to reproduce the many fascinating debates, discussions, and dialogues, formal and informal, that were a highlight of the conference for many participants. We are, however, pleased to include an introduction to each thematic section; these are original to this volume, although several of the distinguished "introducers" had served as section discussants in Macapá.

Acknowledgments

We wish to thank several institutions and many individuals who helped make both this book and the symposium possible. The conference received the generous sponsorship of the Ford Foundation, the Conselho Nacional de Desenvolvimento Cientifico e Tecnológico (CNPq), the Wildlife Conservation Society, and the Institute of Economic Botany, The New York Botanical Garden. The conference also enjoyed the support of several Amapá institutions and we especially would like to thank the Honorable Annibal Barcellos, then Governor of the State of Amapá; Raimundo Brabo, the Chief of CPAF, the research arm in Amapá of EMBRAPA; Fernando Guimarães, the President of IRDA; Antonio Carlos da Silva Farias, the Chefe de Gabinete da CEMA; Professora Zaíde Soledade, the Director of the Teatro das Bacabeiras, where the Conference's formal meetings took place; and Armando Alves Jr., the Chefe do Gabinete Militar do GEA.

The organizers of the conference were Márcio Ayres, Deborah Lima, Ana Rita Alves (Projeto Mamirauá); Christine Padoch, Andrew Henderson (The New York Botanical Garden); and Miguel Pinedo-Vasquez (Columbia University); many other persons, however, contributed generously of their time and other resources.

In Macapá our special thanks go to Jaime Rabelo and his entire family for their constant support and help with innumerable tasks. We also wish to thank Andrea Quong, Jennie Hawkins, Jeanne Haffner, Dawn Ward, and Robin Sears for their expert help in communicating with our far-flung authors, as well as their aid in editing, finding references, converting text and graphic files, and combatting the fierce computer viruses that arrived on some diskettes.

C. Padoch
J. M. Ayres
M. Pinedo-Vasquez
A. Henderson

Section 1:
Fish and Fisheries

Introduction

Michael Goulding

An understanding of the Amazon fish resource is one of the greatest challenges facing tropical ecology. Amazon fish represent our planet's greatest diversity of freshwater vertebrates. There are an estimated 2500–3000 fish species in the Amazon basin. Somewhere between 200 and 300 species are exploited for food or the aquarium trade. At present the retail value of the wild fish resource is worth at least US$200 million; when secondary and tertiary activities are added, the total value might be double the retail estimate. Considering the aquatic resources available, the potential value of aquaculture in the Amazon basin is very high and could one day surpass that of wild stocks exploited in the commercial fisheries. Perhaps 200,000 people living in the Amazon basin depend on fish for their main or partial income. Finally, locally caught fish are one of the principal protein staples for at least 1 million people in the Amazon basin, most of whom live in towns and cities.

It is now generally recognized that fish is the single most valuable sustainable resource exploited in Amazonian rivers and their floodplains. Earnings from floodplain logging can be greater than fisheries in some local areas, such as in parts of the Amazon estuary and in the Mamirauá Ecological Reserve, but there are no large-scale replanting programs, and most of the Amazon River floodplain downriver of Tefé has been depleted of its timber resources. Although the economic and nutritional value of Amazon fish has been recognized, there is still no clear consensus on what types of models should be experimented with to manage it. The conservation of floodplain biodiversity is even more problematic because of our ignorance about it and about large-scale environmental degradation that has already taken place along much of the Amazon River in the last 50 years. Most of the tributaries, except in headwater regions, have been less severely affected, although mercury pollution from gold mining operations could possibly have serious negative effects far downstream of contamination sites. Any serious mercury contamination would also destroy export markets for Amazonian fish in general.

Authorities working in the Amazon have traditionally attempted to manage fisheries mostly through closed seasons and size limits. A visit to any major Amazon fish market will reveal that size limits, closed seasons, and other such measures cannot be enforced. Even if laws directed at fishermen could be enforced without their agreement, it would still be unclear how effective these tools would be for managing the fish resource. As Ronaldo Barthem points out in this volume, the Amazon fish resource in general is not being overexploited.

As of 1996, only three food species were being overexploited: piramutaba, tambaqui, and pirarucu. Life histories and management models for piramutaba and tambaqui have recently been completed by Ronaldo Barthem, Carlos Araujo Lima, and Michael Goulding, and it should be possible to prevent the commercial extinction of these important species. Neither the fisheries nor the biology of the giant pirarucu, however, have been studied in detail. Though the species is widespread in the Amazon basin, there are insufficient data to know exactly where and to what extent it is being heavily exploited. In a chapter dealing with pirarucu fishing inside the Mamirauá Ecological Reserve, Helder Lima de Queiroz points out that the species is being heavily exploited and that immature fish make up most of the catch taken by local residents.

The fact that fisheries in general are not being overexploited does not mean that they are not endangered in the long run. The main problem, from both a conceptual and an ecological point of view, is habitat protection. The three main habitats for food fishes in the Amazon are open waters, floating meadows, and flooded forests. Humans have had little effect on the open waters of floodplain lakes or river channels associated with the Amazon River. The huge volume of water transported by the Amazon River has mitigated against local pollution, though this could change with any large-scale use of pesticides for floodplain agriculture. The life histories of most fish species are intimately linked to floating meadows and flooded forests. These habitats provide shelter, food, and reproductive retreats for most of the species. Floating meadows are relatively easy to study because they are formed of usually no more than three or four grass species. Seines and other sampling gear can be used effectively to determine what types of fish communities live in these habitats. For example, Henderson and Robertson in this volume show that about 80 fish species might be expected to be found in a 400 m^2 area of floating meadow, but only a few species account for most of the biomass. Total biomass falls within the range of about 17–195 g/m^2, the highest values recorded during the low-water season, when the floating meadow habitat is greatly reduced in size. It will probably be many years before we can make a similar fish biomass estimate for flooded forests.

We still know very little about how fishes use flooded forests. Fruit- and seed-eaters have been the most investigated because of their economic importance and peculiar ecology in comparison with most other fish in the world. Flooded forest is very difficult to sample because seines, the main gear used in the Amazon to quantify floodplain communities, cannot be employed effectively in this habitat. Most researchers have dealt with this problem by ignoring it, and thus our biomass information is all from floating meadows. William Crampton points out in his chapter on the discus how difficult it was to catch this fish in flooded forests, though the beautiful cichlid was suspected to migrate to this habitat during high water. My own collections suggest that nearly all species migrate to flooded forests during the high-water period, though parts of many populations in the muddy river floodplains also inhabit floating meadows. When studying the life histories of floodplain fish, much more attention needs to be given to flooded forests. This will require the development of new collecting methodologies.

Neither local communities nor federal, state, or municipal authorities in the Amazon have done much to protect fish habitats; nor do they seem to be alarmed by floodplain deforestation. This is an ironic situation, considering the great importance of fish in the economy and diet. The destruction of fish habitats is part of the boom-and-bust historical process that has dominated Amazonian economies. Recently the "boom" is being driven by cattle and water buffalo ranching whose large-scale staging ground was originally the island of Marajó. Livestock ranching is now literally driving westward along the Amazon River floodplain and adjacent areas of terra firme. Large parts of the Rio Amazonas floodplain downriver of the Rio Negro were deforested historically for cacao and jute. Even with these activities, however, significant remnants of floodplain forest remained and floating meadows were only minimally affected. As livestock ranching began to expand rapidly in the 1980s and 1990s along the Rio Amazonas, more floodplain was deforested for pasture development. In turn, the floating meadows that colonized areas previously shaded by flooded forest began to be destroyed on a larger scale each year as livestock populations increased. The two main habitats on which fish depend are thus being removed—either permanently, by deforestation, or prematurely on a seasonal basis, in the case of floating meadows—in order to provide grass for cattle and water buffalo.

There is no evidence that the economy of the Rio Amazonas floodplain or associated urban centers has improved as a result of livestock ranching. Productivity is relatively low because livestock must be removed to upland pastures during the floods. By the time cattle and water buffalo are reintroduced onto the floodplain after the annual floods, they are emaciated because of the poor quality of grass on nutrient-poor terra firme soils. Absentee "landowners" have nevertheless been attracted to floodplains for livestock ranching because the fish resource, though much more productive, cannot be corralled and controlled, as it is found in open waters during much of the year. Furthermore, subsistence and commercial fishermen do not cede rights to the fish resource to large landowners. Local floodplain communities have also begun to buy livestock, either for investment or in imitation of large ranches. Floodplain communities, however, rarely kill cattle or water buffalo for food; rather, they depend on fish as their main protein source. Other than the factors mentioned above, there is also a certain romanticism and machismo associated with livestock ranching.

If most of the Amazon River floodplain is turned into livestock pasture, the fish resource as we now know it will be radically changed. McGrath, Castro, and Câmara suggest in their chapter that local communities could better manage floodplain fisheries than could federal or state agencies. The problem with this approach is that it focuses attention in the wrong place—that is, on fishing rights rather than on habitat protection. The main question for both fisheries and floodplain biodiversity in general is this: What levels of management would be most effective in reducing deforestation and the destruction of floating meadows? As mentioned above, neither local communities nor government agencies have faced up to the ecological consequences of the destruction of flooded forests and floating meadows. From an ecological point of view, it makes little difference whether local communities or urban fishermen exploit the fish resource. Of more importance is the degree to which each does something to protect the habitats on which their main resource depends. Otherwise both groups are only competing for the spoils left after livestock ranching activities. The community approach to fisheries management will be effective only once local peoples are convinced to leave floodplain forests and floating meadows intact. That will be the main challenge of community-oriented projects.

The most important fisheries in the Amazon are based on migratory species that use huge areas for their life histories. There is occasionally the suggestion that these fisheries cannot be managed because the areas they cover are too large and no single local community has the means to enforce protective measures. Along much of the Amazon River, however, local communities often depend more on migratory species captured in river channels for cash income than on catches made in floodplain waters. An example is the seasonal exploitation of upstream migrating dourada and piramutaba catfishes, species that rarely enter floodplain waters. Ruffino in this volume points out that about 50% of the total Santarém catch consists of migratory catfishes. With characins added, the migratory total is perhaps 90%. Many of these fish are captured by floodplain dwellers who then sell their catches in Santarém.

Both the migratory catfishes and the characins have foodchains linked to flooded forests and floating meadows. Although dourada and piramutaba are not commonly captured in floodplain waters, they depend on prey whose foodchains are linked to flooded forests and floating meadows. All of the important migratory characins captured in commercial fisheries spend the high-water period in flooded forests. The two principal examples are tambaqui (*Colossoma macropomum*) and jaraqui (*Semaprochilodus* spp.), traditionally the two most important fish groups for Manaus, the largest market in the Amazon.

There is no single model that can be used to manage Amazonian fisheries. At present, too much attention is being given to managing local social conflicts rather than finding ways to protect habitats. The fish are seen but not the forests and floating meadows that sustain them; all too often, the larger picture of a wealth of wildlife—birds, mammals, reptiles, amphibians, and countless invertebrates—that depends on these habitats is all but forgotten. To focus myopically on single players is to ignore the fact that there is a great diversity of interests concerned with the development and conservation of the Amazon floodplain. The floodplain is not an isolated entity but, rather, tied ecologically to river channels and even adjacent terra firme forest and streams. We must not make the mistake of seeing Amazonian ecology and resource management as an artificial reflection of the geographical limits of local communities. There are many types of communities, including local floodplain villages, fishers' co-operatives, scientific institutions, conservation organizations, and government agencies. It will take a creative mixture of all of these, adapted to specific cultural conditions and ecologies, to bring about the long-term management of the world's greatest freshwater fish resource. May we not forget the trees and meadows in the process.

Várzea Fisheries in the Middle Rio Solimões

Ronaldo B. Barthem

Introduction

The aquatic environment plays a central role in the Amazon, by virtue of both its great biological and habitat diversity and its auspicious conditions for human settlement. Rivers, estuaries, furos, and lakes form the basis of this region, which sustains vegetation made up of high and low restingas, chavascal, igapó, and matupá (Ayres, 1993). Water-level fluctuations, especially flood and drought cycles, define the structure of animal and plant communities and determine the abundance and composition of the aquatic biota. Furthermore, a positive correlation between fish production and flooded areas (Welcomme, 1976) indicates that the regular flooding of this environment is the aquatic fauna's main source of energy (Junk et al., 1989).

The human population in the Amazon region has learned to live in this flooded environment, called "várzea," and to obtain not only the raw materials necessary for building houses and boats but also the fish required for subsistence (Smith, 1979, 1981). Fish consumption by the population along the banks of the large rivers in the Amazon region, particularly the Solimões-Amazonas, is higher than the mean consumption in other regions of Brazil or of the world. Manaus, for example, consumes seven times more fish than developed countries (Shrimpton & Giugliano, 1979). Fishing is economically important in Amazonia. For instance, the "artisanal" fishing industry accounts for an annual revenue of US$100–200 million, if we consider that the annual catch is more than 200,000 tons, and that fish prices vary from US$0.50 to US$1.00/kg (Petrere, 1992). This amount is greater than the hard currency from shrimp and piramutaba exports from the States of Pará and Amazonas, which are almost US$50 million. There are 70,000 fishing jobs in the northern region, 30,000 of which are found in "fishing colonies" (SUDEPE, 1985).

The middle Solimões region, which includes the confluences of the Rios Solimões-Japurá and Solimões-Juruá, is one of the largest floodplains in the Amazon region, and its flood cycle varies considerably. The region is fished by fleets from several municipalities and supports a complex aquatic community that strongly interacts with the várzea. The largest municipality in the region is the city of Tefé, but boats from Manaus and Manacapuru (Brazil) and Leticia (Colombia) supply their own respective markets and fish-processing plants with fish from this area.

To date, there is no evidence of a general overfishing in the Tefé region, or even in the Amazon as a whole. Nevertheless, conflicts between different groups of fishermen have been observed and described by various authors, such as Penner (1980), Melo

(1985, 1989), Loureiro (1985), and Hartmann (1989). The conflicts are usually the result of disputes among fishermen for a common and accessible resource. Fishing along the Amazon River developed very quickly with the introduction of new fishing and storage technologies, and some groups were unable to keep pace or even benefit from the development (Barthem, 1992; Barthem et al., 1995). Increased pressure on the aquatic environment has brought about a decrease in the total catch of highly profitable commercial species, such as tambaqui (*Colossoma macropomum*) and pirarucu (*Arapaima gigas*), and has had a remarkable impact on the lives of the people living in the várzea, whose traditional market product was dried pirarucu.

The disproportional development of the fishing industry, together with a lack of efficient legislation or knowledge about the biology of the species fished, may hamper the recovery success of fishery resources and give rise to the "tragedy of the common," that is, the failure to manage a common resource (Hardin, 1968; Chapman, 1989). Such failure would be extremely undesirable, because of both the importance of fish as a source of protein and the industry's importance as a social function. Fishing is a "buffer profession" in society—that is, when there is unemployment in the cities, the fishing industry provides an opening for some of the unemployed, thus decreasing the number of jobless people.

This paper is part of the Mamirauá Project, conducted at the Mamirauá Ecological Station (MES) in the confluence of the Rios Solimões and Japurá. It includes a preliminary analysis of the fisheries in the region, based on local observation and unloading statistics from Tefé, which is the main city of the region.

Material and Methods

The Tefé municipal market supplies fish to nearly 30,000 people and is the main fish-consuming center in the region of the middle Solimões. Fishing boats of various sizes arrive at the Tefé port, where workers transport the catch, in wooden boxes over their heads, to the Municipal Market. Each boxful can hold 50–100 kg of fish. The fish are unloaded twice a day, from 6:00 to 10:30 A.M. and from 3:00 to 5:30 P.M., and are delivered to the fish vendors who hold counters at the Municipal Market.

The information on fishing was obtained at the time of unloading in the port of Tefé and also when the fish was delivered for sale at the Municipal Market. During unloading, a data collector interviewed fishermen from all boats bringing fish to be sold at the market. At that point in time, information was obtained about the location fished, the type of environment, the fishing period and gear, and the total catch, as estimated by the interviewed fisherman. A second data collector recorded the weight of the fish delivered for sale at the Municipal Market. For this purpose, a scale was placed at the market entrance where the porters climbed when they entered the market with their boxes of fish. To facilitate control, each box was numbered. The weight of the fish was calculated by subtracting the weight of each porter with an empty box from the weight of each porter with a full box. At the market, fork-length measurements were made of four target species: pirarucu, tambaqui, pirapitinga, and large catfish.

Results

From October 1991 to August 1993, 3400 tons of fish were unloaded at the Tefé Municipal Market. In 1992, total unloading was 1983 tons, with a monthly average of 153 tons. The data set the mean annual per capita consumption of fish at 66.1 kg, or a daily 181 g/person, for 1992, considering the population of Tefé to be 30,000 inhabitants. These statistics do not include dried salted fish or fish marketed by the fish-processing plants in the region, which are unloaded outside the market and usually supply consumers in other regions.

Unloading in Weight

The estimated linear regression for the unloading estimated by the fishermen and that weighed at the market was highly significant ($P < 0.0001$; $r = 0.97$; $r^2 = 0.94$). The resulting equation is as follows:

$$\text{scale information} = 1.89 + 0.9687 \times \text{information from fisherman}$$

Figure 1 shows the monthly distribution of unloading and the water-level fluctuations. The period encompasses two low-water seasons, for the years 1991 and 1992, and two high-water seasons, for the years 1992 and 1993. The monthly unloading shows

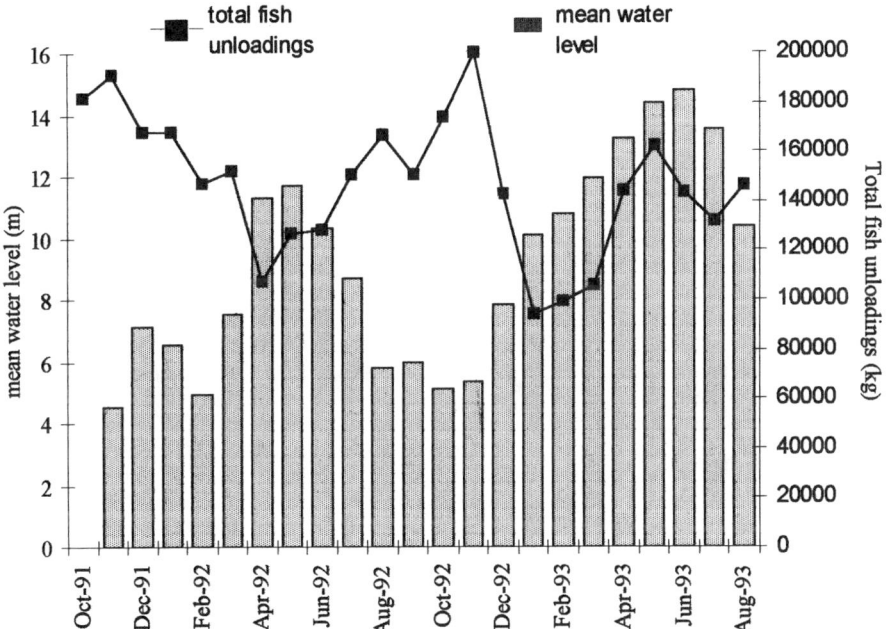

Figure 1. Monthly water-level means for the Rio Solimões and monthly total fish unloadings at the port of Tefé, from October 1991 to August 1993.

a significant negative correlation ($r = -0.60$; $P < 0.01$) with regard to water level, clearly indicating that a catch is inversely proportional to water level. During the period studied, the average water level of the river was at its highest value (14.8 m) in June 1993, and at its lowest (lower than level zero of the Tefé ruler) in the months of September and October 1991. The largest catches were obtained in November 1991 and 1992, during the dry season, when 191.1 and 199.8 tons were fished and the average water level was 4.5 and 5.3 m, respectively. The smallest catch was obtained in April 1992 (107.1 tons), when the monthly water level averaged 11.3 m, close to the highest water level for that year. In 1993, the smallest catch was obtained in January (94.2 tons), at the beginning of the flood season and when the average water level was 10.1 m.

There was also a significant negative correlation ($r = -0.75$; $P < 0.01$) between the number of unloadings per month and the water level; that is, the lower the water level, the higher the number of fishing boats unloading their catch at the port of Tefé. Unloadings were more frequent in the 1991 dry season than in that of 1992, since the drought was more intense in the former than in the latter year (Figure 2).

Average catch per boat from October 1991 to August 1993 was 430 kg/trip. There is a significant positive correlation ($r = 0.63$; $P < 0.01$) between the average catch of each boat and the water level: in the high-water season, boats return from fishing trips with more substantial catches than they do in the low-water season. In the months of May 1992, June 1992, and April 1993, during the high-water season, the average catch exceeded 500 kg/boat trip (Figure 3). Coincidentally, the highest variation coefficient values (>3) were obtained during these periods, which indicates that the increased

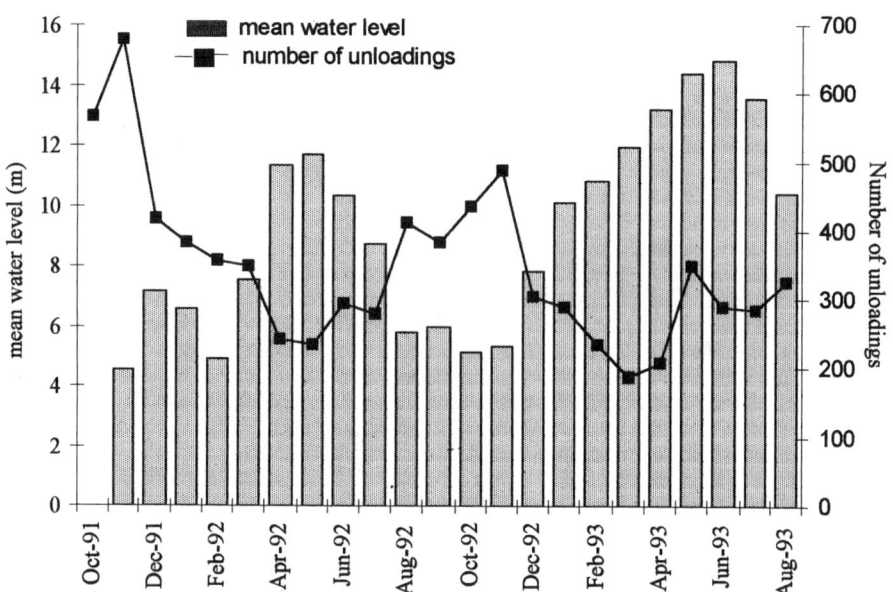

Figure 2. Monthly water-level means for the Rio Solimões and monthly frequency of unloadings by the Tefé fishing fleet, from October 1991 to August 1993.

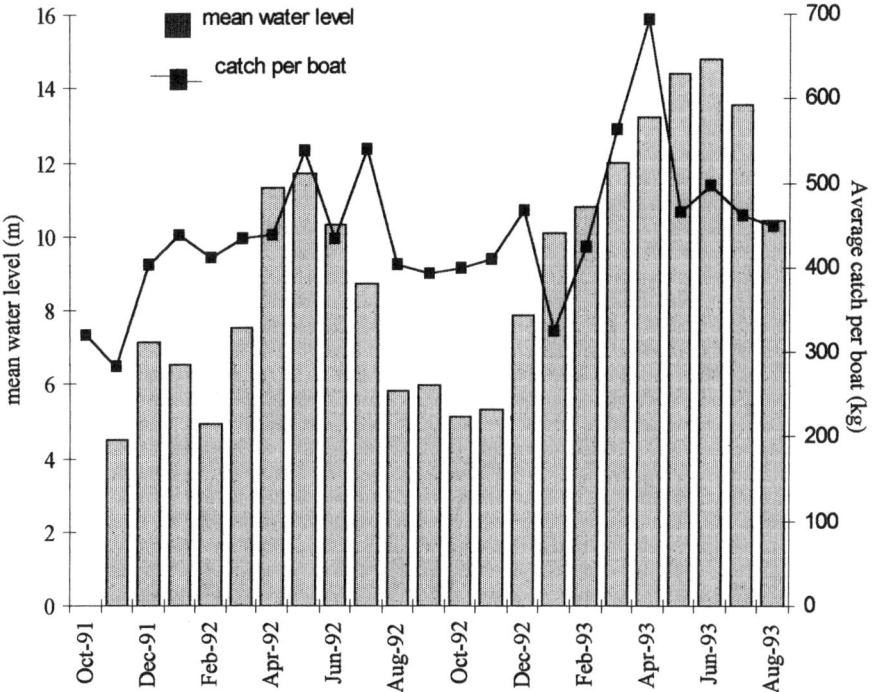

Figure 3. Monthly water-level means for the Rio Solimões and monthly catch per boat unloading at the port of Tefé, from October 1991 to August 1993.

productivity per trip goes hand-in-hand with an increased uncertainty about the success of a fishing trip (Figure 4).

The unloading data were grouped according to the type of fishing craft, type of fishing gear, species marketed, and type of fishing environment.

TYPES OF FISHING CRAFT   The craft unloading fish in Tefé were grouped into six categories:

1. canoes with oars: wooden-hull canoes propelled by oars;
2. canoes with rabetas: wooden-hull canoes propelled by outboard motors (rabetas);
3. fishing boats: boats propelled by a center-mounted diesel engine, equipped with a fixed ice chest, and manned by professional fishermen; deadweight varies from 0.3 to 9 tons;
4. boats-without-boxes: multipurpose boats propelled by a center-mounted diesel engine. Since it is used for purposes other than fishing, this type of boat has a precarious box lined with styrofoam (which is removed when the boat is used for other purposes) rather than a fixed ice chest; deadweight varies from 0.1 to 9 tons, 200 kg mode;
5. regular-line boats: boats used on the regular passenger-transportation lines, which

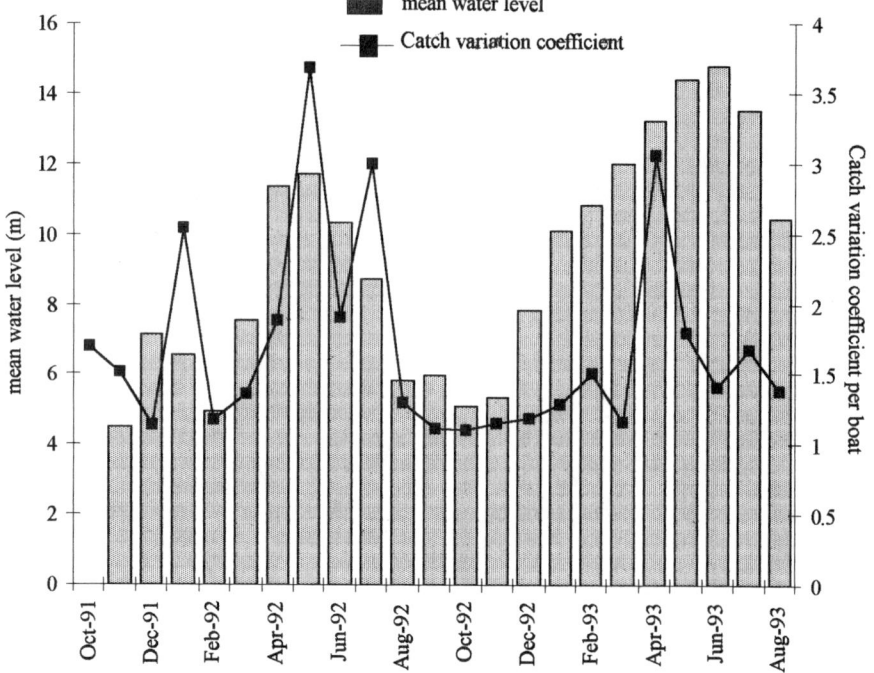

Figure 4. Monthly water-level means for the Rio Solimões and monthly catch variation coefficients per boat unloading at the port of Tefé, from October 1991 to August 1993.

may purchase and store the fish in their cold-storage chambers; deadweight varies from 0.2 to 3 tons;

6. middlemen boats: boats that buy fish from local fishermen rather than manning boats with their own fishermen; deadweight varies from 0.1 to 6 tons.

The fishing boats can be considered really specialized or commercial enterprises. They account for 68% of all fish unloaded. Canoes propelled by outboard motors catch close to 18% of total fish production, and the boats-without-box catch close to 12%. Unloadings from other boats were negligible for the Tefé market (Figure 5).

The distribution of the monthly catch per type of craft shows that the fishing boats maintain a regular production throughout the year, and that the canoes propelled by outboard motors and the boats-without-boxes are more active during the low-water season (Figure 6). There is an inverse correlation between the catch of the fishing-boats and that of the boats-without-boxes ($r = -0.46$; $P < 0.05$), and no significant correlation between any of the other craft categories.

The increase in average production achieved by boats during the high-water season is associated with a decrease in the activities of smaller and less specialized boats (Figure 3), such as the rabetas and the boats-without-boxes. Average production per boat increases during the high-water season due to the decreased activity of these boats. The opposite occurs during the low-water season, when a larger number of small boats with

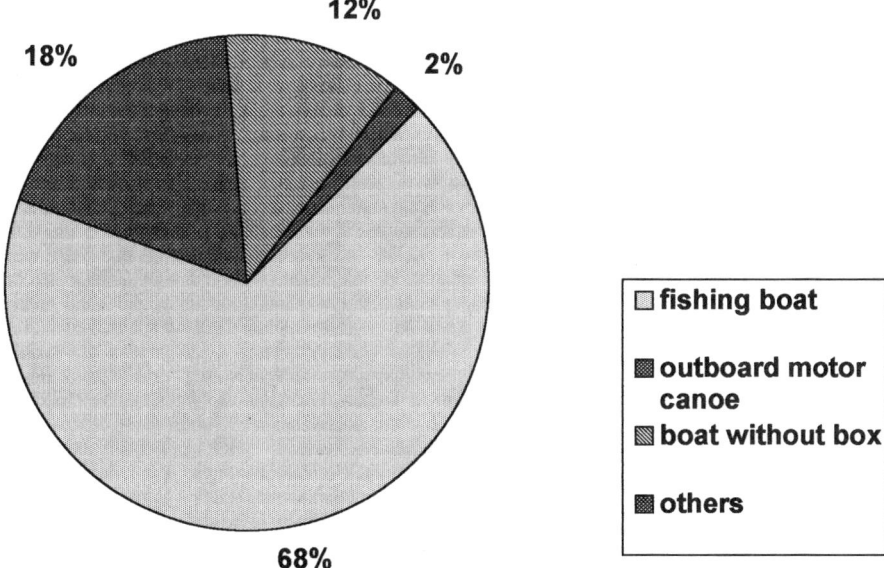

Figure 5. Composition of the Tefé fishing fleet by type of craft.

minimal deadweight begin fishing and unload their catches at the port, thus decreasing the catch-per-boat rate.

Another factor that may be associated with the negative correlation between fishing boats and boats-without-boxes is the increased "risk of success of the catch," which may be observed in the fluctuation of the variation coefficient (Figure 4). During the high-water periods, when fish disperse throughout the flooded area and make fishing more difficult, the risk of not finding the schools increases. The possibility of failure discourages less specialized fishermen with little capital, experience, or equipment from investing in this activity. On the other hand, fewer fish in the market means higher purchase prices, and a corresponding increase in the number of fishing boats, which are better prepared to take on such risks. The opposite occurs during the summer, when the number of boats fishing and the number of active fishermen increase, as does the fish supply in the market. Consequently, fish prices fall, and this discourages investment in high-cost fishing expeditions, which involve long distances and high fuel and ice expenditures.

Types of Fishing Gear   A wide range of fishing gear is used in the Amazon, and each one is adapted to a given environmental context and to a set of target-species. In the middle Solimões, 15 fishing-gear categories have been identified:

1. cast net (*tarrafa*), which is used in shallow waters with little or no vegetation;
2. gill net (*malhadeira*), widely used in environments with no water current;
3. harpoon (*arpão*), specialized gear for the pirarucu, although it can also capture other large animals;

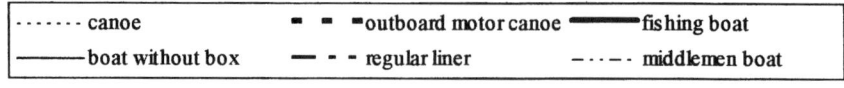

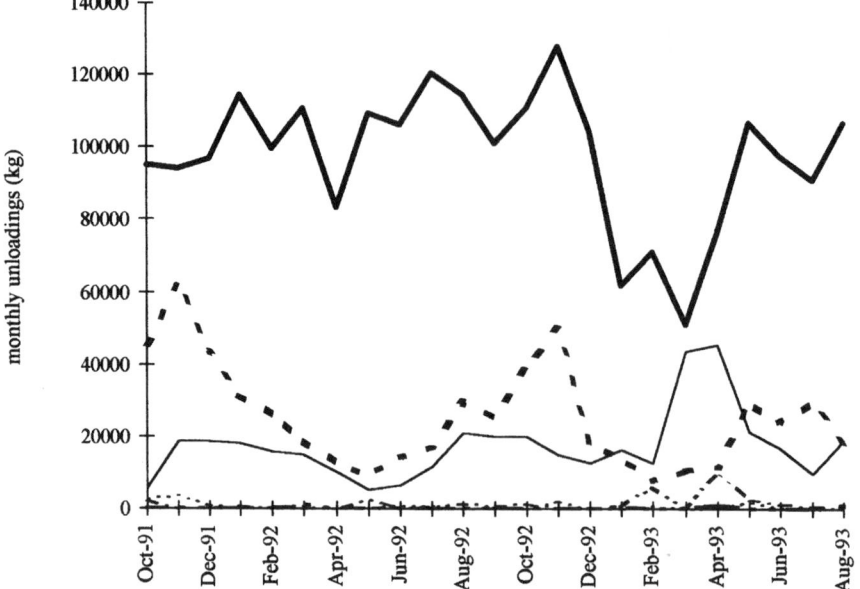

Figure 6. Distribution of the monthly unloadings at Tefé, with regard to the type of craft.

4. baited handline (*linha de mão*), a long line with a baited hook placed midway, held by the fisherman, primarily used for catfish;
5. long line with several baited hooks (*poita-estiradeira*), which is thrown into the river channel and tied to the bank;
6. long line with one baited hook (*curumim*), which is tied to the bank;
7. long line with several medium-size baited hooks and both ends tied to the bank (*espinhel*);
8. cane, line, and baited hook (*caniço*), which is important for catching small fish during the low-water season;
9. cane, line, and unbaited hook (*pinauaca*), with a red cloth tied to the line so that the movement of the cloth attracts the fish;
10. unbaited hand line with a spoon at the hook (*currico*); the shining metal attracts the fish as the boat moves;
11. trident (*zagaia*), a cane with a trident at the end, used for night fishing by the light of lanterns;
12. arrow with or without bow (*flecha/flechão*), a hollow cane with a harpoon or trident at the end, which is either thrown by hand or using a bow;
13. beach seine (*arrastadeira*), a beach net for areas without vegetation;

14. purse seine (*redinha*), used in deep waters or areas where the bottom is free of obstacles;

15. gill net used as beach seine (*rede malhadeira*), a mixture of gill net and beach seine used for capturing catfish.

Fishing fleets unloading at Tefé use most of these devices throughout the year. Nevertheless, in almost 90% of trips from October 1991 to January 1992, only one type of fishing gear was used, accounting for 81% of total unloaded weight (Table I) .When two types of gear were used, the unloaded weight accounted for 17% of the total unloaded weight; the catch of boats which used more than two types of gear (2%) is not representative in terms of the local production.

The total catch from the fishing trips in which only one type of gear was used shows the productivity of the gill and purse seine (96%), followed by gill net used as beach seine (1.2%) and the cast net (1.9%) (Table II). The use of one or more fishing devices in each fishing trip and the high productivity of gill nets and purse seines indicate that commercial fishing in the middle Solimões region is becoming increasingly specialized in the use of these types of gear. This, in turn, requires greater skill of the younger fishermen, who become specialists in the use of the equipment and are no longer interested in less productive fishing modes such as harpoons, arrows, or hooks.

COMMERCIALIZED SPECIES   Based on the names used by the fishermen and fish vendors during the unloading operation, 57 categories of fish were identified. Out of these, 41 are common names given to a single species; 10 are common names given to more than one species in the same genus; 5 are a grouping of various species from the same family; and 1 category is called *salada* (salad). The latter includes fish from various species, which, because of either their small size or their unpopularity with local consumers, have a lesser commercial value (Table III).

Table IV shows the composition of the fish caught and unloaded at Tefé, by species, from October 1991 to January 1992. The data are based on the information supplied by the scales at the market, and from estimates made by the fishermen. The curimatá

Table I. Distribution of total capture by number of different types of fishing gear used in each fishing trip between October 1991 and January 1992

Number of types of fishing gear used	Frequency	Total capture (kg)
1	1416	394,532
2	152	83,043
3	7	6171
4	1	Not weighed
5	0	0
6	1	100
7	1	970

Table II. Percentage distribution of total capture registered on balance, by fishing gear type

Fishing gear	Balance weight (percent)
Harpoon	0.04
Beach seine	0.61
Cane	0.17
Baited hook	—
Long line with several hooks	0.03
Hand line	0.06
Gill net	61.54
Gill net and beach seine	1.16
Purse seine	34.29
Castnet	1.88
Trident	0.22

(*Prochilodus nigricans*) was the most important species during the period, followed by the aruanã (*Osteoglossum bicirrhosum*) and tucunaré (*Cichla* spp.). The monthly unloading of the 10 most important species and of pirarucu is shown in Table V.

Each type of fishing gear has a targeted set of species (Table VI). During the low-water season, gill nets, purse seines, and cast nets captured the largest number of fish species. The gill nets caught mainly curimatá, aruanã, tucunaré, acará-açu, and tambaqui;

Table III. List of common names of commercialized fish in Tefé

Acará-açu[a]	Bacu pedra[a]	Matrinchã[a]	Piramutaba[a]
Acará-tucunaré[a]	Bico-de-pato[a]	Mapará[b]	Piraíba[a]
Acará-bararuá[a]	Curimatá[a]	Mandí[b]	Pirarara[a]
Other acarás[c]	Cubiu[a]	Orana[c]	Pacamom[a]
Aruanã[a]	Caparari[a]	Pacu comum[b]	Peixe-lenha[a]
Apapá or sardinhão[b]	Cuiu-cuiu[a]	Pacu galo[a]	Piranambu[a]
Aracu comum[a]	Dourada[a]	Pacu jumento[a]	Sardinha comprida[a]
Aracu cabeça gorda[a]	Filhote[a]	Pirapitinga[a]	Sardinha chata[a]
Other aracus[c]	Gogóta[a]	Pescada[b]	Surubim[a]
Acari bodó[b]	Jaraqui grossa[a]	Piranha caju[a]	Salada[d]
Branquinha comum[a]	Jaraqui fina[a]	Piranha preta[a]	Tucunaré[b]
Branquinha cabeça lisa[a]	Jatuarana[a]	Other piranhas[c]	Traíra[a]
Branquinha cascuda[b]	Jejú[a]	Pirarucu[a]	Tamoatá
Bacu liso[a]	Jundiá[b]	Peixe cachorro[b]	Tambaqui[a]

[a] Common names representing only one species.
[b] Common names representing more than one species of one genus.
[c] Several species of one family.
[d] Mixed, representing one fish category unloaded including several species not weighed separately.

Table IV. Composition of fish species, weighed on balance and estimated by fishermen in Tefé, between October 1991 and January 1992

Fish	Weight on balance (kg)	Estimated weight (kg)
Curimatá	124,307	135,713
Aruanã	77,863	86,099
Tucunaré	46,426	51,206
Salada	45,671	45,259
Acari-bodó	34,851	20,753
Pacu comum	23,611	28,875
Sardinha comprida	18,947	21,135
Acará-açu	17,862	19,167
Jaraqui escama fina	13,668	13,543
Tambaqui	13,516	19,466
Pirapitinga	12,747	14,552
Matrinchã	7126	8000
Caparari	6347	8808
Pescada	3340	3586
Jaraqui escama grossa	3208	3005
Piranha caju	3052	3310
Surubim	2942	5311
Branquinha comum	2154	2400
Cuiu-cuiu	1942	2230
Acará-tucunaré	1590	1969
Aracu comum	1492	3445
Acará (others)	1310	1440
Pirarucu	1015	1834
Dourada	773	1055
Peixe-lenha	694	816
Sardinha chata	489	1472
Jaraqui	287	300
Tamoatá	273	290
Piranha preta	222	160
Apapá or sardinhão	164	140
Traíra	157	128
Aracu	111	40
Branquinha cabeça lisa	83	200
Acará-bararuá	60	70
Cubiu	59	50
Filhote or piraíba	58	50
Jejú	54	45
Branquinha cascuda	52	50
Aracu cabeça gorda	50	100
Piranambu	35	52
Pacu jumento	30	20
Bacu-liso	20	20
Pacamom or jaú	19	
Mandi	12	
Pirarara	5	29
Pacu		145
Pacu galo		314
Branquinha		10
Piramutaba		20

Table V. Monthly capture (kg) of more important species unloaded at Tefé market

Species	October 1991	November 1991	December 1991	January 1992
Curimatá	49,036	24,958	34,107	16,206
Aruanã	12,485	18,471	36,619	7288
Tucunaré	8406	5194	23,881	8945
Acari-bodó	13,191	11,972	8993	695
Pacu comum	2526	8521	7799	4756
Sardinha comprida	5728	3951	7619	1649
Acará-açu	4479	4736	6469	2178
Jaraqui escama fina	1121	3259	9288	0
Tambaqui	4668	7392	0	208
Pirapitinga	7147	2284	2211	1105
Pirarucu	293	722	0	0

more-specialized purse seines caught curimatá, common pacu and pirapitinga; and cast nets basically caught acari-bodó.

ENVIRONMENTS   Nine types of fisheries were identified in the region of the middle Solimões: rio (river), *lago* (lake), *igapó* (seasonal blackwater swamp), *praia* (beach), *paraná* (channel between a river and a lake, or a river and another river), *ressaca* (an old canal, with a silted end), *enseada-pauzada* (coves formed by the curve of the river, where bank erosion is most intense), *boca de rio ou lago* (mouth of a river or lake), and *capim-matupá* (areas with large concentration of aquatic macrophytes). Figure 7 shows the distribution of the various environments along the river or *várzea*. During the 1991 low-water season, the lake was the environment most frequently fished by the fishermen unloading their catch at Tefé, followed by river, beach, and enseada-pauzada (Table VII).

INTERACTION AMONG THE CATEGORIES (FISHING GEAR, ENVIRONMENTS, FISH SPECIES)   The distribution of the fish caught, sorted by environment and type of fishing gear, is shown in Table VIII. The largest number of fishing devices were used in the lakes, particularly gill nets, followed by purse seines and cast nets. The river ranked second in terms of fish abundance, and purse seines and gill nets were the most frequently used fishing gear types in this environment. Beaches and enseada-pauzadas are special environments, since water currents limit the use of gill nets and cast nets. In these environments, active nets, such as the *rede malhadeira* (gill net used as beach seine) in the beaches and purse seines in the enseada-pauzada are much more productive. The ressacas are very similar to lakes, and thus fishermen favored the use of gill nets and purse seines in this environment.

Table IX shows the highest catch results, sorted by fish species, fishing gear, and environment. It should be noted that lakes are important for many types of fishing gear

Table VI. Distribution of the capture (kg) by fishing gear and species

Species	Harpoon	Beach seine	Cane/line/baited hook	Long line with baited hook	Baited hand line	Gill net	Gill net as beach seine	Purse seine	Cast net	Trident
Acará-açu		65	475			15,006		30	80	60
Acari-bodó						1629		508	26,247	
Aruanã		292	250			51,077	54	5401	1065	946
Curimatá		228			97	52,165	1033	46,991	1410	
Jaraqui gross						450		2143		
Pacu comum		20				6366	243	10,461	245	
Pirapitinga			51			2626		8347	120	
Pirarucu	230			18		578			25	
Tambaqui						8794	568	3136	43	
Tucunaré		120			255	30,727	82	3929	167	261

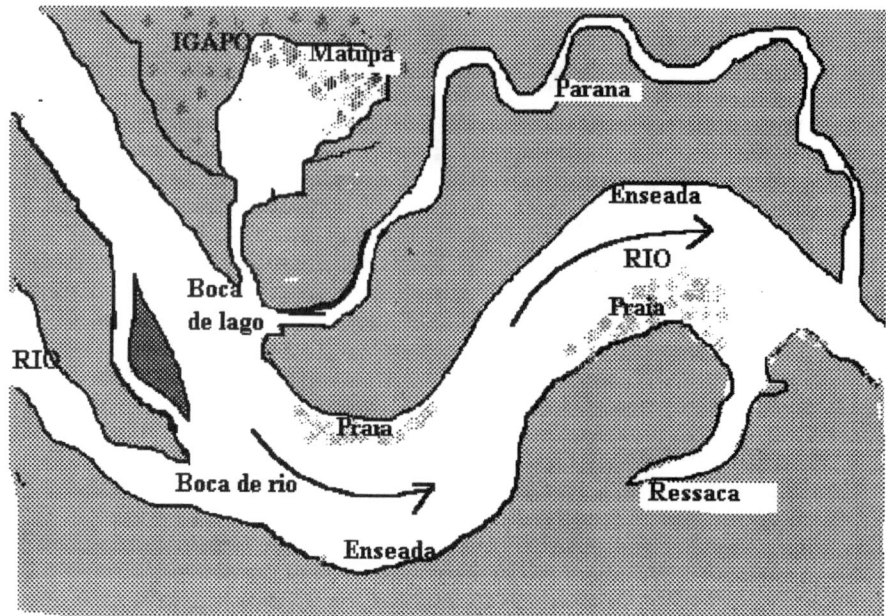

Figure 7. Distribution of the types of environment in the rivers and floodplain of the middle Solimões (rio, river; lago, lake; boca de rio, river mouth; boca de lago, lake mouth; praia, beach).

and species of fish. Consequently, lakes are extremely difficult to manage and are a highly disputed environment.

A preliminary analysis of the catch distribution in regard to the Mamirauá Ecological Station shows that the most important lakes for fishing in the municipality of Tefé are outside the station. Because only 7.4% of the recorded fish caught came from the

Table VII. Distribution of capture by habitat

Habitat	Total capture (kg)
Mouth of river or lake	8123
Capim matupá	406
Enseada-pauzada	11,846
Igapó	120
Lake	329,397
Parana	1833
Beach	24,071
Ressaca	19,661
River	28,520

Table VIII. Distribution of capture (kg) by fishing gear and habitat

Fishing gear	Mouth of river/lake	Capim matupá	Enseada-pauzada	Igapó	Lake	Paraná	Beach	Ressaca	River
Harpoon					80		50		100
Beach seine							278	772	
Cane					1040				
Baited hook					18				
Long line with several hooks							142		
Hand line					352				
Gill net	256		84	84	185,702	900	5597	8319	7648
Gill net & beach seine		406			2471	1032			36
Purse seine	7146		11,707		47,273		13,145	6100	15,869
Castnet					23,028	238	699	3970	764
Trident					1173				94

Table IX. Distribution of capture of the main species unloaded in Tefé in relation to habitat and most productive fishing gear

Species	Fishing gear	Habitat	Capture (kg)
Acará-açu	Gill net	Lake	14,125
Acari-bodó	Cast net	Lake	21,001
Curimatá	Gill net	Lake	43,256
Aruanã	Gill net	Lake	44,030
Jaraqui fina	Purse seine	Lake	3982
Pacu comum	Purse seine	Lake	8544
Pirapitinga	Purse seine	Enseada	3795
Pirarucu	Gill net	Lake	350
Sardinha comprida	Purse seine	Lake	8148
Tambaqui	Gill net	Lake	8057
Tucunaré	Gill net	Lake	28,631

reserve (Table X), it seems that our setting up the reserve did not especially affect the fish supply for the city of Tefé.

Unloaded Fish Lengths

In August 1993, we began gathering the fork length data of fish being sold at the Tefé and Alvarães markets. Nine species were regularly measured, and Table XI shows the basic biometric statistics of such species. Larger species belong to the Pimelodidae family, whose mean fork length is more than 55 cm (pirauaca, filhote, surubim, caparari, and dourada). The tambaqui was the species most frequently studied because of the large number of specimens sampled in the market. This was also an opportunity to monitor the commercial fishing of this species.

Table X. Distribution of capture by area or lake systems near Tefé

EEM Lakes	Capture (tons)	Lakes outside EEM	Capture (tons)
Jaraua	114	Tefé	812
Mamirauá	61	Cubuá-copeá	273
Preguiça	24	Capivara-repartimento	217
Putiri	15	Janamã	208
Tijuaca	12	Jacaré	200
Maguari	9	Tambaqui-pantaleão	123
Aiucá	9	Tarara	35
Cauaçu	7	Mapixari	11

Table XI. Basic statistics of length of commercialized species in Tefé

| Species | Number of individuals | Length (cm) | | | S.D. |
		Minimum	Maximum	Average	
Caparari	578	28.5	119	59.6	11.071
Dourada	26	37	75	58.8	10.402
Filhote	45	39.5	76	55.3	9.211
Jandiá	11	36.5	66.5	50	8.555
Pirauaca	21	40	124	69.8	20.029
Piramutaba	377	23.5	85.5	47.4	7.567
Pirapitinga	847	10	93	41	8.574
Surubim	513	10.5	99.5	55.3	11.220
Tambaqui	1465	16	89.5	46	8.855

SUMMARY OF THE BIOLOGY OF TAMBAQUI   The tambaqui (*Colossoma macropomum*) is the largest caracidae in the Amazon region. It grows to more than 1 m in length and weighs in excess of 30 kg (Goulding, 1988). In the late 1970s, this species accounted for more than 40% of unloaded fish at the port of Manaus (Petrere, 1978), and the tambaqui stock did not show any signs of overfishing (Petrere, 1983). However, in the 1980s, the situation changed and there was a decline in the total catch of tambaqui. Merona and Bittencourt (in press) presented evidence of overfishing in the stock fished by the Manaus fleet.

The life cycle of the tambaqui is reasonably well known. The species inhabits the main rivers of the Amazon region and migrates along almost the whole extension of the Solimões-Amazonas River and some of their tributaries; its life cycle is associated with sediment-laden waters. Juveniles (length <55 cm; weight <4–6 kg) are abundant in the várzea lakes, where they feed on zooplankton, fruits, and seeds (Carvalho, 1981; Goulding & Carvalho, 1982). Though larger specimens are considered to be adults and migrate along the main river, little is known about their natural cycle of reproduction (Goulding, 1988). At the beginning of the high-water season, the tambaqui spawn in the main sediment-laden river (in a place still unknown). Their fertilized eggs and larvae are then carried downstream, until they reach the recently flooded várzea, where the fish grow to maturity (Goulding, 1979; Lima, 1984).

The unloading data obtained by Petrere (1978, 1983) shows that commercial fishing in Manaus focuses on adult tambaqui (>55 cm), whereas commercial fishing in the lower Amazon (although no unloading records are available for this area) targets young tambaqui (<55 cm).

UNLOADING TAMBAQUI AT TEFÉ   Data obtained in 1991 and 1992 show that only 11% of specimens unloaded at the Tefé market met commercialization regulations, namely, IBAMA Administrative Ordinance 1534, dated 20 December 1989 (>55 cm). If the existing legislation were enforced (Figure 8), 89% of the unloaded fish should have been confiscated by IBAMA.

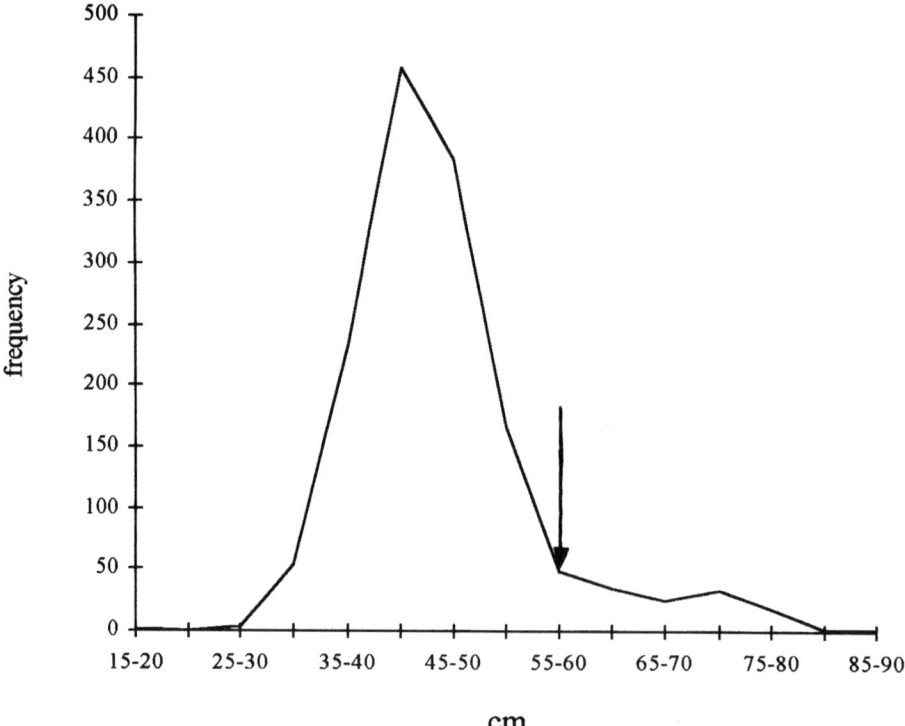

Figure 8. Length distribution of tambaquis unloaded in the market (arrow, minimum capture size allowed).

The origin of small-size tambaqui was investigated during the low-water season, when it was possible to accompany some commercial fishing trips. The average length of specimens captured in the lakes (38.9 cm; $n = 23$) is less than that of specimens caught in the river (65.1 cm; $n = 6$) ($P < 0.001$). Figure 9 shows the separate environments of juvenile and adult tambaqui: juveniles stay in the lake, even under stressful conditions, whereas adults meet in the river, in shoals, preparing to spawn as soon as the flood season begins.

The prevalence of juvenile and small tambaqui in the unloadings at the Tefé market during the 1991 low-water season was due to the fact that most of the fishing was done in the lakes (Table IX).

Discussion

Fishing in Tefé

In 1992, the total catch unloaded at Tefé, per inhabitant (considering a population of 30,000), was equivalent to 180 g/day. This value is very similar to the 1980 estimate for Manaus (with a fish production of approximately 30,000 tons and a population of

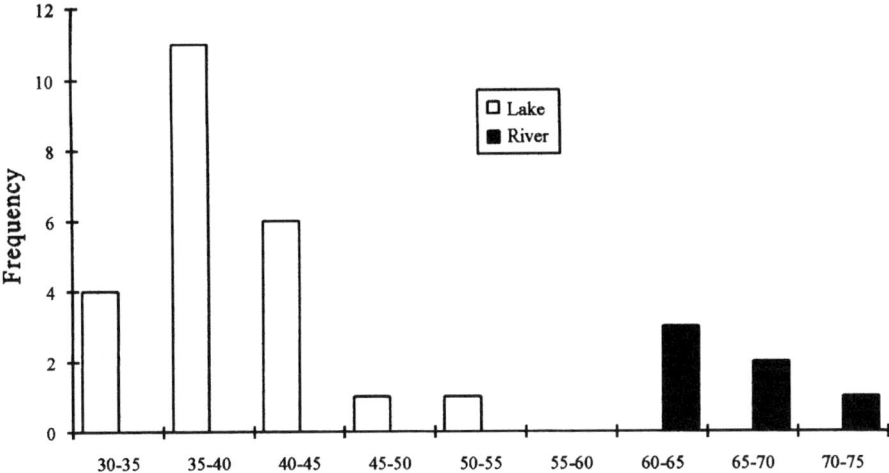

Figure 9. Length distribution of tambaquis caught by the Tefé commercial fishing fleet by type of environment.

700,000 inhabitants), which was 117 g/day. Moreover, these numbers are close to the daily per capita food consumption for the 1973–1974 period, which was 139 g of clean fish, which was considered rather high at the time (Shrimpton & Guigliano, 1979). The data indicate that Tefé does not have a fish supply problem.

From October 1991 to August 1993, the daily unloading:inhabitant ratio varied from 104 to 222 g for the Tefé population. The periods of scarcity are harder to predict than those of abundance, since the latter occurred in October and November, during the low-water season, and the former occurred in different months in 1992 and 1993 but always during the high-water season. The phenomenon must be related to the intensity of the flood and the speed at which the water rises, since the 1992 and 1993 floods were quite different.

As is evident in the large number of species caught, no single species is targeted by the Tefé fishing fleet. Nevertheless, the activity is becoming increasingly specialized, as the description of fishing gear used indicates. Petrere (1978) mentions the importance of gill nets and purse seines in the Manaus unloading in the mid-1970s, and the same is true for Tefé in the 1990s. Thus, the fishing and stock management regulations for the region should focus on these types of equipment.

The species caught most frequently in the region during the 1991 low-water season were the curimatá, aruanã, tucunaré, and acari-bodó, the latter three of which do not migrate. The lakes were the most productive environment in that period, where the largest number of specimens from the above-mentioned species were most often caught. The várzea lakes are the most disputed type of environment, not only because they are more productive, with the more valued fish species caught there, but also because they allow for the use of a wider assortment of fishing gear.

Tambaqui fishing in Tefé focuses on juvenile fish that live in lakes and do not migrate. It is more difficult to catch adult tambaqui, since they have a great ease of locomotion, hide themselves in the pauzadas in order to escape from the nets, and leave the lakes during the low-water season, thus becoming less vulnerable to gill nets. Pirarucu are also quite vulnerable in the lakes, where they build nests for spawning during the low-water season. During the parental care activities, the couples become easy targets for fishermen. Consequently, the management of the tambaqui and pirarucu should focus on the lake phase, when they are at their most vulnerable.

Fishery Management Prospects in the Middle Solimões Region

The development of fishing in the Amazon region occurred in a random manner and created a series of problems related to the decrease in the production of some stocks and conflict between the different categories of fishermen. These problems result from a crisis arising from three important factors:

1. The Amazon region lacks a policy for the conservation and use of renewable resources, so that fishing activities are uncontrolled.
2. The potential of tropical fisheries relies on very many species, but the size of the stock of each species is relatively small.
3. Human activities have brought about important changes in the environment—through dams, through hydroelectric power plants, such as Tucuruí on the Rio Tocantins (Ribeiro et al., 1995), and through deforestation of *várzea* forests (Goulding, 1993)—that impair the recovery of stocks.

In view of the dependence of the population on fish consumption, the possible inability of fish stocks to recover their natural abundance is a source of concern for the future. The maintenance of this renewable resource depends on the conservation of the environment and appropriate management of fished stocks. At the present time, the control of fisheries in Brazil is based on the prohibition of certain fishing gear and on other restrictions concerning specific periods, certain species, delimited fishing areas, and maximum catch size (Isaac et al., 1993; Barthem et al., in prep.).

A possible way to manage fisheries made up of numerous species, where many different types of fishing gear are used, is to restrict the use of certain gear to given areas. At the Mamirauá Ecological Station, the project contemplates the preservation of the forest beyond the establishment of a system of water uses. An inland lake may be preserved in one of three categories:

1. totally protected areas
2. restricted-use areas for the survival of the communities
3. areas for small-scale commercial fishing.

Since most commercially fished species are migrating species, this type of reserve model apparently does not meet the needs of the main species. Nevertheless, as the Tefé unloading data have shown, part of the migrating population remains in the flooded areas during part of their life cycle. Thus, by restricting fishing in preservation areas, it is hoped that there will be a decrease in the mortality of the whole population as a result of fishing activities; that is, the benefit derived from protecting one area will make up for the high catches in neighboring areas. How setting up such reserves will affect the recovery of the whole stock is still uncertain, however, because there is scant knowledge available concerning the patterns of exploitation by commercial fishing fleets and of migration of most species.

Literature Cited

Ayres, J. M. 1993. As matas de várzea do Mamirauá: médio rio Solimões. Estudos do Mamirauá. CNPq, Sociedade Civil Mamirauá, Brasília DF.

Barthem, R. B. 1992. Desenvolvimento da pesca comercial na bacia amazônica e consequências para os estoques pesqueiros e a pesca de subsistência. Documentos Básicos da Conference on Environmentally Sound Socio-economic Development in the Humid Tropics, organizada pela AB/UNESCO, UNAMAZ e Academia de Ciências do Terceiro Mundo; Manaus, de 13 a 19 de junho de 1992.

————, **H. Guerra & M. Valderrama.** 1995. Diagnostico de los recursos hidrobiologicos de la amazonia. Ed. 2. Tratado de Cooperacion Amazonica.

Carvalho, M. L. 1981. Alimentação do tambaqui jovem (*Colossoma macropomum* Cuvier, 1818) e sua relação com a comunidade zooplanctônica do Lago Grande-Manaquiri, Solimões, AM. Master's thesis, INPA, Manaus.

Chapman, M. D. 1989. The political ecology of fisheries depletion in Amazonia. Environmental Conservation 16: 331–337.

Goulding, M. 1979. Ecologia da pesca do Rio Madeira. INPA/CNPq, Belém.

————. 1988. Ecology and management of migratory food fishes of the Amazon basin. Pages 71–85 *in* F. Almeida & C. M. Pringle, eds., Tropical rainforests, diversity and conservation. California Academy of Sciences, San Francisco.

————. 1993. Flooded forest of the Amazon. Scientific American 268(3): 114–120.

———— & **M. L. Carvalho.** 1982. Life history and management of the tambaqui (*Colossoma macropomum*, Characidae): An important Amazonian food fish. Revista Brasileira Zoologica, São Paulo 1(2): 107–133.

Hardin, G. 1968. The tragedy of the commons. Science 162: 1243–1248.

Hartmann, W. D. 1989. Conflitos de pesca em águas interiores da Amazônia e tentativas para sua solução. Pages 103–118 *in* A. C. Diegues, ed., III Encontro de Ciências Sociais e o Mar. Pesca artesanal: tradição e modernidade, 3 a 5 de abril, São Paulo.

Isaac, V. J., V. L. C. Rocha & S. Mota. 1993. Considerações sobre a legislação de "Piracema" e outras restrições da pesca da região do Médio Amazonas. Pages 187–211 *in* L. Furtado, W. Leitão & A. F. de Mello, eds., Povos da águas: Realidade e perspectiva na Amazônia. Museu Paraense Emilio Goeldi, Belém.

Junk, W. J., P. B. Bayley & R. E. Sparks. 1989. The flood pulse concept in river-floodplain systems. *In* D. P. Dodge, ed., Proceedings of the International Large River Symposium (LARS). Canadian Special Publications in Fisheries and Aquatic Science 106: 110–127.

Lima, C. A. R. M. A. 1984. Distribuição espacial e temporal de larvas de caraciformes em um setor do rio Solimões-Amazonas, próximo a Manaus, Amazonas. Master's thesis, INPA, Manaus.

Loureiro, V. R. 1985. Os parceiros do mar. Natureza e conflito social na pesca da amazônia. Museu Paraense Emílio Goeldi, Belém.

Melo, A. F. 1985. A pesca sob o capital; a tecnologia a serviço da dominação. Universidade Federal de Pará, Belém.

————. 1989. Contribuições para uma teoria dos conflitos pesqueiros no Brasil: partindo do caso "Amazônico." Pages 63–75 *in* A. C. Diegues, ed., III Encontro de Ciências Sociais e o Mar. Pesca artesanal: tradição e modernidade, 3 a 5 de abril, São Paulo.

Merona, B. de & M. M. Bittencourt. 1988. A pesca na Amazônia através dos desembarques no mercado de Manaus: Resultados preliminares. Memoria Sociedad de Ciencias Naturales, La Salle 48: 433–453.

Penner, M. E. S. 1980. A pesca no nordeste paraense. Raízes, Belém 1(1): 47–56.

Petrere, M., Jr. 1978. Pesca e esforço no estado do Amazonas. II. Locais, aparelhos de captura e estatística de desembarque. Acta Amazônica Suppl. 2(3): 1–54.

————. 1983. Yield per recruit of the tambaqui (*Colossoma macropomum* Cuvier, 1818) in the Amazonas State, Brazil. Journal of Fish Biology 22: 133–144.

————. 1992. Pesca na Amazônia. Pages 72–78 *in* PARÁ—Secretaria de Estado de Ciência, Tecnologia e Meio Ambiente. SIMDAMAZÔNIA. Belém, PRODEPA.

Ribeiro, M. C. L. B., M. Petrere Jr. & A. A. Juras. 1995. Ecological intregrity and fisheries ecology of the Araguaia–Tocantins river basin, Brazil. Volume 10 *in* Regulated Rivers: Research and Management 10: 31–45.

Shrimpton, R. & R. Giugliano. 1979. Consumo de alimentos e alguns nutrientes em Manaus. 1973–1974. Acta Amazonica 9(1): 117–141.

Smith, N. 1979. A pesca no rio Amazonas. Conselho Nacional de Desenvolvimento Científico e Tecnológico (INPA), Manaus.

————. 1981. Man, fishes, and the Amazon. Columbia University Press, New York.

SUDEPE. 1985. Relatório do grupo permanente de estudo da piramutaba. Sub-Grupo Economia Pesqueira. Março, Base de Operações do PDP, Belém.

Welcomme, R. L. 1976. Some general and theoretical considerations on the fish yield of African rivers. Journal of Fish Biology 8: 351–364.

The Impact of the Ornamental Fish Trade on the Discus *Symphysodon aequifasciatus*: A Case Study from the Floodplain Forests of Estação Ecológica Mamirauá

William G. R. Crampton

Introduction

No formal studies of the impacts of commercial exploitation on discrete discus populations have been conducted to date. My intention here is to fill this gap through the study of a recent case of overfishing of the discus *Symphysodon aequifasciatus* Pellegrin in Estação Ecológica Mamirauá, near Tefé, Amazonas. The area, hereafter referred to as the "Mamirauá Reserve," is now protected as an ecological sanctuary. Since the commercial exploitation of discus must be seen in the context of the Amazonian ornamental fish trade and the way the trade works, this paper begins with a summary of the trade and what is known about its impact on stocks of ornamental fish. It then describes the variety of *S. aequifasciatus* found in the Mamirauá Reserve and presents a general description of the habitats represented within the reserve. Then follows the kernel of the study: an evaluation of the effects of recent fishing activity in the Mamirauá Reserve. This is based both on an investigation of the nature and history of ornamental fish catching in the area and on direct assessments of discus stocks.

Since there are no published accounts of the ecology and natural history of wild discus, I describe here aspects of their natural history that are of importance to their conservation. An investigation of the site fidelity and seasonal movements of discus is presented separately. Of particular importance is the fact that discus, like many cichlids, have distinct intraspecific color varieties. These may form an important part of the genetic diversity of discus species. In any assessment of the impact of commercial exploitation, such diversity issues must be considered. I summarize here the nature of the color varieties of Amazonian discus species and their relevance to the conservation of the genetic diversity of discus.

The Ornamental Fish Trade in the Amazon Basin

More than 150 million freshwater and marine ornamental fish are sold annually, and the sale of fish and aquarium accessories supports an industry of over US$7 billion a year (Andrews, 1990). Most of the demand is from the United States, western Europe, and Asia. Exports of Amazonian fish began in the 1930s but boomed in the late 1950s with increasing demand, improved overseas air connections, and better-organized suppliers. Brazil exports the majority, although Peru, Colombia, and Venezuela also have substantial industries (McGrath, 1990). Between 15 and 20 million ornamental fish, around 90% of Brazil's annual total, are exported through Manaus (Andrews, 1990; Leite & Zuanon, 1993). Export revenue from Manaus averaged US$1.2 million in the late 1980s and surpassed US$2 million in 1991 (see Chao, 1993). Many fish destined for export die during capture or transit. Chao (1993) believes that the number of ornamental fish removed annually from the Brazilian Amazon is likely to be between 30 and 40 million.

Around 150 species are regularly exported from the Amazon as ornamentals. Demand, however, focuses on a small number of species. In exports of ornamental fish from Manaus in 1980, the cardinal tetra *Paracheirodon axelrodi* accounted for 81.6% of the total number of fish exported and 53% of the export revenue. Discus, *Symphysodon* spp., represented just 0.7% of the total number but 22% of the revenue (Junk, 1984). Today these two fish constitute approximately the same respective proportions of the market as they did in 1980.

Recent years have seen an increasing number of cases of overfishing involving ornamental fish. Coral reef fish, all wild caught, are particularly vulnerable (see Wood, 1985). Rare freshwater ornamentals are also sensitive, as illustrated by local extinctions of species in parts of Sri Lanka and Malaysia (Banister, 1989). At present, the cultivation of ornamental fish remains uneconomical in Amazonia, with the consequence that all fish for export are wild caught. Can the ornamental fish trade of the Amazon basin continue this exploitation without compromising diversity? The aquarium trade has long argued that the exploitation of ornamental fish in the Amazon basin has had little impact on wild stocks. The artisanal nature of the trade and the limited periods of the year suitable for collecting are, Axelrod (1988) argues, sufficient to protect stocks. Chao (1992, 1993) has investigated the sustainability of upper Rio Negro ornamental fish using the cardinal tetra as an indicator species; he concludes that, for the most part, the trade has had little or no impact on stocks. There is evidence, in spite of this, to suggest that intensive fishing has caused the commercial extinction of discus (*Symphysodon* spp.) and cardinal tetra (*Paracheirodon axelrodi*) in the middle Rio Negro (Bayley & Petrere, 1990). Leite and Zuanon (1993) also document a one-year ban on discus fishing imposed by governmental decree following crashes in discus populations in the Rios Jufarí and Tea, and near Barcelos and Santa Isabel on the Rio Negro.

Discus Species of the Amazon Basin

Two species of discus are recognized from the Amazon basin. *Symphysodon discus* Heckel occurs principally in the Rio Negro and in north-bank tributaries around the Rio Trombetas. The distribution of the other species, *S. aequifasciatus*, mostly does not overlap with that of *S. discus* but is more widespread. All the main Amazonian affluents across the entire basin other than those inhabited by *S. discus* are known to contain *S. aequifasciatus* (for discussions of geographical range, see Kullander, 1986; Schultz, 1960;

Schulze, 1988; Mayland, 1994). Both species of discus have distinct color forms. *Symphysodon aequifasciatus*, according to Schultz (1960), occurs as three subspecies based on color differences: subsp. *axelrodi* Schultz (the "brown discus"), subsp. *aequifasciata* Pellegrin (the "green discus") and subsp. *haraldi* Schultz (the "blue discus"). Subspecies of *S. discus* based on color forms have also been proposed (e.g., Burgess, 1981). In the aquarium literature, many attempts have been made to further subdivide these color forms and to claim the existence of other unique forms.

Although the division of *Symphysodon* species into subspecies has been treated with scepticism (see Kullander, 1986), there remains the important question of whether the color forms are genetically based, reflecting diversity within the species. The alternative explanation is that water chemistry and diet are more important determinants of color (see Burgess, 1981). The definitive test would be a genetic study of discus populations around the Amazon basin. This has yet to be done, but there are two reasons to believe that the color varieties of *S. aequifasciatus* have a large genetic component. First, as is well known in discus breeders, the three main color varieties of this *S. aequifasciatus* are recognizable in second-generation thoroughbred fish in captivity. Second, commercial breeders have long known the predictable heritability patterns of traits explored through crossing the different varieties (see Mayland, 1993).

For the most part, the distribution of brown, green, and blue discus in the Amazon basin does not overlap. Brown discus are found mainly in the easternmost affluents such as the Tocantins, Xingu, and Tapajos. Green discus are found in the Rios Tefé and Japurá and in the Peruvian and Colombian Amazon. Blue discus are found in the affluents around Santarem and in the Manacapuru lakes. In some areas the forms coexist sympatrically, as is believed to occur in the Brazilian territory between the Rios Solimões and Japurá. Here, distinct green, blue, and possibly brown discus are believed to coexist (Mayland, 1994). The green and blue forms of *S. aequifasciatus* are rarer than the "brown" form and are also more strikingly colored. They are very popular amongst aquarists and form the parental stock of many artificial strains created by discus breeders. As a consequence of their popularity and rarity, green and blue strains command very high export prices and are much sought after by ornamental fish catchers. Some populations such as the "Tefé green discus" and "Manacapuru royal blue discus" are, for reasons unknown, even more colorful than the usual green and blue forms (e.g., Mayland, 1994; Schulze, 1988; Bleher, 1986).

Symphysodon species are almost exclusively confined to the margins of seasonally flooded rivers and to the lakes and inundated forests of floodplains. They live in colonies amongst submerged branches or tree crowns. A breeding colony can contain many thousands of fish. Little is known and nothing published about the ecology of wild discus. On the other hand, the mate choice, breeding behavior, and parental care of captive discus are well documented in the aquarium literature.

Discus of the Mamirauá Reserve and Tefé Region

The study, to establish the impact of fishing on wild stocks of *S. aequifasciatus* and other ornamental fish, focused on a network of lakes known locally as the Jarauá system. This area was chosen for two reasons: (1) because it had an active fishery at the time the study began, in 1992, and (2) because it showed all the features of a white-water flood-plain system (*várzea*) including the characteristic mosaic of channels (*paranás*), lakes (*lagos*), and levees. The entire Mamirauá Reserve is subject to flooding by the annual ebb and flow of the Amazon, so that most of the forest is inundated for three to seven months a year. Ayres (1993) presents a general description of the habitats represented within the reserve. During the low-water season, the discus in the area are found in *galhadas*, the submerged crowns of fallen trees at the edge of lakes. Although the inter-mittently flowing paranás that drain the várzea plain do contain discus, most are found in the lakes. Most várzea lakes are shallow and ribbon shaped and have drainage channels connecting them to the paranás. Some lakes are isolated during low water. Nearly all the várzea lakes contain substantial areas of floating meadow, an important nursery for young cichlids. Table I presents physicochemical data and latitude/longitude for the lakes involved in the study.

Morphometric measurements were taken from eight adult and three juvenile discus from lakes in the Jarauá system, six adult discus from the Mamirauá lake system, and one adult from the blackwater ria Lake Amanã, outside the reserve. All of these spec-imens were identified as *Symphysodon aequifasciatus* on the basis of a comparison of morphometrics with measurements presented by Kullander (1986). Color photographs of the discus were compared to a large number of published photographs of freshly caught specimens. On the basis of coloration, all of the specimens encountered could be assigned to the "blue discus" variety proposed as the subspecies *S. aequifasciata haraldi* Schultz. The discus from Rio Tefé and Lago Tefé, some 40 km south on the other side of the Rio Solimões, are of another variety: the "green discus" of the proposed sub-species *S. aequifasciata* subsp. *aequifasciata* Pellegrin (see Bleher, 1986).

Assessing the Impact of Recent Exploitation on the Discus of the Mamirauá Reserve

The study took a two-pronged approach. First, informal interviews were conducted with local residents from communities involved in the ornamental fish catching, the aim being to get an estimate of the numbers of fish being removed annually from different areas of the reserve. The second part of the approach was to conduct direct stock assessments.

Interviews: History and Nature of Exploitation

1989: Commercial exploitation of the Jarauá area began in earnest in 1990. However, local residents recall that, during the mid- to late 1980s, some boats entered the Jarauá lake system and extracted discus and angelfish (*Pterophyllum scalare*). At this time there were known to be large discus colonies in many of the lakes. In 1989 a boat owned by an ornamental fish–exporting company in Manaus visited briefly to collect information about the potential of the area. No fish were removed. The adjacent Mamirauá lake

Table I. Sampling sites of *Symphysodon aequifasciatus*: locations and physicochemical parameters

Location, lake system	Lat, long. (S, W)	Depth (m)[a]	Temperature (°C)[b]		Conductivity[c]	Surface pH[a]	Flow[d]	Oxygen[e]	Transparency[f]
			Surface	Bottom					
Lago Sumaumarinho, Jarauá (Jarauí)	2°48'40", 65°04'35"	1.5/8.0	31.3/29.2	31.0/28.7	79.4/67.8	6.4/6.2	0/0	3.7/0.8	0.45/1.0
Lago Cedrinho, Jarauá	2°48'30", 65°05'24"	2.0/9.0	31.5/29.0	30.0/28.1	143.4/71.2	7.0/6.5	0/0	3.2/1.2	0.4/1.1
Lago Jaraqui, Jarauá (Jarauí)	2°42'09", 65°06'05"	5.5/11.5	31.8/28.9	30.8/26.9	191.5/69.5	6.6/6.6	0.01/0	2.4/0.4	0.75/1.3
Lago Promessa, Mamirauá	3°06'05", 64°46'37"	3.5/9.5	26.9/29.6	26.9/27.1	83.0/86.1	6.8/6.8	0/0	2.6/0.1	0.75/0.8
Lago Bocinha, Mamirauá	3°03'51", 64°49'51"	1.5/7.0	31.3/32.0	31.3/31.8	82.8/84.2	6.5/6.6	0/0	3.7/0.3	0.4/1.23
Lago Içé, Ilha Içé (Solimões)	3°14'50", 64°42'17"	3.0/10.0	29.7/28.4	29.6/28.1	131.5/91.0	6.9/7.0	0.06/0.30	2.5/1.0	0.55/0.48
Lago Amanã, Lago Amanã	2°30', 64°40'	—/3.0	—/31.3	—/28.5	—/7.3	—/5.3	—/0	—/3.0	—/2.2

[a] During low water/high water.
[b] At midday, during low water/high water.
[c] During low water/high water; conductivity = μS cm^{-1} at 25°C.
[d] During low water/high water; flow = ms^{-1} at surface.
[e] Surface dissolved oxygen (mg/l) at midday.
[f] Transparency (m) with secchi disc.

system has not, to the recollection of any of the residents, been regularly visited by ornamental fish catchers. For unknown reasons it does not hold any sizeable colonies of discus.

1990: The same boat returned in 1990. Men from the communities of Nova Colômbia and Jarauá were hired to provide manual labor and to guide the fish catchers to lakes holding discus. Discus were then removed intensively during the low-water months, from October to December, and shuttled to large submerged holding cages (*viveiras*) near the communities. Here the fish awaited transportation to Manaus. Estimates of how many discus were removed in the first year range from 25,000 to 40,000. Most estimates are close to 30,000. The figures are probably fairly accurate, since the local people were paid by the number of discus they caught. Around 4000–5000 discus died in the holding cages. The fish were taken to Manaus in the modified hold of the boat. A large number of fish are likely to have died in transit (see, e.g., Mayland, 1994, and Axelrod, 1978, for problems of transporting discus). The submerged branches holding discus colonies were surrounded by fine mesh seines and the branches cut and removed. This method is almost completely effective, allowing the fish catchers to systematically remove the discus from any lake. Several people remember a catch of 2000 discus from one submerged tree crown alone. The figures mentioned do not appear to be out of the ordinary in comparison to accounts of the number of discus removed from Lago Tefé in the mid-1980s. (Details of the main lakes exploited are presented in Crampton, 1993.) A small number of *Pterophyllum scalare* was removed along with the discus, but otherwise no other fish were exploited.

1991: For two months in the low-water season, the same boat returned and continued to fish in the Jarauá area. This time, however, the boat brought its own crew and made no use of local manpower or expertise. Reports for this period are essentially second hand and hence anecdotal. Estimates ranged from 10,000 to 20,000.

1992: This year marked a turning point in the catch figures. Representatives of both Nova Colômbia and Jarauá report that the boat was able to remove only between 300 and 500 fish in six weeks. This would suggest that the point of commercial extinction was approaching.

1993: The boat returned in September 1993 to test whether it was worth continuing fishing. Despite the recently introduced protection order, an estimated 200 fish were removed. The scarcity of fish seems to have been more instrumental than the protection order in hastening the boat's departure.

1994: The protection order was being effectively maintained by late 1994. The boat visited the reserve briefly in November en route to the Paraná Juamí, further up the Rio Japurá. As far as I know, no fish were caught, but the company appears to be monitoring the stocks in Jarauá. The strength of the protection order will prevent exploitation from outside companies for the foreseeable future.

In summary, one boat alone is reported to have caused a serious decline in the stocks of discus from within the Jarauá area, where previously there existed large and healthy discus populations. The pattern resembles that of the discus fishery of Lago Tefé in the mid- to late 1980s. The collapse of the Tefé fishery is thought to be the main explanation for why a Manaus company moved to the Jarauá area. The discus stocks in Lago Tefé have not yet recuperated and are still considered commercially extinct.

Stock Assessment

Stock assessments were conducted by comparing present populations of *S. aequifasciatus* in selected lakes to other cichlid fishes that were not harvested. A more appropriate method would have been to compare the disturbed lakes to an area that held intact stocks of discus. Unfortunately, the fish catchers had been so efficient in locating lakes holding discus that I could find no such unexploited locality in the Mamirauá reserve or within a reasonable distance of Tefé. Comparisons of the numbers of discus to other cichlids was done by direct fish counts or by netting. Another useful assessment of stock quality was the proportion of the discus in breeding condition in comparison to other cichlids. This was possible only by netting and was conducted at the end of the low-water period, when most cichlids, including *S. aequifasciatus*, are normally in breeding condition.

FISH COUNTS   Density estimates were made by slowly patrolling the edge of a lake by canoe for 4–9 hours (beginning at 7:00 P.M., on one night or two successive nights) in November 1993 and again in November 1994. A halogen light was employed to locate fish as each galhada was visited sequentially until the lake was circled. Three lakes in Jarauá (Sumaumarinho do Jaraquí, Cedrinho, and Jaraquí) and one in the Mamirauá system (Lago Promessa) were selected as representative sites. Table II presents the results of these observations.

It was clear from the counts presented in Table II that discus are not present in large numbers. Most other cichlids, on the contrary, are relatively abundant. Local residents claim that before the overfishing, discus in the Jarauá lakes were the most plentiful of the cichlids found in submerged branches. Lago Promessa has never been commercially exploited for discus but does not hold any sizeable colonies. In all of the Jarauá lakes, the relative densities of discus appeared to have decreased from 1993 to 1994. This deduction should be treated with caution, however, since the water level at the time of the 1994 observations was some 2 m higher than it was in 1993. This discrepancy could have influenced the amount of habitat available for observation.

GILL NETTING   Monofilament nylon gill nets (40 mm stretched mesh), 40 m high, were pegged around an area of submerged branches, and a pole was used to disturb the branches and thus flush out discus and other fish. In November 1994, nets were set around these habitats in two lakes in the Jarauá system (Sumaumarinho and Cedrinho) and two lakes in the Mamirauá system (Promessa and Bocinha). The objective was to compare the numbers caught and the breeding condition of discus in comparison to other cichlids. The catches are nonquantitative and represent approximately one morning of netting for each lake. Table III gives the numbers, sizes, and breeding conditions of all cichlids captured from the four lakes.

As was discovered with the direct counts, discus are present in much lower densities than most other cichlids in Jarauá lakes, and are rare in the Mamirauá lakes. One important observation is that no *Symphysodon aequifasciatus* in these or in any other samples in 1994 were in breeding condition. The size values of all the individuals were nev-

Table II. Counts of *Symphysodon aequifasciatus* and all other cichlids at low water 1993 and 1994

| | | Low water 1993 | | | | | Low water 1994 | | | | |
| | | Per lake | | | | | Per lake | | | | |
Lake	Perimeter (km)[a]	t[b]	N/total[c]	%[d]	Km (n/total)[e]	Hr (n/total)[f]	t[b]	N/total[c]	%[d]	Km (n/total)[e]	Hr (n/total)[f]
Sumaumarinho	2.60	4	16/262	6.1	6.2/100.8	3.2/52.4	6	11/289	3.8	4.2/111.2	0.2/48.2
Cedrinho	5.85	8	28/356	7.9	4.8/60.9	3.5/44.5	9	24/501	4.8	4.1/34.3	2.7/55.7
Jaraqui	10.0	9	11/312	3.5	1.1/31.2	1.2/34.7	9	4/259	1.5	0.4/25.9	0.4/28.8
Promessa	4.60	5	0/40	0	0/8.7	0/8	4	1/61	1.6	0.2/13.3	0.3/15.3

[a] Measured from 1:50,000 ground survey maps.
[b] Hours of observation to survey entire perimeter of lake.
[c] Number of discus/total number of cichlids.
[d] Percent of cichlids represented by *S. aequifasciatus*.
[e] Number of discus/total number of cichlids per km of lake edge.
[f] Number of discus/total number of cichlids per hour of observation.

Table III. Relative numbers, weight, and breeding condition of cichlid fish captured by unquantified gill net sampling in low water 1994

Species	N[a]	Mass[b] X	S.D.	Breeding status[c] NR	PRM	RM	PRF	RF	%R
Lago Sumaumarinho: Jarauá									
Astronotus ocellatus	6	295.8	84.0	5	—	—	—	1	16.6
Caquetaia sp. 1	11	202.7	37.4	11	—	—	—	—	0
Chaetobranchus flavescens	1	400.0	0	1	—	—	—	—	0
Cichla monoculus	1	298.0	0	1	—	—	—	—	0
Geophagus cf. proximus	2	86.0	11.3	2	—	—	—	—	0
Heros appendiculatus	11	180.2	29.8	5	—	—	—	6	0
Hypselecara temporalis	1	146.0	0	—	—	—	—	1	100
Mesonauta insignis	3	78.3	9.4	—	—	1	—	2	100
Satanoperca sp. 1	6	163.0	13.3	1	—	4	—	1	83.3
Symphysodon aequifasciatus	3	180.7	13.0	3	—	—	—	—	0
Lago Cedrinho: Jarauá									
Caquetaia sp. 1	8	194.5	16.4	8	—	—	—	—	0
Chaetobranchus flavescens	3	420.7	83.2	2	1	—	—	—	0
Heros appendiculatus	3	145.3	19.0	—	—	—	—	3	100
Mesonauta insignis	2	85.6	9.5	—	—	1	—	1	100
Satanoperca sp. 1	7	152.3	12.3	4	—	—	—	3	42.9
Symphysodon aequifasciatus	4	175.5	18.0	4	—	—	—	—	0
Lago Promessa: Mamirauá									
Astronotus ocellatus	6	311.3	48.9	3	—	1	—	2	50.0
Caquetaia sp. 1	10	204.5	15.6	1	1	1	1	6	70.0
Heros appendiculatus	4	178.9	12.4	3	—	—	—	1	25.0
Mesonauta insignis	2	78.9	12.5	—	—	1	—	1	100
Lago Bocinha: Mamirauá									
Astronotus ocellatus	8	247.3	50.9	2	—	1	2	3	50.0
Caquetaia sp. 1	12	177.0	24.2	8	—	—	—	4	33.3
Geophagus sp. 1	1	94.2	0	—	—	—	—	1	100
Heros appendiculatus	2	137.5	3.5	1	—	—	—	1	50.0

[a] Number of indidviduals.
[b] X, mean mass; S.D., sample standard deviation of mass.
[c] NR, nonreproductive; PRM, pre-reproductive male; RM, ripe male; PRF, pre-reproductive female; RF, ripe female; %R, percent of individuals that are ripe (full breeding condition).

ertheless those of potentially mature adults. Furthermore, the end of the low-water period is the expected time of the year for discus breeding. Only two other species were not represented by breeding adults: *Cichla monoculus* and *Chaetobranchus flavescens* (Table III). At the same time of the year, however, I observed many ripe female and male individuals of both these species in the catches of local fishermen. Since only a small number of discus were caught here, a larger sample size might have been necessary to support the conclusion that they were not breeding; but, due to time restrictions and the need to avoid further disturbance of the discus stocks, this further sampling was not possible.

SUMMARY OF STOCK ASSESSMENT   Both direct fish counts and gill net samples indicate that in lakes where discus were previously common, they are now a relatively scarce part of the cichlid fauna. In no single locality was a large colony of discus found. These conclusions appear to confirm the recent history of exploitation of discus in the area. The apparent lack of breeding discus may not be significant, due to the small sample sizes presented here, but it is an observation that extends to nonquantitative samples of discus made in other lakes. The monogamous behavior of *Symphysodon* spp. and the time and effort discus expend in finding a compatible mate are well known. It is also acknowledged by many breeders that fish that are constantly disturbed or agitated fail to pair-bond. It is plausible that the breakup and reduction of discus colonies, combined with a regime of disturbance, have not only depleted stocks but may also have severely disturbed the normal breeding behavior of the discus in the Mamirauá Reserve.

Site Fidelity and Movements

SITE FIDELITY   No large-scale migrations of cichlid fishes are known in the Amazon. On the contrary, the territorial nature of most cichlids means that their home ranges are usually very restricted. Local people say that a resident discus colony will remain indefinitely in the same area, always breeding in the same submerged tree at low water. Presumably, as a colony grows, it expands into satellite colonies in the same or nearby lakes. To investigate whether or not discus do remain in the same proximity from year to year would require an extensive tagging program for a large colony. This was impossible, given the absence of intact colonies, but 11 discus were nevertheless tagged in November 1993: 6 from a submerged tree crown at the east end of Lago Cedrinho and 5 from a similar habitat at the west end of the lake. These fish were captured by night with a halogen spotlight and small dip net, weighed, measured (standard length only), and tagged with a Floy tag in the caudal region. No individuals were successfully sexed (see Silva & Kotlar, 1980, on the difficulties of sexing live discus). Furthermore, due to the small scale of the experiment, the scales of recaptured fish were not removed for age determination.

A remarkable 3 out of 11 fish were recaptured, all at or near the original sampling point. Table IV summarizes the capture and recapture data. These results do provide some evidence for the hypothesis that discus exhibit high site fidelity.

Table IV. Capture and recapture data for three tagged *Symphysodon aequifasciatus*

Tag	Original location, 1993	Standard length (mm) 1993	1994	Increase (%)	Weight (g) 1993	1994	Increase (%)	Distance moved (m)[a]
1	Lago Cedrinho (east)	121	138	14.0	135	174	28.9	<5
2	Lago Cedrinho (east)	131	147	12.2	151	203	34.4	<5
3	Lago Cedrinho (west)	125	140	12.0	140	198	41.4	40

[a] Approximate distance between original sampling point and recapture site.

MIGRATION INTO THE FLOODED FOREST AT HIGH WATER   The seasonal flooding of the várzea floodplain forests opens up a vital habitat to fish. Many of the resident and nonresident fish of the floodplains move in to feed on the allochthonous and autochthonous plant and invertebrate food supplies (e.g., Goulding, 1980; Table V). There are no published accounts of discus entering flooding forest, although several local fishermen I spoke to reported seeing them, often in pairs, in the submerged branches of trees. All of these fishermen remarked that the discus were never more than 20–30 m from a lake edge.

To investigate this matter, gill nets (40 mm stretched mesh) were set between trees in the flooded forest bordering on Lago Sumaumarinho (Jarauá), at 15 m away from the lake edge and at 40 m away from the lake edge. A total of 40 m of gill nets were set for nine hours on each of two consecutive nights (16–17 May 1993, between 5:00 P.M. and 2:00 A.M.). A camp was set up and the nets investigated on an hourly basis in order to prevent damage to the catch or nets by piranhas. Over the two nights the nets yielded 6 *Cichla monoculus*, 4 *Astronotus ocellatus*, 2 *Heros appendiculatus*, 1 *Chaetobranchus flavescens*, and 1 *Symphysodon aequifasciatus*. The presence of this single specimen confirms that discus do, at least periodically, enter the flooded forest. The possibility that discus use flooded forests at high water highlights the need to conserve várzea forest around lakes.

Discussion

The Discus of Jarauá: Management Considerations

Several important points arise from the study. First, commercial exploitation in recent years appears to have precipitated a large decline in discus populations and to have disrupted their breeding. The ease with which entire colonies of discus can be captured at low water and the high value of the "blue discus" variety of the Mamirauá Reserve accounts for the rapidity of the depletion.

Second, the high site fidelity of discus, as suggested by this study, and its attendant restriction on the dispersal of populations render the recolonization of an overfished area much slower than would be the case for a fish with a higher dispersal rate such as the cardinal tetra.

And finally, no breeding stock remained in the discus sampled at low water in

Table V. Dietary preferences of cichlids

Mean percentage volume of an individual's stomach contents represented by a given food type (standard deviation)

Species	MVC[a]	n[b]	Detritus[c]	Plant periphyton	Plankton[d]	Ostracods	Choncostracans	Chironomidae	Macrocrust[e]	Au Aql[f]	Al Aql[g]	Fish[h]
Low water, submerged branches (21–29 Nov 1994/10–15 Oct 1993)												
Acaronia nassa	102	10	—	—	2.0 (2.7)	39.0 (16.7)	31.0 (13.4)	—	—	28.0 (16.4)	—	—
Astronotus ocellatus	45	12	—	—	—	—	—	—	2.1 (7.2)	49.6 (46.2)	2.5 (8.7)	45.8 (49.8)
Caquetaia sp. 1	99	10	—	—	60.0 (16.9)	32.5 (14.7)	7.5 (8.6)	—	—	—	—	—
Chaetobranchus flavescens	42	12	—	—	67.5 (16.6)	25.8 (17.3)	3.3 (4.9)	—	—	—	—	—
Cichla monoculus	76	12	—	—	—	—	—	—	1.3 (3.1)	2.5 (4.5)	—	97.1 (6.2)
Crenicichla sp. 2	17	6	—	—	—	—	—	—	—	1.7 (3.9)	—	75.0 (38.9)
Geophagus cf. proximus	257	6	30.0 (25.6)	3.8 (7.4)	—	39.4 (39.8)	—	5 (7.6)	—	25.0 (38.9)	—	—
Heros appendiculatus	41A	12	—	10.8 (26.1)	—	—	—	—	—	16.9 (25.2)	0.8 (2.9)	11.7 (19.1)
Mesonauta insignis	44	10	—	57.5 (34.4)	—	4.2 (7.9)	—	10.4 (14.5)	4.6 (12.0)	67.9 (31.1)	4.2 (10.0)	—
Pterophyllum scalare	73	2	—	90.0 (14.1)	—	—	—	10.0 (14.1)	—	22.9 (34.1)	—	—
Satanoperca sp. 1	100	6	—	10.8 (16.3)	—	48.3 (39.2)	4.2 (4.9)	—	—	2.5 (4.2)	2.5 (4.2)	48.3 (53.0)
Symphysodon aequifasciatus	72	12	—	90.8 (8.7)	—	—	—	5.0 (4.3)	—	4.2 (7.0)	—	—
High Water - Flooded forest (16–17 May 1994)												
Astronotus ocellatus	45	10	—	—	—	2.0 (4.2)	—	—	16.0 (32.4)	7.0 (14.9)	35.0 (42.5)	40 (51.6)
Chaetobranchus flavescens	42	4	—	—	—	10.5 (17.1)	—	—	6.5 (11.1)	8.5 (10.6)	43.5 (36.1)	31.0 (42.8)
Cichla monoculus	76	8	—	—	—	—	—	—	12.5 (35.4)	—	21.3 (36.4)	66.3 (44.4)
Heros appendiculatus	41A	5	—	—	—	—	—	—	13.5 (18.6)	5.0 (8.5)	62.5 (25.5)	19.0 (30.7)
Symphysodon aequifasciatus	72	1	—	73.3 (0)	—	—	—	10.0 (0)	—	—	16.7 (0)	—
Low Water - Floating meadow (17 Nov 1993)												
S. aequifasciatus (postlarval)	72	4	—	—	—	12.5 (15.0)	02.5 (18.9)	47.5 (29.9)	—	27.5 (35.9)	—	—
S. aequifasciatus (immature)	72	2	—	—	—	—	5.0 (9.0)	5.0 (9.0)	—	77.5 (10.6)	—	—

Note: Many stomachs contained fragments of wood; these were not included as a foodstuff.
[a] Accession number in Mamirauá Ichthyological Voucher Collection.
[b] Mean number of individuals. Only adult specimens with stomachs >30% full were included.
[c] Organic detritus and sediment.
[d] Mainly cladocera and copepods.
[e] Amphipods, isopods, shrimps.
[f] Autochthonous aquatic insects.
[g] Allochthonous aquatic insects.
[h] Parts of fish, including scales, or whole fish.

1994. Discus in captivity breed at about 18 months of age, when they attain approximately 120 mm of standard length. Discus in the Mamirauá Reserve appear to reach this standard length as they approach about 24 months. The breeding season for discus, as for most other cichlids of the várzea lakes, is toward the end of the low-water season (pers. obs. in low-water 1993; also see Mayland, 1994). Due to the need for discus to enter into a monogamous relationship with a partner, it is likely, however, that discus reach the peak of their fecundity not immediately upon reaching maturity but, rather, in their third year or after. Even if the discus that survived the recent overfishing were to begin reproducing immediately, the first generation they spawn will not begin to produce a significant year-class until at least 1999. This estimation contrasts to ornamental fish with rapid reproductive turnover, such as the tetras that mature and breed at one year of age.

The estimates for the recovery of the population through breeding, combined with the slow rates with which discus are presumed to recolonize vacant habitats, mean that it will be 5–15 years before discus stocks regain anything like their former levels. Although these estimates are speculative and may seem far from optimistic, we must remember that healthy stocks of discus have never returned to Lago Tefé since the "Tefé green" discus became commercially extinct in the mid- to late 1980s.

A probable development in the community-based management plan for the Mamirauá Reserve is to promote artisanal activities that provide seasonal income to local communities and that consequently remove the pressure on less sustainable activities such as timber extraction. One such potential enterprise is a small-scale ornamental fish–catching enterprise. In light of the monopoly held by Manaus companies, it would be very difficult to organize an export base. A more realistic strategy would be to form a cooperative of local communities to capture and store ornamental fish for sale to visiting boats from Manaus companies. This would need to be coordinated by the nongovernmental organization (NGO) managing the reserve. Some of the main points of a forthcoming management plan are listed below:

1. Although *Symphysodon aequifasciatus* is the commercially most valuable species, it is also the most sensitive. A total ban on commercial fishing of the species until the year 2000 is proposed to allow stocks to recover. During this time it will be possible to assess whether sustainable catches will be possible and at what level of intensity. In the meantime, the recuperation of discus may be encouraged by felling some trees along lake edges to provide new submerged branch habitats.

2. There are, in the Mamirauá Reserve, many other fish of potential value as ornamental fish which will not require complicated stock monitoring and which could be exploited at any time in the future. To diversify the local fishery would be expedient in the long term, as all evidence points to the ornamental fish trade requiring more variety in exports from the Amazon. A provisional list of species can be found in Crampton (1993).

3. The communities will need to be trained in the techniques of catching, transporting, and caring for ornamental fish. A guide book for use by local people is in preparation.

4. Due to the interaction of ornamental fish and forest resources, forests around

lakes used for ornamental fish catching will need to be preserved, at least in the form of a strip.

The socioeconomic management plan for the Mamirauá Reserve renders it ideal for an experimental fish-catching cooperative designed to preserve a lucrative but sustainable ornamental fishery. If successful, it could become a model for cooperatives in other rural areas of the Amazon basin.

Implications for the Conservation of *Symphysodon* Species in the Amazon Basin

The combination of vulnerability, accessibility, and high commercial value of the more colorful varieties of *Symphysodon* species makes them vulnerable to localized depletion. Both species of *Symphysodon* consist of color forms that are likely to reflect an underlying genetic diversity. In the case of *S. aequifasciatus*, the more colorful and also commercially valuable varieties are comparatively rare. Where commercially valuable varieties such as the blue discus of the Mamirauá Reserve or the green discus of Lago Tefé are encountered, companies are prepared to spare no expense in exploiting the populations to the point of commercial extinction. This is emphasized by the fact that the Jarauá area continued to be visited even beyond the point of a clear crash in the stocks.

Does the depletion of local populations affect the species? The fact that the ranges of the different color forms usually do not overlap and that different forms are known to coexist apparently without hybridizing would suggest that there is some degree of assortative mating between color forms. This is supported by the fact that captive discus tend to form pairs with partners of similar coloration (Silva & Kotlar, 1980). Assortative breeding is, furthermore, a common phenomenon in many cichlids (e.g., McElroy et al., 1991). Nonetheless, intermediates between two wild forms may occur naturally, as may have happened with discus intermediate between green and blue forms in the Rio Coarí (Mayland, 1994). The long-term effect of assortative mating between color forms would be to restrict genetic exchange between some populations of a species and hence to maintain the genetic diversity of the species. Speciation based on the divergence of color morphs has been proposed for many cichlid groups such as the haplochromine cichlids of the African Great Lakes (Fryer & Iles, 1972). The removal or depletion of a particular color form of *Symphysodon aequifasciatus* may hence have deleterious effects that extend beyond the direct population losses: genetic diversity of the species may be lost and the continuing evolution of the species interrupted. These last possibilities are made all the more likely by the fact that the commercially most valuable varieties are the rarest.

Is extinction of some color forms possible? The trade often argues that the proportions of *Symphysodon* spp. being captive-bred outside the Amazon is increasing, due mainly to the efforts of large companies in Florida and Asia. While this is certainly true, it should be stressed that there is still a large demand for captive-caught fish, which are generally hardier. Furthermore, because repeated cross-breeding and inbreeding eventually leads to inferior progeny (see Schulze, 1988), there is a constant demand for new wild stock to enrich cross-bred strains and from which new varieties are made. For these purposes the ornamental fish trade is especially interested in the colorful varieties. The famous "turquoise discus," for example, is an interbred and hybridized strain derived from wild green and/or blue discus.

It is also tempting to imagine that the vast size of the Amazon basin offers sufficient protection. This is not necessarily the case, however, since most discus populations are found in the floodplain areas of rivers which are always accessible even if remote. The commercially more valuable strains of *Symphysodon aequifasciatus*, such as the green and blue, and also *S. discus* itself are, moreover, within only a few days' range of Manaus. Perhaps the most powerful brake on an accelerated depletion of *Symphysodon* stocks is the way in which the ornamental fish trade operates. The trade in Manaus is monopolized by a dozen or so companies that have resisted competition by making it difficult for new businesses to set up. The effect has been to prevent an uncontrolled boom in supply.

It is unlikely that either *S. aequifasciatus* or *S. discus* will ever become extinct, because the less valuable strains, such as the "brown discus," are less threatened and because commercial extinction does not rule out an eventual recovery of stocks. Nevertheless, it remains to be seen whether the depletion of color forms may advance to the point of compromising the genetic integrity and natural evolution of *Symphysodon*. The protection of discus populations in conservation areas such as the Mamirathis Reserve may become increasingly necessary in the future.

Future Work

Many unresolved issues were raised during this study. A full-scale study of the ecology, breeding biology, and movement patterns of a large intact discus colony would be immensely valuable, as would an investigation of the genetic basis of discus color varieties.

The discus population in the Jarauá area, and possibly in Lago Tefé, will continue to be monitored in the coming years both before and after the limited commercial exploitation by local communities is allowed to recommence. The species may prove to be a useful indicator of the success of the Mamiraná Reserve as well as a potential source of revenue for the local communities.

Acknowledgments

This work was conducted as a satellite project to a doctoral thesis on the ecology and diversity of gymnotiform electric fish in the Mamiraná Reserve. I thank Márcio Ayres, Director of Projeto Mamiraná, for inviting me to write a management plan for ornamental fish in the reserve. I received some extra funding for field expenses via Projeto Mamiraná from the ODA for this work. Peter Henderson provided constant support and advice. Without my excellent field assistants Jonas de Oliveira, Tito Martins, and Waldecy, the fieldwork would have been impossible. I also thank Helder Queiroz, Marise Reis, and Andrea Pires for advice on some aspects of this work. The following people were especially helpful during the interviews with residents of local communities: Esmeraldo Françamora (alias Pelé), Roberto de Souza Coelho, Arnaldo, Marino,

Antonio, João de Almeida, João Busca, Raimundo, and Dão. Sr. Segeru in Tefé was an important source of information about the ornamental fish trade in the Tefé region. Peter Henderson and Charles Paxton commented helpfully on the manuscript.

Literature Cited

Andrews, C. 1990. The ornamental fish trade and fish conservation. Journal of Fish Biology 37(A): 53–59.

Axelrod, H. R. 1978. All about discus. TFH Publications, Neptune City, NJ.

————. 1988. Editorially speaking . . . conservation? Tropical Fish Hobbyist 36: 6.

Ayres, J. M. 1993. As matas de várzea do Mamirauá: Médio Rio Solimões. MCT-CNPq, Brasília.

Banister, K. 1989. Shoals of fish. Naturopa 62: 20–21.

Bayley, P. N. & M. Petrere. 1990. Amazon fisheries: Assessment methods, current status, and management options. Pages 385–398 *in* D. Dodge, ed., Special publications of the Canadian Journal of Fisheries and Aquatic Sciences.

Bleher, H. 1986. Solving the mystery of the Tefé green discus. Tropical Fish Hobbyist 12: 10–49.

Burgess, W. E. 1981. Studies on the family Cichlidae 10. New information on the species of the genus *Symphysodon* with the description of a new subspecies of *S. discus* Heckel. Tropical Fish Hobbyist 29(7): 32–42.

Chao, N. L. 1992. Ornamental fishes and fisheries of the Rio Negro. Tropical Fish Hobbyist 40(12): 84–102.

————. 1993. Conservation of Rio Negro ornamental fishes. Tropical Fish Hobbyist 41(5): 99–114.

Crampton, W. G. R. 1993. Preliminary report on the conservation of acará-disco and other ornamental fish from the Estação Ecológica Mamirauá. Unpublished report, Projeto Mamirauá Biannual Report (April–October 1993). Sociedade Civil Mamirauá, Belém.

Fryer, G. & T. D. Iles. 1972. The cichlid fishes of the Great Lakes of Africa. Oliver & Boyd, Edinburgh.

Goulding, M. 1980. The fishes and the forest. University of California Press, Berkeley.

Junk, J. 1984. Ecology, fisheries and fish culture in Amazonia. Pages 443–476 *in* H. Sioli, ed., The Amazon: Limnology and landscape ecology of a mighty tropical river and its basin. Junk, Dordrecht.

Kullander, S. O. 1986. The cichlid fishes of the Amazon River drainage of Peru. Swedish Museum of Natural History, Stockholm.

Leite, R. G. & J. A. Zuanon. 1993. Peixes ornamentais—Aspetos de comercialização, ecologia, legislação e propostas de ações para um melhor aproveitamento. Pages 327–332 *in* A. L. Val, R. Figliuolo & E. Feldberg, eds., Bases científicas para estratégias de preservação e desenvolvimento da Amazônia: Fatos e perspectivas. INPA, Manaus.

Mayland, H. J. 1993. Diskus—Wettbewerb der Diskuszuchter auf der "Aquarama 1993" in Singapur. Aquarium (Bornheim) 294: 15–17.

————. 1994. Adventures with discus. TFH Publications, Neptune City, NJ.

McElroy, D. M., I. Kornfield & J. Everett. 1991. Coloration in African cichlids—Diversity and constraints in Lake Malawi endemics. Netherlands Journal of Zoology 41(4): 250–268.

McGrath, D. G. 1990. Preliminary investigation of the trade in extractive products in the Brazilian Amazon. Final report submitted to the World Wildlife Fund–US. WWF, Washington, DC.

Schultz, L. P. 1960. A review of the pompadour or discus fishes, genus *Symphysodon* of South America. Tropical Fish Hobbyist 8(10): 5–17.

Schulze, E. 1988. Discus fish: The king of all aquarium fish. Discus, Bangkok.

Silva, T. & B. Kotlar. 1980. Discus. TFH Publications, Neptune City, NJ.

Wood, E. 1985. Exploitation of coral reef fishes for the aquarium trade. Marine Conservation Society, Herefordshire, UK.

On Structural Complexity and Fish Diversity in an Amazonian Floodplain

P. A. Henderson and B. A. Robertson

Introduction

There is considerable agreement in the ecological literature that an understanding of the factors influencing local species diversity will come only through the consideration of systems dynamics over a wide range of spatial and temporal scales. Any area, irrespective of size, is patchy, and those species which come to occupy a patch are determined by factors such as availability, chance, and pre-adaptation. A fish may move into a lake to feed only if there is an available population in an interconnected lake or channel. Chance clearly plays a role in the arrival of migrants, particularly as many fish live in small populations subject to redistribution due to events such as flood, drought, and river migration. A species adapted to living in the confined spaces of aquatic plants may find itself pre-adapted to life within leaf litter if such a habitat becomes locally available or if it is flushed from the headwater to the floodplain. In this article we describe the major processes active within the floodplain of the Rios Solimões and Japurá (Fig. 1) which we believe are important for the maintenance of fish diversity. This floodplain is now designated a reserve, Estação Ecológica Mamirauá (herein called the "Mamirauá Reserve"). The studies summarized here are part of a multidisciplinary study to collect information for a reserve management plan.

There are many measures and kinds of ecological diversity. It is conventional to think in terms of species diversity both within a habitat (α diversity) and between habitats (β diversity) as ecological diversity (Magurran, 1988). However, an important aspect of floodplain diversity is also linked to the instability of the habitat and the great changes this can cause in fish density, distribution, and feeding. In this brief review we consider ecological diversity in the broadest sense, including feeding niche and population size. Species diversity is certainly an important feature of this region. At present, about 286 species of fish have been identified within the floodplain and its immediate vicinity, an area of approximately 2×10^3 km². The significance of this number can be appreciated from the fact that the same total for the Amazon basin, with an area of 7×10^6 km², is about 2800 species (for a recent review of fish species numbers, see Val & Almeida-Val, 1995). Probable upper limits for species diversity within the Mamirauá Reserve and the Amazon are in the region of 300 and 3000 species, respectively. In almost any lake within the Solimões-Japurá floodplain are to be found more fish species than occur in all of the freshwaters of Northern Europe. Even more impressive than

45

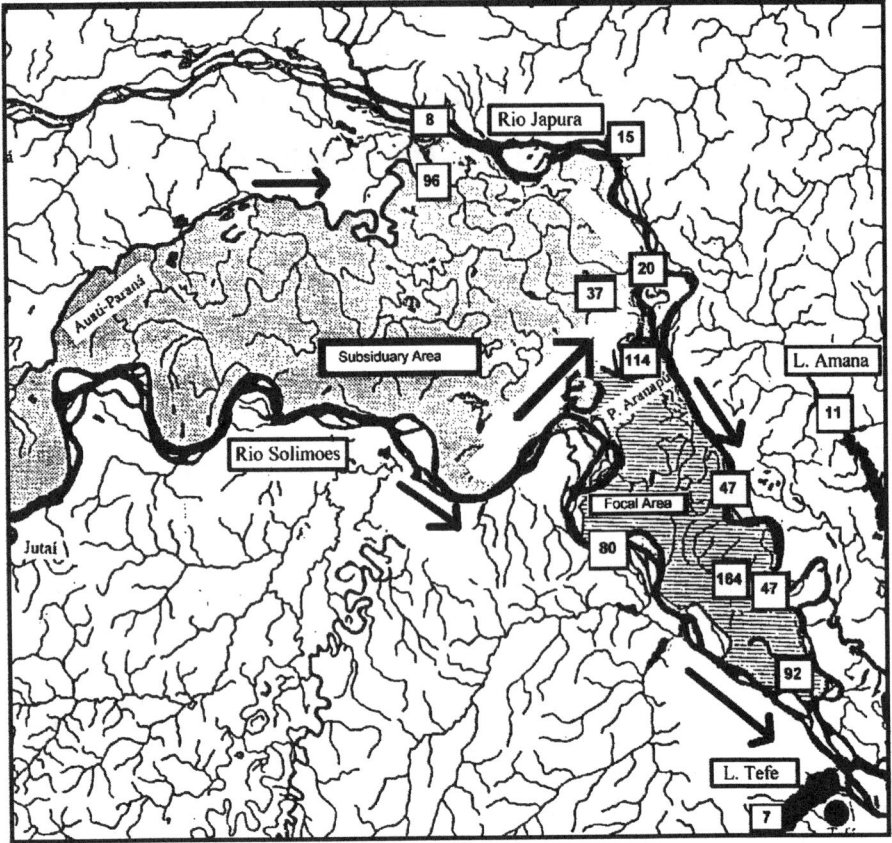

Figure 1. Sketch map of the reserve showing principal rivers and lakes. The arrows show the direction of flow of the main channels. Boxed numbers indicate water conductivity, in μS cm^{-1} at 20°C.

the number of species is the diversity of form and lifestyle shown by floodplain fish. Fish diversity is not based on the rapid speciation of a few ancestral forms such as is found, for example, in the remarkably species-rich cichlid communities of the African Great Lakes (see Ribbink, 1994, for a recent review of cichlid fish speciation).

A Summary of the Fish Fauna

A number of sampling methods have been used within the Mamirauá Reserve to obtain adequate coverage of the fish fauna. The fish within floating vegetation mats were quantitatively sampled using a seine net. Within the forest, qualitative sampling was undertaken using fyke and gill nets. The open water of lakes was sampled with a range of gill nets. Hand nets were also used to sample small and juvenile fish particularly within floating vegetation and leaf litter. Traditional fishing methods including harpoon, bow and arrow, throwing nets, and rod and line were also used when appropriate. Table I lists the families and number of species of fish found within the reserve. The dominant groups in terms of species number are the Curimatidae, Serrasalmidae, and Characidae; the electric fish groups Sternoppygidae, Hypopomidae, Apteronotidae; the catfish groups Doradidae, Auchenipteridae, Pimelodidae, and Loricaridae; and the Cichlidae. The importance of electric fish within the floodplain has been demonstrated by Crampton (1996). However, each of two species-poor families, the Erthyrinidae and Electrophoridae, includes a single dominant species—*Hoplias malabaricus* (Bloch) and *Electrophorus electricus* (L.), respectively—that is widely distributed and abundant in terms of both number and biomass. This species list is typical of Amazonian floodplain habitats (Lowe-McConnell, 1987).

One of the most important habitats within the floodplain, in terms of both species number and standing crop, is floating meadow. A general account of the density and distribution of fish during the low-water season is given by Henderson and Crampton (1995). The availability of shelter and food amongst the roots and rhizomes gives an ideal habitat for small fish species including the juveniles of important commercial

Table I. The number of species subdivided into families recorded from the Rios Solimões–Japurá floodplain

Family	Number of species	Family	Number of species
Potamotrygonidae	2	Electrophoridae	I
Clupeidae	I	Doradidae	16
Engraulidae	2	Auchenipteridae	10
Arapaimidae	I	Aspredinidae	2
Osteoglossidae	I	Pimelodidae	25
Erythrinidae	3	Ageneiosidae	4
Ctenolucidae	I	Cetopsidae	I
Crenuchidae	I	Hypophthalmidae	3
Characidiidae	6	Trichomycteridae	3
Anostomidae	9	Callichthyidae	4
Hemiodidae	3	Loricariidae	17
Lebiasinidae	9	Belonidae	I
Curimatidae	16	Cyprinodontidae	3
Gasteropelecidae	2	Synbranchidae	3
Serrasalmidae	12	Sciaenidae	2
Characidae	48	Cichlidae	21
Sternopygidae	13	Gobiidae	I
Rhamphichthyidae	3	Soleidae	I
Hypopomidae	13	Tetraodontidae	I
Apteronotidae	13	Lepidosirenidae	I
Gymnotidae	7	Total	286

species (Saint-Paul & Bayley, 1979; Goulding & Carvalho, 1982). Studies undertaken within this habitat serve to show some general features of floodplain fish communities. During quantitative seine net sampling, 79 species have so far been captured (see Table II). The species accumulation curve shows an almost linear increase in species number up to a sampled area of 400 m^2 (Fig. 2). Given the lack of asymptotic behaviour, there can be no doubt that the total species number for this habitat will exceed 100. This view is supported because other sampling methods, such as hand nets, have already shown further species to be present. Thus this one habitat holds more than one-third of the total species complement for the region (presently 286 species including forest steams and ria lakes) and about one-half of all the species found within the várzea. However, no single site ever holds such species diversity. Floating meadows with the most species occur along the edge of flowing channels such as the Apará-Paraná, which carries water from the Rio Japurá into Lago Mamirauá. In this channel the maximum species number recorded at one station within an area of 10 m^2 was 23. Probably the most species-rich localities in the várzea have complements of about 25 species per 10 m^2. Rank-abundance of fish within the floating meadow is plotted in Figure 3. This shows that a few species contribute >90% of the biomass. These dominant species— *Hoplias malabaricus*, *Aequidens* cf. *teramerus*, *Hypopomus* spp., and *Parauchenipterus* sp.— can be classified as generalists, as they were found at almost all the stations sampled. As shown in Figure 4, this distribution does not conform to the log normal that is often found to fit large, mature, natural communities (Magurran, 1988). It is closer to a log series that is considered most applicable when only a few factors dominate the ecology of a community. This would seem reasonable, as flow and oxygen concentration have been found to be the dominant variables determining species presence. The majority of species that have been caught in floating meadow are infrequently caught and are never abundant. It is probable that these species use the habitat for only one period within the life cycle or only when other, more favoured areas such as flooded forest are unavailable. While the literature has tended to emphasise the rich range of specialised adaptations shown by Amazonian fish, observations on the floating meadow fauna suggest that many species are facultative in their use and can be viewed as adaptable and unspecialised, particularly with respect to feeding niche.

The general picture is of a high number of species, each patchily distributed, with the patch size for different species ranging in area from that of a large lake to individual fallen trees or a small clump of floating plant material. With this scaling of patch size, we can no longer speak in terms of α (within habitat or patch) and β (between habitat or patch) diversity; the two become indistinguishable because habitat boundaries differ between species. This patchiness of distribution directs our attention to the physical structure of the habitat: Perhaps species diversity can be linked to and explained by structural complexity. However, we cannot expect the physical processes that mould the floodplain structure to act alone to create the diversity of habitat upon which the fish community depends. The biota is also important—for example, the structures used by fish are often living plants. However, insofar as the presence of the plants is determined by the processes of land formation and erosion, the physical processes must create the foundation upon which the diversity is supported. Similar conclusions to those presented above, concerning species abundance patterns and distribution, also hold for

Table II. Fish species and biomass within floating meadow of the Rio Solimões–Japura floodplain. Mass is wet weight and the total area sampled between April 1994 and March 1995 was 395 m²

Species	Total biomass (g)	Mean density (g/m²)	Species	Total biomass (g)	Mean density (g/m²)
Osteoglossum bicirrhosum	6.00	0.015	Hypopomus sp. 6	58.60	0.148
Hoplias malabaricus	4078.60	10.299	Hypopomus sp. 7	1058.45	2.673
Rhtiodus microlepsis	114.00	0.288	Steatogenys elegans	601.90	1.520
Anostomus taeniatus	24.50	0.062	Adontosternarchus sp. 1	131.50	0.332
Anostomus trimaculatus	28.00	0.071	Adontosternarchus sp. 2	197.00	0.497
Anostomus indet.	413.00	1.043	Apteronotus sp. 1	351.85	0.889
Leporinus fasciatus	42.10	0.106	Apteronotus sp. 2	146.25	0.369
Schizodon fasciatum	856.00	2.162	Gymnotus angularis	5.90	0.015
Hemiodopsis sp.	0.45	0.001	Gymnotus carapo	334.50	0.845
Copella nattereri	1.20	0.003	Pseudodoras niger	1040.00	2.626
Pyrrhulina indet.	1.70	0.004	Amblydorus hancocki	14.00	0.035
Nannobrycon eques	0.80	0.002	Anadorus sp. 1	361.80	0.914
Branquinha	8.00	0.020	Anadorus grypus	24.00	0.061
Semaprochilodus taeniurus	2.00	0.005	Doradidae indet.	344.80	0.871
Metynnis sp.	6.20	0.016	Parauchenipterus sp. 1	1688.10	4.263
Mylossoma aureum	236.15	0.596	Pimelodella sp.	0.40	0.001
Mylossoma duriventre	100.80	0.255	Pimelodus cf. blochii	34.90	0.088
Serrasalmus nattereri	635.70	1.605	Pimelodidae indet.	12.00	0.030
Serrasalmus elongatus	5.95	0.015	Hoplosternum thoracatum	2.50	0.006
Serrsalmus indet.	126.50	0.319	Ageneiosus sp.	746.00	1.884
Tetragonopterus chalceus	2.00	0.005	Hypophthalmidae indet.	40.00	0.101
Tetragonopterus indet.	141.10	0.356	Hypopthalmus sp. 3	280.00	0.707
Triportheus albus	22.00	0.056	Hypostomus sp. 2	16.00	0.040
Triportheus angulatus	46.00	0.116	Pterygoplichthys sp.	166.00	0.419
Chalceus erythrurus	1.00	0.003	Sturisoma sp. 1	15.50	0.039
Roeboides sp.	7.50	0.019	Loricariidae indet.	123.25	0.311
Charax sp. 1	34.50	0.087	Synbranchus marmoratus	735.05	1.856
Gymnocorymbus sp.	2.00	0.005	Aequidens cf. tetramerus	2190.80	5.532
Rhabdolichops sp. 1	8.00	0.020	Apistogramma spp.	28.50	0.072
Distocyclus sp.	185.00	0.476	Astronotus ocellatus	678.00	1.712
Eigenmannia trilineata	4.50	0.011	Crenicichla spp.	236.80	0.598
Eigenmannia virescens	21.90	0.055	Cichla monoculus	148.70	0.376
Eigenmannia sp.	233.10	0.589	Mesonauta festivium	375.95	0.949
Sternarchogiton sp. 1	16.30	0.041	Heros appendiculatus	164.00	0.414
Sternopygus macrurus	956.20	2.415	Satanoperca sp. 3	48.40	0.122
Rhamphichthys sp. 1	246.40	0.622	Hypselecara temporalis	758.00	1.914
Hypopomus sp. 1	2.50	0.006	Cichlidae indet.	8.50	0.021
Hypopomus sp. 2	12.00	0.030	Columesus asellus	2.40	0.006
Hypopomus brevirostris	670.54	1.693	Lepidosiren paradoxa	10.00	0.025
Hypopomus sp. 4	179.85	0.454			

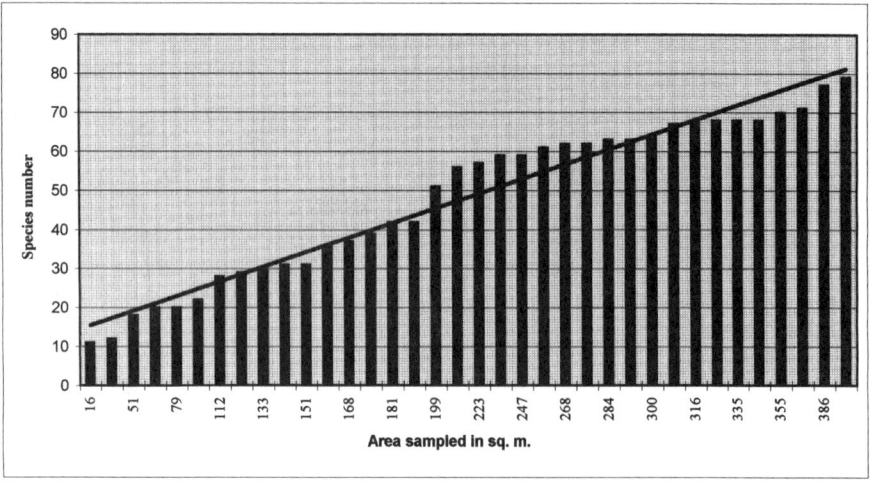

Figure 2. The increase in fish species captured from floating meadow with sampling effort. Samples were from the Rio Solimões–Japurá floodplain. The straight line was fitted by regression.

the flooded forest and open water communities, although they have been less intensively studied.

Factors Moulding Species Diversity

Physical Structure

The landscape processes creating the general structure of the Solimões-Japurá floodplain (Fig. 1) have been reviewed and described by Salo (1990). Below, in order of magnitude, are the landscape structures of this floodplain that are relevant to fish. The processes acting within the reserve create water bodies that differ in size, longevity, depth, flow, temperature profiles, and physicochemical properties such as oxygen concentration and dissolved solid concentrations, thus creating the complex structure within which high fish diversity can be retained.

Megaform processes create structures over time scales of $>10^4$ years. The most important of these events for the Mamirauá Reserve have been eustatic changes. During periods of lower sea water level, channel incision occurred. Over the last 4000 years the rise in sea water level has widened the floodplain and created blocked valley (blackwater) lakes along the border of the reserve. Lago Tefé is the largest of these lakes, whose habitat is characterised by low suspended sediment and conductivity (8–14 μS cm^{-1} at 20°C). They also support typical igapó inundation forest and have little floating meadow. However, they are considerably less acid than typical back waters of the Rio Negro system (pH ca. 6.0 vs. 4.0). These ria lakes are an important fish habitat, holding many species never or infrequently encountered within the várzea lakes of the floodplain. For example, during May 1994, sampling of floating meadow communities in

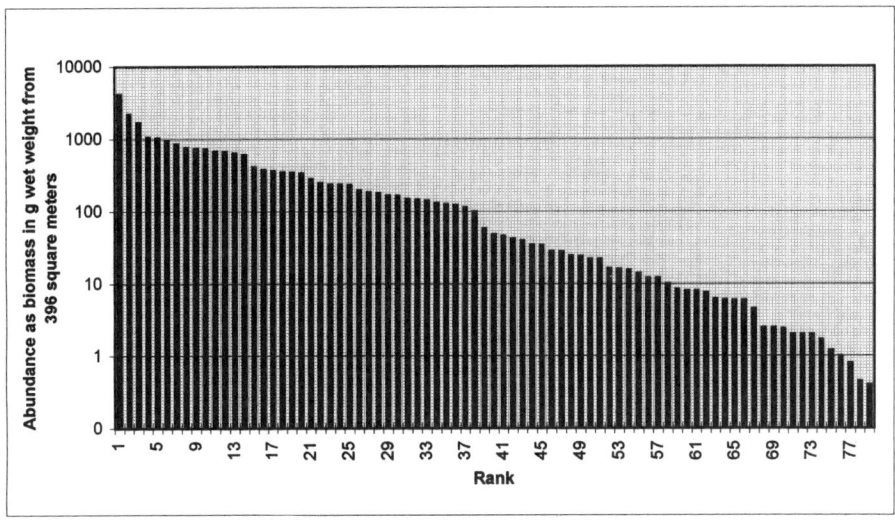

Figure 3. Rank abundance plot for fish from floating meadow in the Rio Solimões–Japurá floodplain.

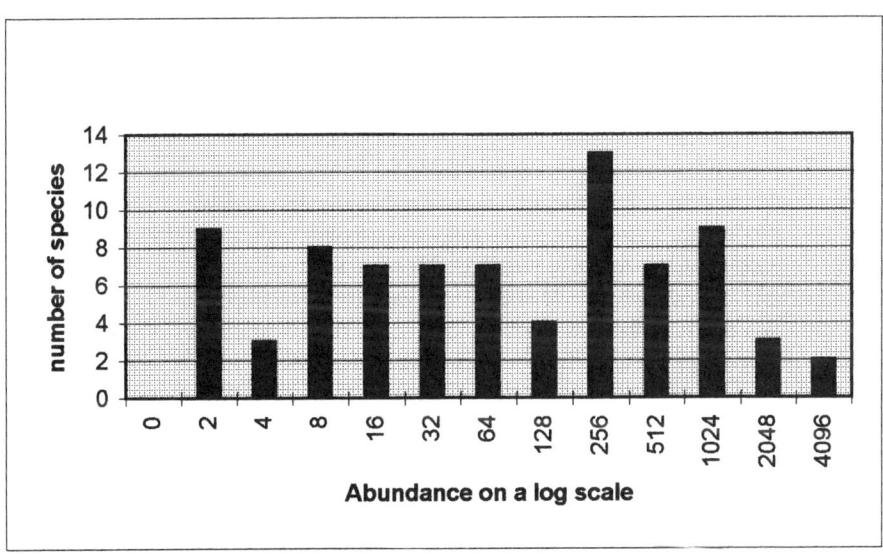

Figure 4. Distribution of species abundance (measured as biomass) within the floating meadow community of the Rio Solimões –Japurá floodplain.

blackwater and whitewater lakes revealed the presence of 20 and 37 fish species, re-spectively (for more detailed comparison of white- and blackwaters, see Henderson & Crampton, 1997). However, six of the blackwater species were not found in the white-water lakes. It is not known if terraces created by eustatic change form large-scale structures influencing fish distribution and speciation.

Macroform processes create structures over periods of 10^2–10^4 years. Main channel migration over these time scales results in the general pattern of lakes and channels within the Mamirauá Reserve. Channel migration is rapid: Mertes (1985) estimated rates of 400 m/year in the border region between Peru and Brazil. A number of different water bodies are formed. Meander migration results in a scoll-swale topography within meander loops. The low-lying region between the meander bar and the inside bank initially forms a crescent lake which later becomes a back swamp because of overbank sedimentation. There are many crescent lakes within the reserve which offer series of successional habitats for fish. Larger oxbow lakes are formed by the cutting of meanders. Within the floodplain, cut-off lakes vary considerably in age and size. The largest ones result from the abandonment of a main river channel and may include a number of meanders. Within the reserve it is likely that Lago Mamirauá is an abandoned main river channel. This extensive lake system has a maximum depth of about 40 m, which is similar to that of the adjacent main channel of the Rio Solimões. The patchy, often laminar, structure created by macroform processes may be the most important element in the maintenance of floodplain fish diversity.

Mesoform processes occur over time scales of 1 year to 10^2 years. The most im-portant of these processes are shear stress and sediment deposition. Many of the char-acteristic channel structures such as undercut banks, bars, and levees are the result of these processes. Erosion and sedimentation create, change, and destroy fish habitats. For example, bank erosion creates fallen tree habitats which are used by important com-mercial fish such as the tucunaré (*Cichla monoculus*) and tambaqui (*Colossoma bidens*). Sedimentation changes flow patterns, and hence well-oxygenated flowing channels are transformed into languid, oxygen-poor waters. This, in turn, causes radical changes in the fish community.

Microform processes occur at time scales of <1 year. By far the most important is the seasonal change in river flow which creates the annual inundation of the forest (for seasonal variation in water depth, see Fig. 6). This cycle creates flooded forest and floating meadow habitat, both of which, because of their spatial complexity and scale, can support many fish species. This cycle also greatly effects fish density and standing crop, at low water the fish are concentrated in the permanent lakes. For example, a floating meadow site within the reserve, sampled at low water, had an average fish standing crop of 195.4 g/m^2, whereas at high water the value was only 17.3 g/m^2. Order-of-magnitude changes in fish density are probably typical. Seasonal flooding also changes the physicochemical environment of the lakes, which, in turn, determines phyto- and zooplankton production. Zooplankton availability influences the reproduc-tive timing of many fish.

From the above it can be appreciated that the floodplain holds water bodies ranging in size from a few square centimeters to abandoned river channels 40 km long and 200 m wide. These may range in age from 1 year to >10^4 years. It is this physical structure generated by flow and sedimentation which is the template upon which fish biodiversity is formed. However, it is necessary to consider habitat physicochemical variability and vegetational structure to obtain an understanding of the forces determining diversity.

Physicochemical Variability

It is inevitable that the Rios Japurá and Solimões, because they collect waters from different geographical regions, will differ in chemical content, flow, and suspended sediments. Above the Auati-Paraná, which is the first channel via which the Rio Japurá receives Rio Solimões water, the Rio Japurá is generally described as a blackwater river. Thus the floodplain has the interesting large-scale feature of being bounded by both a blackwater and a whitewater river. The effect of this on water conductivity (Fig. 1) is that the Rio Japurá side of the reserve receives water of considerably higher conductivity than the Solimões side. Conductivity gives a general measure of dissolved nutrient availability and thus the ability of the water to support plant growth. Lakes in the reserve, then, differ in the amount of nutrients and sediment they receive from the rivers, and this, in turn, affects their productivity. Main river inputs are not the only factor determining conductivity. Some lakes within the reserve have during the low-water period conductivities >180 μS cm^{-1} at 20°C, compared with <80 μS cm^{-1} for the Rio Solimões. These exceptionally high conductivities are probably linked to decomposition of floating meadow.

Other important determinants of fish distribution—flow, turbidity, temperature, oxygen concentration—also vary temporally and spatially over all scales. More detail of this variation is given in Henderson et al. (in press). Possibly one of the most important features are small-scale frontal systems. These occur at the intersection of lake and channel water and result in quite different aquatic conditions in localities only 20 m apart. Figure 5 shows the situation at the intersection of the Apara-Paraná and Lago Mamirauá. The intersection of the two waters is often clearly visible, as the lake water has lower sediment and is thermally stratified so that the surface waters do not mix. Vertical stability within the lake is often sufficient to cause water deeper than 1 m to be almost anoxic. In contrast, water flowing along the Apara-Paraná is well mixed, with oxygen levels favourable to fish available at all depths. Sites on either side of this front differ greatly in fish community structure (Table III), due probably to the fact that many of the species found in the Apara-Paraná avoid oxygen-poor waters. Further, because of the concentration of detritus and zooplankton in the frontal system, such zones are much favoured by juvenile fish. Open water fish densities and fish predators such as dolphins also tend to be more abundant at these intersections (Fig. 5).

It was stated above that one of the most important microform processes determining fish distribution and diversity was the seasonal variation in water depth linked to changes in river flow. This pattern of inundation shows considerable between-year variation. Figure 6 shows water depth at Tefé for the years 1990–1995. Marked on the graphs is the depth at which the forest surrounding Lago Mamirauá becomes completely inundated. As can be seen, the timing and duration of inundation vary considerable between years. Forest-dwelling fish must adapt to this variation in habitat availability. No research has yet focused on the influence of this variability. It can be expected that year-class

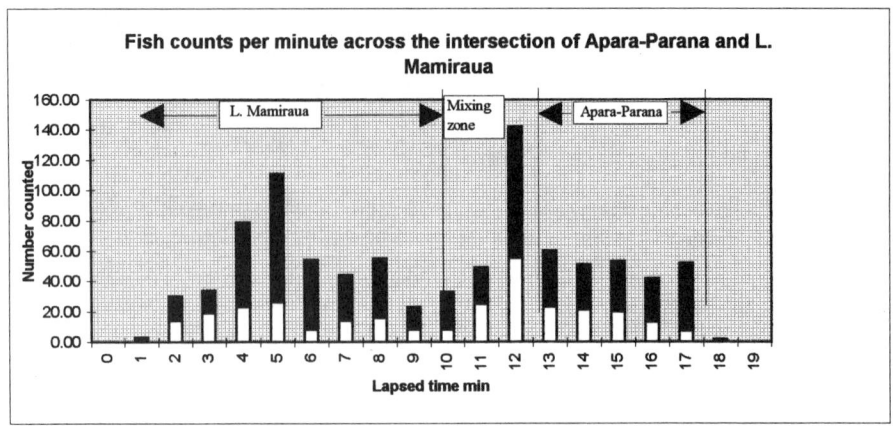

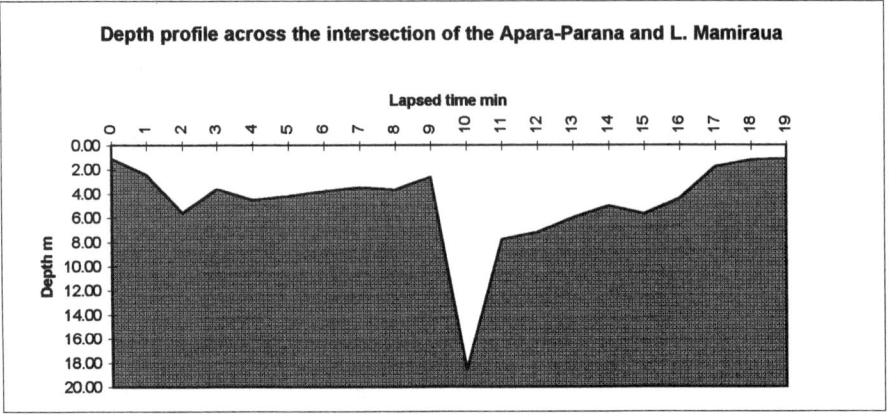

Figure 5. Fish density and water depth across the intersection of the Apara–Paraná and Lago Mamirauá, October 1994.

strength will show between-year variation due to differences in the floodplain carrying capacity between years.

Plant Communities

Higher plants within the floodplain offer two principal habitats to fish, namely, floating meadow and forest. Flooded forests are a characteristic feature of all Amazonian flood-plains. Whitewater inundation forest, called *várzea*, is typically richer in tree species than *igapó*, the blackwater equivalent. It seems plausible that the spatial complexity of the floodplain, in which lakes and channels of differing age are separated by bands of forest, could be a factor in the generation and maintenance of diversity. It might seem that, during the high-water season, fish could move unhindered between different lakes via

Table III. Community structure within the floating meadow at two localities only 200 m apart. For the 20 species capture the two localities had only 6 in common. Samples were take during high-water season of April 1994.

Species	Number of individuals captured	
	Lago Mamirauá: still water	Apara Paraná: flowing water
Hoplias malabaricus	4	9
Anostomus trimaculatus	1	
Hemiodopsis		2
Mylossoma aureum	1	
Serrasalmus nattereri	3	
Triportheus angulatus		1
Hypopomus sp. 1	2	
Hypopomus sp. 2		3
Hypopomus sp. 3	2	2
Hypopomus sp. 6		2
Gymnotus angularis	1	3
Gymnotus carapo	1	2
Amblydorus hancocki		1
Doradidae indet.		1
Pimelodella sp.		1
Hoplosternum thoracatum	1	
Synbranchus marmoratus	10	14
Aequidens cf. tetramerus	5	4
Apistogramma spp.		12
Lepidosiren paradoxa	1	
Number of species	12	14
Total area sampled	28	51

the forest. However, many of the infaunal species of the meadows such as electric fish and small cichlids are reluctant to enter open water or flooded forest, perhaps because of the threat of predation. This results in lakes that, though geographically in close proximity, are effectively isolated for infaunal species. Such isolation would tend to maintain diversity, as it reduces the chances of a dominant form competitively excluding others. It is interesting to note that fish groups that hold members that move easily through forest are often species poor—e.g., Synbranchidae, Arapaimidae, Osteoglossidae, Electrophoridae, and Erythrinidae. The structure of the forest changes with age, degree of inundation, and nutrient and sediment load of the floodwater. Forest type has an influence on the distribution of fish insofar as it determines the availability of shelter and food.

The floating meadow habitat of várzea lakes is dominated by two grasses, *Paspalum repens* Berg and *Echinochloeta polystachya* Hitchcock; during the low-water season these

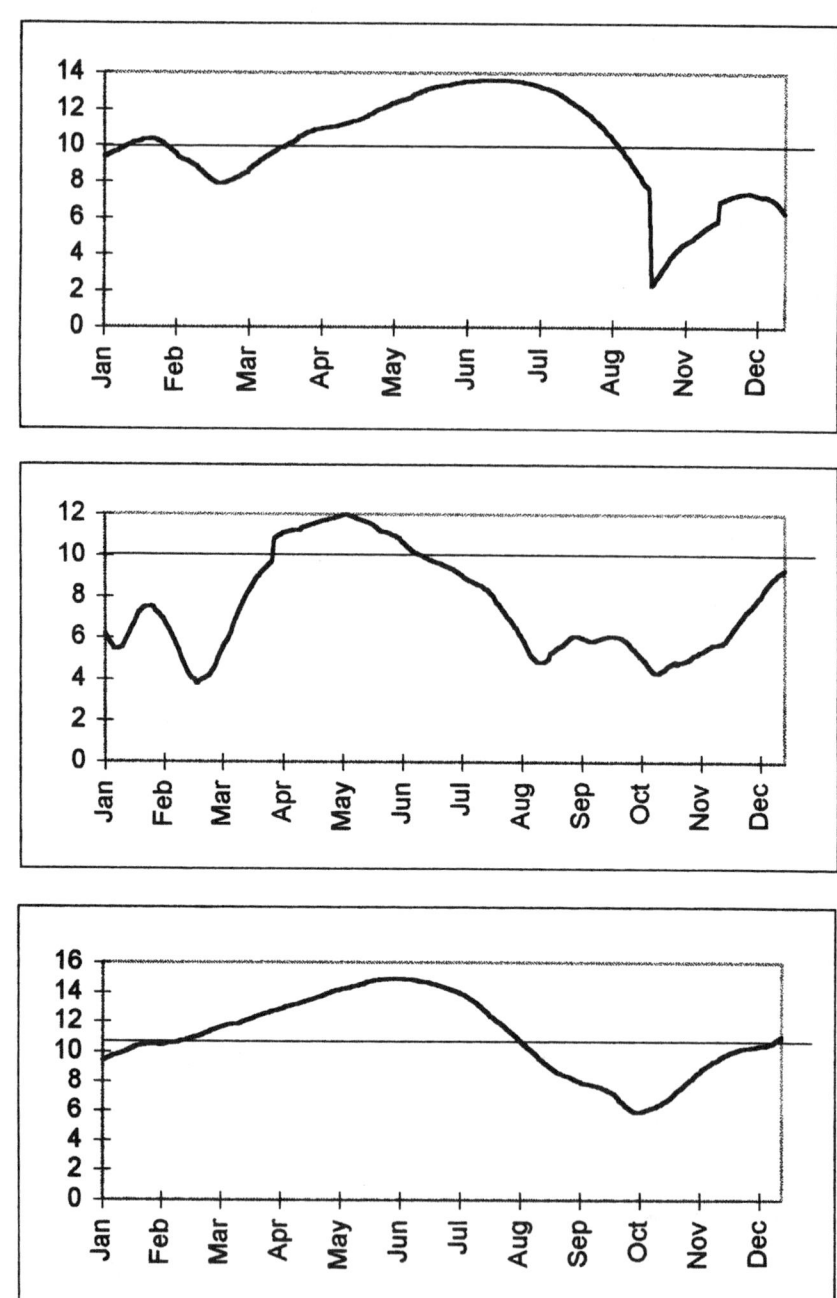

Figure 6. Water depth recorded at Tefé, 1991–1993. The 10 m line is the depth at which large areas of várzea forest begin to flood.

are rooted on the exposed muddy banks. The grass grows with the rising water but may eventually break free, forming a floating mat with a submerged mass of rhizomes and roots. As the water falls, this mat collapses and rots, releasing nutrients that are used by the next cycle of growth. Other aquatic plants include *Oryza* sp., *Salvinia* sp., *Azolla* sp., *Pistia stratiotes*, *Utricularia* sp., and *Eichhornia crassipes*. Junk (1970) gives a detailed account of the ecology and productivity of the várzea floating meadows. In black waters the rafts are dominated by *Cyperus* and *Oryza* with fewer plant species; however, because of the accumulation of dead plant material, they can be denser than those within the várzea. The total area of floating habitat per unit area of lake is very much smaller in the black than white waters. The fish communities of these meadows has been described above. As in the case of forest, the age, diversity, and physical structure of a meadow determine species composition and richness.

Species richness per unit area is enhanced by the mosaic structure of meadow and forest. Many fish species prefer to live or feed within the transitional zones between different vegetation, such as the forest/open water or forest/meadow interfaces. The importance of these ecotones in land-water systems has been recognised previously (Naiman & Decamps, 1990). Species diversity may be linked to the variety and quantity of these transitional zones offered by the floodplain.

Concluding Remarks

Given the spatial and temporal complexity of the floodplain, and some of the ways in which this influences the diversity of its fish community, it seems likely that any theory hoping to explain fish diversity will require an understanding of the functional relationship between species number and structural complexity on scales from >100 km to <1m. Much work is required before we understand how to describe and quantify structural complexity in terms of its meaning for fish. The task is made difficult by the fact that the importance of a given structural element is different for different species.

In terms of fish species diversity, probably the most damaging changes man undertakes within floodplains is the simplification of the structural complexity. This may take many forms, from the dredging of strength-sided canals to land reclamation to the removal of fallen trees that obstruct boat movements. Successful conservation may require a much improved knowledge of the types and amounts of change that are acceptable. However, it seems clear that conservation will require the maintenance of three things: (1) structural complexity over a wide range of spatial scales, (2) spatial dynamics (i.e., lakes and channels must be created and destroyed), and (3) connectivity between headwater, ria lake, main stream, and várzea lake. This will require the establishment of conservation plans covering areas of forest greater than those thus far achieved. In the case of the Mamirauá Reserve, consideration needs to be given to extending the reserve to include some of the surrounding blackwater (ria lakes) and headwater streams.

Literature Cited

Crampton, W. G. R. 1996. Gymnotiform fish: An important component of Amazonian floodplain fish communities. Journal of Fish Biology 48: 298–301.

Goulding, M. & M. L. Carvalho. 1982. Life history and management of the tambaqui (*Colossoma macropomum*, Characidae): An important Amazonian food fish. Revista Brasileira de Zoologia Sao Paulo 1: 107–133.

Henderson, P. A. & W. G. R. Crampton. 1995. Standing crop and distribution of fish in drifting and attached floating meadow within an Upper Amazonian várzea lake. Journal of Fish Biology 47: 266–276.

———— & ————. 1997. A comparison of fish diversity and abundance between nutrient-rich and-poor lakes in the Upper Amazon. Journal of Tropical Ecology 13(2): 175–198.

————, **W. D. Hamilton & W. G. R. Crampton.** In press. Evolution and diversity in Amazonian floodplain communities. *In* D. Newbury, ed., Speciation in tropical systems. Blackwells, London.

Junk, W. J. 1970. Investigations on the ecology and production-biology of the floating meadows (*Paspalum–Echinochloetum*) on the Middle Amazon. Amazoniana 2: 449–495.

Lowe-McConnell, R. H. 1987. Ecological studies in tropical fish communities. Cambridge University Press, Cambridge.

Magurran, A. E. 1988. Ecological diversity and its measurement. Croom Helm, London.

Mertes, L. A. K. 1985. Floodplain development and sediment transport in the Solimões-Amazon River, Brazil. University of Washington, Seattle.

Naiman, A. & D. Decamps. 1990. The ecology and management of aquatic-terrestrial ecotones. UNESCO, Paris.

Ribbink, A. J. 1994. Biodiversity and speciation of freshwater fish with particular reference to African cichlids. *In* P. S. Giller, A. G. Hilldrew & D. G. Raffaelli, eds., Aquatic ecology, scale, pattern, and process. Blackwell Scientific, London.

Saint-Paul, U. & P. Bayley. 1979. A situacão da pesca na Amazonia central. Acta Amazonica 9: 109–114.

Salo, J. 1990. External processess influencing origin and maintenance of inland water-land ecotones. *In* A. Naiman & D. Decamps, eds., The ecology and management of aquatic-terrestrial ecotones. UNESCO, Paris.

Val, A. L. & V. M. F. Almeida-Val. 1995. Fishes of the Amazon and their environment. Physiological and biochemical aspects. Springer-Verlag, Berlin.

Community Management of Floodplain Lakes and the Sustainable Development of Amazonian Fisheries

David McGrath, Fábio de Castro, Evandro Câmara, and Célia Futemma

Introduction

The Amazon floodplain is one of the world's last major inland fisheries to remain thus far little exploited. Over the last 30 years, however, intensification of the commercial fisheries has greatly increased pressure on floodplain fish stocks, affecting not just fisheries but also the populations that exploit these resources (Furtado, 1990; Goulding, 1983; Smith, 1985; Junk, 1984b). While Amazonian fisheries have undergone major changes, the development of these fisheries is still in its early phases (Bayley & Petrere, 1989). As they develop, two management strategies are emerging: one based on the conventional model of state-centered fisheries management, and the other, community lake reserves, on community management of floodplain fisheries (McGrath et al., 1993a).

Development of Amazonian fisheries represents both a problem and an opportunity. Fish are a highly productive and renewable resource. If fisheries are managed sustainably, integrating the local smallholder populations that now exploit them, they can contribute significantly to the development of the floodplain. If the fishery is exploited unsustainably, and without integrating floodplain smallholders, intensification of the fishery could lead to the degradation of várzea ecosystems and the marginalization of the ribeirinho population, those who inhabit the floodplain (McGoodwin, 1990; Weber, 1994). The purpose of this paper is twofold: (1) to evaluate these two management models in terms of their impacts on floodplain smallholders, fisheries, and ecosystems and (2) to assess the degree to which the emerging community management model could serve as the basis of a regional strategy for development of floodplain fisheries. This paper is divided into three parts. In the first, the main features of the two models are presented; in the second, the potential of each model for developing Amazonian fisheries is explored; and in the third, the main barriers to the implementation of the community management model are discussed.

Background: The Intensification of Floodplain Fisheries and the Emergence of the Community Management Model

The present status of Amazonian fisheries is the result of the interaction of three sets of factors. First, changes in fishing technology, combined with increasing demand (regional and export) for Amazonian fish products have substantially increased pressure on the resource (Bayley & Petrere, 1989; Meshckat, 1960). Second, with the decline of jute production, once the base of the floodplain economy, agriculture has lost its traditional role in the regional economy, and ribeirinho labor has shifted from farming to fishing (Gentil, 1988; Furtado, 1990). Today, with fishing the mainstay of the ribeirinho economy, the great majority of the ribeirinho population depend on fishing for at least part of their annual income (Gentil, 1988; McGrath et al., 1993a). It is estimated that Amazonian fisheries presently involve some 230,000 fishers, the great majority of whom are smallholders living on the várzea (Bayley & Petrere, 1989; Furtado et al., 1990). The third factor is the expansion of cattle and water buffalo ranching to take advantage of the natural grasslands of the várzea. Ranching is now the dominant land use on the Amazonian várzea (particularly for the Amazon below Manaus; this is probably not the case on the Solimões). The expansion of cattle ranching has contributed to deforestation of floodplain forests and the overgrazing of the natural grasslands, significantly modifying floodplain environments and contributing to the degradation of the productive capacity of aquatic ecosystems (Goulding, 1996).

Although fish stocks are not yet overexploited, the increase in fishing effort has considerably reduced the productivity of fishing effort for artisanal fishers (Bayley & Petrere, 1989). Concerned with the decline in fishing productivity, ribeirinho communities have sought to prohibit the entry of nonlocal commercial fishers into local lakes. As a result, conflicts between communities and outside commercial fishers have proliferated throughout the basin, leading to human injury and the destruction of gear and boats (Hartman, 1989; Junk, 1984b; McGrath et al., 1993a). Thus far the state has proven incapable of monitoring and managing fisheries effectively or of mediating conflicts between ribeirinho communities and outside commercial fishers. While existing fisheries legislation is quite comprehensive, with the exception of the industrial fisheries of the estuary, an open-access management regime prevails, encouraging fishers to exploit the resource with little concern for maintaining long-term productivity.

The efforts of ribeirinho communities to assume control of local lakes represent an attempt to fill the vacuum left by the absence of state control. Throughout the region, communities are taking control of local lakes and forming lake reserves in which the community defines and implements rules regulating lake fisheries, frequently with the support of fisher's unions such as the Colônia de Pescadores and Catholic organizations such as the Comissão Pastoral da Pesca (C.P.P.) and Comissão Pastoral da Terra (C.P.T.)(C.P.T., 1992a, 1992b; McGrath et al., 1993a, 1993b, 1994; Pinedo Vasquez et al., 1992; Lima, 1994). Given the mismatch between the size and ecological complexity of the Amazon floodplain and the limited resources of IBAMA, the government agency responsible for managing fisheries, effective user-group participation in the management of local fisheries is probably essential if management control is to be extended to the local level. In this context, the efforts of ribeirinho communities to manage local fisheries represents a promising alternative to the sustainable development of Amazonian fisheries.

Management Models for Fisheries Development in the Brazilian Amazon

Floodplain fisheries can be characterized as a "common property resource," defined, following Feeny et al. (1990), as a class of resources from which it is difficult to exclude others (excludability) and for which use by one individual reduces the amount available to other users (subtractability). Researchers frequently distinguish four property regimes under which common property resources can be managed: open access, communal property, private property, and state property (Bromley & Cernea, 1989; Feeny et al., 1990).

Open access property regimes are characterized by the absence of rules. With access to the resource unregulated, anyone has the right to exploit it. The open access regime has often been confused with the communal property regime, as in Hardin's (1968) classic paper, "The Tragedy of the Commons," in which he uses the term "commons" to refer to an open access management regime (Feeny et al., 1990).

By contrast, in a communal property regime, rights to the resource are held by a distinct group of users who exclude outsiders and regulate use by members of the user group. Members usually have equal access to and use of the resource. These communal property rights may or may not be formally recognized by the state. Hardin (1968) denied the possibility of functioning communal property regimes and argued that the solution to the environmental destruction caused by open access was either a private property or a state property regime.

In a private property regime, rights to exclude others and regulate use of the resource are vested in individuals. Rights to the resource are transferable and are usually recognized and enforced by the state.

In a state property regime, rights to access and use of the resource belong exclusively to the state. In some cases the state may cede access to and use of the resource to individuals, and in others the state may permit free access to all while regulating use of the resource.

To varying degrees, all four models are operating in Amazonian fisheries. Official fisheries management policy is based on the state property regime. However, the developing community-based management model is more consistent with the communal property regime. At the same time, the state's limited regulatory presence in the interior has led to a situation of open access over much of the floodplain, whereas in other areas, such as eastern Marajó Island, large landowners have taken control of local lakes to establish de facto private property regimes. However, neither open access nor private property regimes are regarded as legitimate management options for Amazonian fisheries, and the debate over fisheries management policy has tended to focus on variants of the state property and communal property regimes, here referred to respectively as the "technocrat" and the "community" management models.

The technocrat model has served as the basis for fisheries development throughout the world (McGoodwin, 1990; McEvoy, 1990). The model takes as its starting point the assumption that the resource belongs to society and that the state has the responsibility to manage the resource so that society obtains the full benefit that can be sustainably extracted from it. Consequently, the overriding concern of management under the technocrat model is for efficiency, specifically, how much of the productive potential of the resource is being exploited. It is assumed that optimal levels of harvesting can be determined scientifically and management policy adjusted to achieve this optimal level of catch (Larkin, 1977, 1978; Nielsen, 1976). Over the course of the twentieth century, an impressive body of theoretical and empirical work has accumulated, much of it concerned with developing models and methodologies for determining optimal levels of catch (Anderson, 1986; Scott, 1979; Welcomme, 1985).

In the technocrat model, fishers are typically full-time professionals moving between fishing sites over the course of the year. The state assumes responsibility for managing the fishery and fishers. Management policy is developed by government technocrats with varying degrees of input from researchers and organized groups of fishers and industry, and is implemented by regional field staffs. Fishers are considered to be motivated by self-interest and concerned primarily with maximizing the value of their catch. They have no effective motivation to conserve local fish populations, since they know that the fish they leave in the water will simply be caught by other fishers. Thus, the state must design and implement regulations to insure that the level of pressure on the resource does not exceed its productive capacity. This model requires a comprehensive regionwide institutional infrastructure for monitoring the status of fisheries and enforcing fisheries regulations, including sophisticated programs of research to collect basic data on the biology, ecology, and economics of the fishery and a regular system of patrols to monitor fishing activity on the major fishing grounds.

The community model of fisheries management has been the focus of considerable attention over the last 10–15 years, especially among anthropologists and other social scientists (McCay & Acheson, 1987; Berkes, 1989; Chernela, 1985; Cordell & McKean, 1986; Stocks, 1987; McGoodwin, 1990). Though long scorned as unscientific and therefore as incapable of taking advantage of the productive potential of local fisheries, the community model is gaining respectability among fisheries researchers and managers, largely as a result of the crisis in many of the world's major fisheries (McGoodwin, 1990; Fairlie et al., 1994; Mathews, 1994). In this model, a definite group of fishers, either members of one or more communities or members of some form of collective organization, control access to and use of a well-defined fishing territory. Fishers in this model are "sedentary" in the sense that they are confined to a distinct fishing territory. Rules regulating use of the fishery are defined by community members or the local user group, with varying degrees of input from other organizations/institutions. Enforcement is mainly the responsibility of the community or members of the local user group. The organization usually enjoys at least tacit support from the local governmental agency responsible for fisheries management.

Fisheries Management Policy in the Brazilian Amazon

Fisheries management in the Brazilian Amazon follows the technocrat model. Instituto Brasileiro de Meio-Ambiente e Recursos Renováveis (IBAMA) is the federal agency responsible for fisheries management. By Brazilian law, waterways belong to the state, and fishers have the right to fish in any body of water they can reach by boat. Ownership of water bodies is formally recognized only in the case of permanently isolated lakes and ponds that are entirely surrounded by private property. Management policies are defined by IBAMA's fisheries management staff with varying degrees of input from independent researchers and representatives of fishers' organizations and industry. Although IBAMA's regional offices have some flexibility to adapt regulations to local conditions, all major management decisions must be formalized in decrees (*portarias*) signed by the president of IBAMA. This requirement places a major constraint on IBAMA's ability to address local management issues.

Fisheries management policy draws on the range of conventional management tools including restrictions on gear types, minimum size requirements, and prohibitions during spawning seasons (Isaac et al., 1993). In some Amazonian states, trawls and seines are prohibited from interior waters and only gillnets, fixed and floating, are permitted. Several species are protected during their three-month spawning season, when fishing for them is prohibited. Minimum sizes are also specified for some commercially important species. With the exception of parks and other forms of biological reserves, there are few territorial restrictions, so that the entire inundated area of the floodplain is theoretically available to fishers throughout the year. Although licenses are required for all commercial fishers, with the exception of the industrial trawl fishery in the estuary, there are no effective restrictions on the number of fishers who can exploit the fishery, and those caught fishing without a valid license are rarely punished. For all intents and purposes, Amazonian fisheries are open access.

As noted earlier, the technocrat model places great demands on the capacity of the state to monitor fish stocks and fishing activity and enforce fisheries policies. This model is difficult to implement effectively even in areas where fisheries ecology is much simpler and better understood than is the case in the Amazon (McGoodwin, 1990; Fairlie et al., 1994). IBAMA does not have the minimum personnel, funds, or equipment needed to implement this model effectively. There is no official register, for example, to provide a reliable estimate of the number of fishing boats operating in the region. Even in the major urban centers, monitoring of fish markets is minimal, and in the interior, IBAMA is unable to maintain the regular patrols necessary for monitoring fishing activity and fish stocks. It is only in the last decade that research groups collaborating with IBAMA have developed programs to monitor fish landings in the major urban centers, so that only a fraction of the data needed to assess fish stocks is available. The result is that fishing activity is regulated more by natural conditions than by the state (Goulding, 1983).

Community Management of Floodplain Lakes: An Evolving Model

The lake reserve model that is developing in the Amazon is based on the community model described earlier (McGrath et al., 1993b, 1994). Here, rather than regarding the floodplain as a single management unit within which fishers circulate freely, management is organized around individual floodplain lake systems and their local user-group population. Fishers are largely sedentary, fishing in the lake system closest to their own community. While fishers are also considered to be self-interested operators, in this case there are mechanisms for protecting and reconciling both individual and collective interests. In this sense, then, the emerging lake reserve model has the potential to satisfy the basic conditions necessary for collective management: a distinct territory controlled by a well-defined user group with exclusive access to the resource (Ostrom, 1990; McCay & Acheson, 1987).

Although the lake reserve movement is a relatively recent phenomenon, it draws on traditional notions of land tenure and fisheries ecology. To a certain extent, the notion of lake ownership is related to the manner in which an individual property is defined. Várzea property is usually defined not in terms of its area but in terms of the length of frontage along the river or river channel (*metros de frente*). Normally, the property is considered to extend inland to the center of the island or lake, where it meets properties extending inland from the opposite shore. One practical result is that it is very difficult to know the total area of an individual property because inland boundaries are only vaguely estimated.

While vague in terms of area, this land-use system guarantees each family access to all of the major environmental zones of the várzea (see Denevan, 1984). The river is used for transportation and seasonally for fishing. House sites are located on the higher ground of the natural levee which parallels the river or side channel, and it is here that most agricultural activity takes place. Cattle are pastured on the seasonally inundated grasslands of the transition zone between lake and natural levee, and fishing activity over much of the year focuses on lake fisheries.

Although individual properties incorporate the major várzea habitats, these ecological zones are treated differently within the land tenure system. There is a gradient from individual to collective use as one proceeds inland from the river (Figure 1) (the river in front of one's house is more or less considered open access, though residents frequently reserve the right to object to activities taking place there which they regard as detrimental to their interests). Along this transect, the levee is considered to be private property with boundaries clearly defined and often fenced. In general, the natural grasslands inland from the levee are regarded as a commons. Cattle are allowed to roam freely throughout this zone, though individual property owners are within their rights to fence off their grassland property. The lake is also considered to be common property, and efforts to restrict access are generally resisted unless the entire lake lies within one or a few properties. The major distinction in terms of land tenure, then, is a distinction between the front levee, which is exploited individually, and the interior grasslands and lake system, which are treated as a commons open to all community members. Ownership of river frontage provides access to the pasture and lake commons behind.

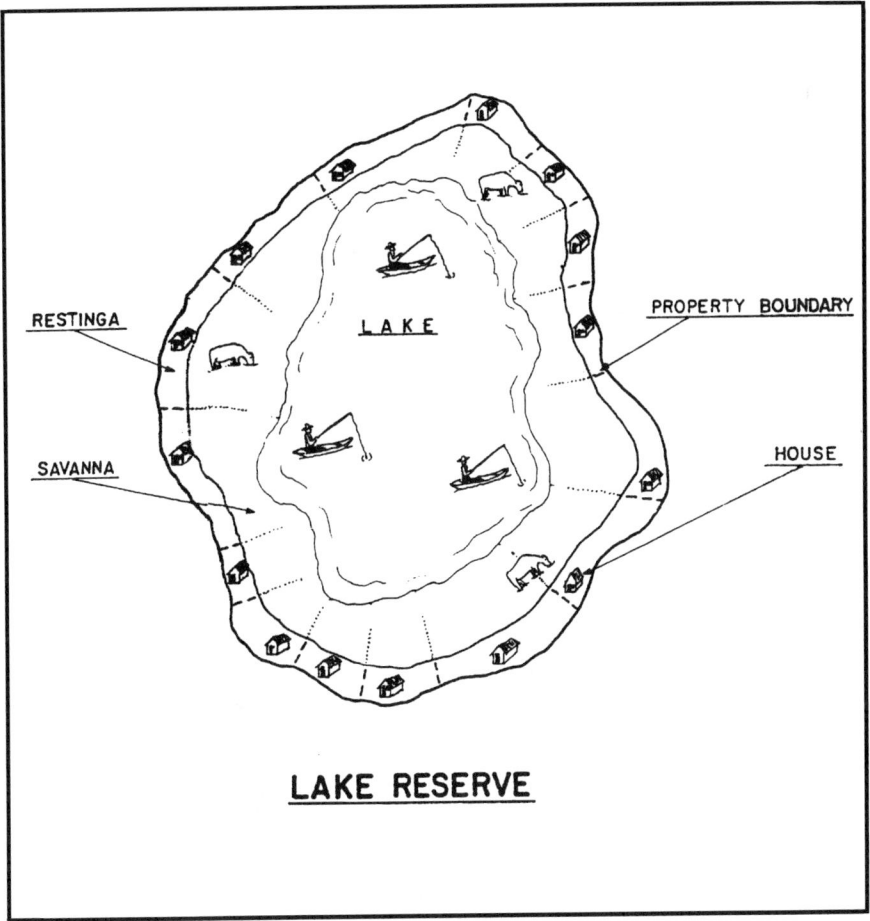

Figure 1. Land tenure patterns in community lake reserves.

Following this logic, community ownership or control of lakes is generally based on the ownership by community members of the property surrounding the lake, though a tradition of use by neighboring communities may also be recognized. In this system the community, as a collective property owner, has a right to the fish in the lake in much the same way that upland landowners would claim ownership of game on their property. This view of lake fisheries provides the basis for the collective management of the resource by defining the group of individuals who have access to the resource and receive benefits from it.

Management Objectives and Measures

Community management of lake fisheries is based on community agreements called "*acordos de pesca*" (fishing accords), which specify measures to be taken and sanctions to be levied against infractors. These documents are usually drafted in community meetings and signed by those present who are in agreement. The document is then typed and the signatures annexed to form a petition, or *baixo assinado*, which is presented for formal recognition to IBAMA, the Colônia, and municipal authorities. Although this recognition implies no formal legal support, it serves to legitimize the agreement in the eyes of the community, providing moral, if not legal, support for actions taken to enforce lake accords.

The principal aim of fishing agreements is to stabilize or reduce pressure on local fisheries. These agreements usually attempt to achieve this objective indirectly through restrictions on gear and storage capacity rather than directly by placing limits on catch size. In addition to regulating fishing activity, fishing agreements frequently include measures intended to conserve habitat considered to be important to lake fish populations. They may also include measures designed to regulate exploitation of other species, especially river turtles. These measures may or may not be consistent with existing state regulations. Typically these agreements include a half dozen types of measures (briefly dicussed below), concerning access to lake fisheries, commercial uses of fish, fishing gear, storage capacity, habitat conservation, zonation of lake systems, and collective organization and enforcement.

The most common measure is to restrict access to lake fisheries, by prohibiting fishers who are not part of the community from fishing in the lake. Where fishing is primarily subsistence oriented, this measure may be sufficient to restore or maintain the productivity of the fishery. Where the majority of the fishers sell their catch, this measure is unlikely to be sufficient to maintain productivity.

One of the easiest ways to control fishing pressure is to restrict commercial use of the catch. In some cases, commercial fishing is prohibited altogether, while in others it is restricted to certain species or groups of species, or permitted only during part of the year. In other cases, sale of fish is permitted only within the community, in effect limiting total catch to the subsistence needs of the community as a whole.

Frequently communities prohibit the use of one or more types of gear during all or part of the year. The main target of these measures is the gill net, which is considered the principal cause of the excessive pressure on lake fisheries. Restrictions on use of gill nets is considered the most effective way of limiting catch size. Gill nets may be prohibited altogether (frequently the case in subsistence fisheries) or during part of the year, usually the low-water season, when fish are concentrated in small bodies of water. In addition, lake accords may also prohibit a range of gear types and fishing techniques, including trawls and seines, which are illegal in some Amazonian states.

Another means of limiting catch size is to limit storage capacity. There are several measures that communities may adopt. First, the community may place restrictions on the size or type of boat that may enter the lake. Motorized boats or boats over a certain size, for example, may be prohibited from entering the lake. Second, they may limit the size of ice boxes used to store fish. Finally, ice may be prohibited altogether, and only salt, the traditional means for preserving fish, permitted.

Fishers have an intimate knowledge of the relationship between lake vegetation and the productivity of lake fisheries, and communities often define measures to protect vegetation that is considered important to local fisheries. For example, some fishing agreements specify rules to preserve the mat of floating macrophytes and protect seasonally flooded fruit trees. Throughout the Amazon, fishers complain of the deleterious effect of water buffalo on aquatic habitats, and in some cases communities have prohibited buffalo from community territory.

Várzea lakes are, in reality, systems of lakes that vary considerably in their physical and ecological characteristics. Some communities distinguish different types of lakes and adapt fishing regulations to the characteristics of each type. For example, fishers distinguish between shallow, seasonal lakes that may dry out in the lower-water season and deeper permanent lakes, often referred to as "*lagos de criação*" (loosely translated as "nursery lakes") where fish tend to concentrate during the dry season. Commercial fishing may be permitted in the shallow lakes during the dry season, since the fish in these lakes may die anyway, whereas fishing in the deepest lakes may be restricted to subsistence needs or be prohibited altogether during the dry season.

Collective organization and enforcement is the most problematic aspect of community lake management. Accords are usually vague with regard to the organizational structure for collective management. Typically, responsibility for enforcing measures falls to the leadership of the local chapter of the Colônia de Pescadores or other community leaders. Since acordos de pesca have no formal legitimacy, individuals responsible for organizing collective management have no formal authority to enforce community measures.

In addition to these types of measures, lake accords usually specify graduated sanctions to be imposed on those who break rules. For example, a first-time infractor may receive a warning; the second time, his nets may be confiscated for a period or until a fine is paid; and the third time, his gear may be confiscated and destroyed. Sometimes the confiscated gear is given to IBAMA or the Colônia and a formal complaint is lodged against the individual or group. This action is largely symbolic and serves more to legitimize actions taken by the community than to punish the offender, since IBAMA rarely takes any action. While punishments are often specified, mechanisms for apprehending those breaking community rules rarely are. While some communities undertake irregular nocturnal patrols to catch those fishing illegally, most seem to rely on chance discovery by members of the community.

The Logic of Community Management

The community model is based on the logic of the várzea smallholder economy. Artisanal and subsistence fishers are typically part-time, with fishing just one of several activities contributing to the household economy. As a result, fishers have limited time to devote to fishing and prefer a relatively high-productivity fishery so they can meet basic needs with minimal investment of time. Artisanal fishers are, in effect, trading production for productivity. This is evident in the results of a study comparing fishing productivity in managed and unmanaged lakes (Table I; McGrath et al., 1994). In the managed lake, management is oriented toward subsistence, although some commercial fishing is permitted. In the unmanaged lake, fishing is commercially oriented, with no effective regulations on local fishing activity. Fishing productivity in the managed lake was 23% higher, and fishing income 30% lower, than in the unmanaged lake. The higher-productivity fishery required by artisanal fishers and smallholders contributes to total production in several ways. Fishing provides a regular source of animal protein for

Table I. Income and labor productivity in managed and unmanaged systems

Activity	Managed regime	Unmanaged regime
Annual household income (US$)		
Fishing	600	950
Agriculture	350	—
Total	950	950
Labor productivity (US$/person/hour)		
Fishing	1.16	0.94
Agriculture	0.74	—
Combined	1.03	0.94

Source: McGrath et al., 1994.

family subsistence. The sale of fish products provides cash income for basic household needs during the growing season, and any surplus income can be used to pay for day labor, seeds, and other agricultural expenses. As várzea fishers say, "Two days fishing and one day farming."

Although labor productivity in agriculture is considerably lower than that in fishing in both communities, it is more than compensated for in the managed lake by the much higher productivity of labor in fishing. As a result, combined labor productivity in the managed lake is 10% higher than that of fishing in the unmanaged lake (Table I). The economy of the managed community is also more diversified, with one-third of total income derived from agriculture. In addition, since total catch is limited by community rules, the system as a whole is theoretically sustainable, an assumption that cannot be made for the unregulated fishery of the unmanaged lake.

The example illustrates the contributions of the high-productivity fishery and economic diversification to the overall efficiency of production in the community model. Here, fishing and farming are interdependent. The higher productivity of fishing in the managed lake is made possible by the lower total catch, which, in turn, is made possible by income from agriculture. Because fishing is an extractive activity, any attempt to increase fishing income beyond present levels (with existing technology) would lead to a decline in labor productivity below that of the diversified system. Thus, the high-productivity fishery essential to the community model is made possible by diversifying the household economy, combining a high-productivity extractive activity with lower-productivity productive activities such as farming and animal husbandry. The result is a more or less stable equilibrium of fishing and farming, extraction and production, that is maintained by the community enforcement of lake fishing agreements (Hecht et al., 1988).

Fisheries Management and Rural Development on the Várzea

In the preceding section we described the two management models that are emerging as Amazonian commercial fisheries develop. In this section, we evaluate the potential of these two models with regard to two broad questions: the relative effectiveness of the two models for managing floodplain fisheries and the impacts of each model on the course of floodplain development.

Várzea Lakes as Management Units for Fisheries and Other Várzea Resources

Perhaps the first issue that must be addressed in evaluating the potential of community lake reserves is the degree to which an individual floodplain lake system constitutes an effective management unit for Amazonian fisheries and for the várzea resource base as a whole. Várzea lakes are open systems, components of an immense and highly dynamic fluvial system that encompasses nearly 40% of the South American continent (Junk, 1984a; Junk et al., 1989; Goulding, 1989; Welcomme, 1985). During half the year, individual lakes disappear as the várzea is inundated by the river, forming, in effect, one huge lake 50–100 km wide and more than 4000 km long (Figure 2). At this time of year, only treetops and houses on stilts remain visible to indicate the contours of individual lake systems. During the other half of the year the flood waters recede, exposing hundreds of thousands of water bodies varying in size from mere ponds to lakes many kilometers in diameter (Fig. 2). At the same time, fish are mobile, with the potential to move from lake to lake throughout the system over the course of the year.

Given these characteristics of floodplain lake systems, it is not at all clear that there is any enduring relationship between local fish populations and individual lake systems. If fish circulate between lake systems spread out over the floodplain, there is no reason to expect that the effect of management policies applied to a given lake will have an effect that endures beyond the next flood season. Lake fisheries depleted during the low-water season would be simply restocked by the circulation of fish during the following flood season. If these conditions prevail, then, the itinerant model would be the more appropriate management strategy, since it would permit fishers to follow fish populations circulating between lake systems. Some variant of the pulse fishing man-

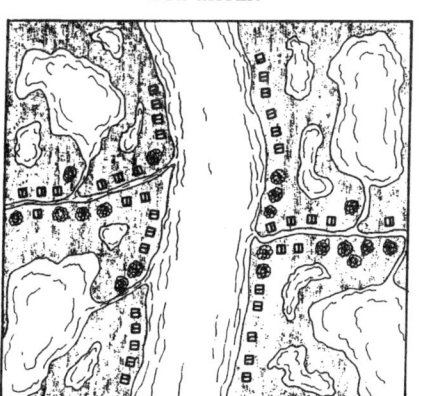
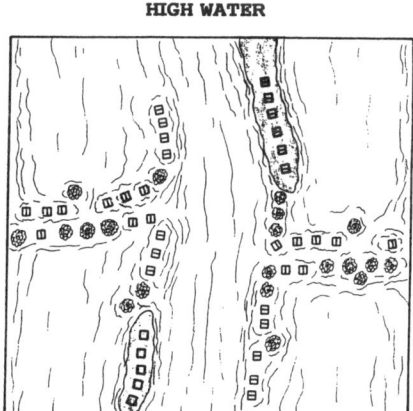

Figure 2. Seasonal variation in várzea lake boundaries.

agement regime proposed by Bayley (1995b) would perhaps be the most appropriate way of managing lake fisheries.

However, the results of Bayley's (1995a) comparative study suggest that the management regime can have an enduring effect on the productivity of lake fisheries. While differences in overall productivity of fishing are difficult to evaluate, when we controlled for gear type, productivity in the managed lake tended to be twice that in the unmanaged lake (McGrath et al., 1994). This is a preliminary finding, and we were unable to control for environmental factors that might account for all or part of this difference in productivity. However, the results are supported by the accounts of fishers managing the managed lake. The community decided to prohibit use of gill nets and commercial fishing in general, in response to the precipitous decline in the productivity of lake fisheries caused by excessive commercial fishing pressure. Since instituting the subsistence management regime, productivity of the lake fishery has increased significantly, indicating that the local effects of local management practices endure beyond the annual flood season.

While this study suggests that individual lake systems are potentially viable management units for floodplain fisheries, the degree to which floodplain lakes can serve as effective management units will vary from species to species, depending on migratory and reproductive behavior and the extent to which floodplain lakes provide significant habitat for the species. Presumably, lake management will be most effective for species that reproduce in lakes and do not undertake regular long-distance migrations, such as the pirarucu (*Arapaima gigas*) and tucunaré (*Cichla* spp.). This is supported by a comparison of pirarucu fishing in the managed and unmanaged lakes described earlier. Pirarucu production in the managed lake, which maintains a closely controlled six-month pirarucu fishing season, was twice that of the commercial lake where pirarucu are fished year-round with no effective restrictions (McGrath et al., 1994). Conversely, management of individual lakes is probably less effective for species that undertake seasonal migrations and that reproduce in rivers rather than in lakes, such as the tambaqui (*Colossoma macropomum*). While probably irrelevant for long-distance migrators such as the dourada (*Brachyplatystoma flavicans*) and piramutaba (*Brachyplatystoma vaillantii*), a regional system of lake reserves extending along the entire length of the Amazon floodplain would help protect medium-distance migrators over much of their range while also providing local centers for restocking surrounding lake systems (Barthem & Petrere, 1991; Goulding, 1980).

The second point with regard to the role of lake reserves as management units is their potential for managing other components of the floodplain ecosystem. In this regard, one of the most important characteristics of the lake reserve is that it integrates simultaneously both the main components of the várzea ecosystem and the principal resources and associated economic activities of the ribeirinho economy (Figure 3). The three main environments of floodplain lakes—the surrounding *restinga* (natural levee) forests, seasonally inundated grasslands, and permanent lakes and channels—are also the focus of the main economic activities of the floodplain economy. Ribeirinho settlement is concentrated on restingas, as are most agricultural activities, whereas cattle are grazed on the seasonally inundated grasslands. Most fishing activity focuses on floodplain lakes rather than the river. Thus, fishing is one of a complex of activities exploiting the main

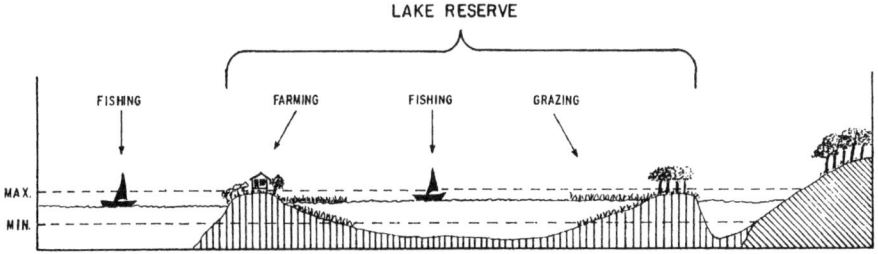

Figure 3. Lake reserve and floodplain land use.

habitats of várzea lakes. Since the productivity of lake fisheries is closely associated with the status of lake habitats, the lake reserve internalizes many of the positive and negative consequences of local land-use practices. As a management model, then, the lake reserve provides a framework for evaluating the impacts of individual activities on other components of the lake reserve system, and in this fashion facilitates the development of management systems that take into account the complex interactions among natural resources, habitats, and economic activities (Junk, 1989; Bayley, 1995a; Goulding, 1993, 1996; Sparks, 1995).

Fisheries and Rural Development

The key underlying theme in this paper is that lake fisheries are the strategic resource of the várzea. In addition to being critical to the economic well-being of the várzea smallholder, lake fisheries, being a highly productive and renewable resource, are capable of generating a large and continuous stream of economic benefits for those who are in a position to take advantage of them. Thus, who gains control of floodplain fisheries and of the way they are exploited will have profound consequences for the course of floodplain development. The two management models described here offer fundamentally different strategies for developing the fishery, and are likely to result in very different outcomes not just for the fisheries but also for floodplain settlement and land use in general. These different scenarios are summarized in Figure 4: Figure 4a is a schematic representation of the probable development trajectory under the technocrat model; Figure 4b, of the potential development trajectory under a successful community model. It should be noted that the two scenarios presented here are not intended to represent the only two possible outcomes for várzea development; rather, they should be taken as the end points of a range of possible outcomes, depending on the specific policy measures adopted, local conditions, and the manner in which the policies are implemented.

Technocrat Model

Fisheries development means transforming the fisheries sector. From the perspective of the technocrat model, artisanal fisheries are inefficient. Their small scale and limited

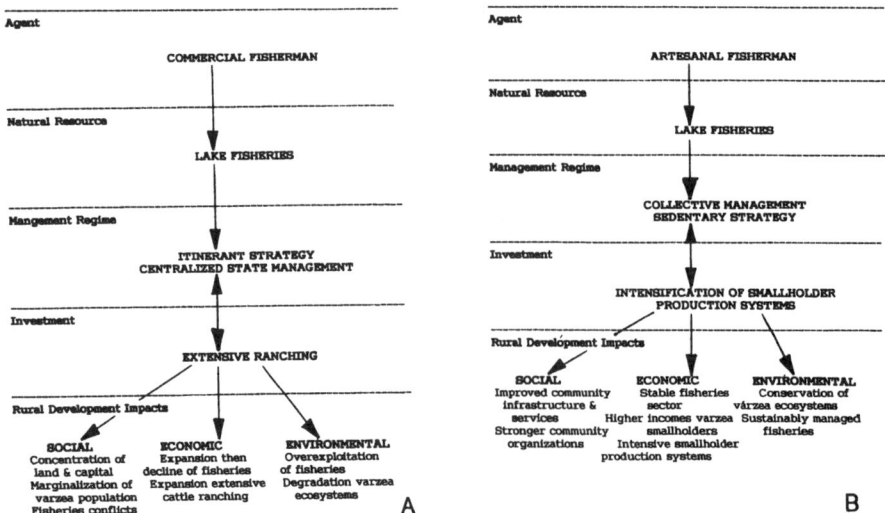

Figure 4. A. Technocrat management model development trajectory. **B.** Community management model development trajectory.

technology, and the fact that they are typically dominated by part-time fishers, combine to limit their ability to efficiently exploit the productive potential of the resource. The solution from the technocrat perspective is to modernize the fisheries sector, increasing the capacity of fishing boats and turning fishers into full-time professionals (McGoodwin, 1990).

In the Amazon, a policy of open access, combined with programs to modernize the commercial fleet, would have the effect of directing access to floodplain fisheries to the relatively small group of professional itinerant fishers in a position to benefit from state loan programs (Cordell & McKean, 1986). Through investment in fleet capacity, total catch would increase, with the modern sector accounting for most of this growth. In this way the modern sector would obtain an increasing proportion of the total value appropriated from the resource. This increase would come at the expense of the artisanal sector, which, lacking the resources to increase effort, would see its share of total catch decline.

This trend would have devastating consequences for artisanal fishers. As the productivity of effort in the artisanal sector declined, so would both subsistence and cash income from fishing. Rural smallholders, for whom fishing is a critical component of the household economy, would find themselves in an increasingly difficult situation (Kurien, 1994). Without viable alternatives to replace the lost income, survival on the várzea would become increasingly difficult. Rural smallholders are likely to respond by attempting to close lakes to outside fishers, resulting in the proliferation of fisheries conflicts (C.P.T., 1992a). However, with government policy supporting open access, they would not be likely to succeed, and the result would be more and more smallholders selling out to neighboring ranchers and moving to the upland frontier or to regional urban centers, swelling the ranks of poor on the urban periphery (McCully, 1991; Furtado, 1990; McGrath et al., 1993a).

This trend would be reinforced by the fact that fishers in the modern sector invest their profits from fishing in ranching, generally regarded as one of the most secure and liquid investments available (Hecht, 1993; Mattos & Uhl, 1994). Through this mechanism, surplus appropriated from the fishery is transferred to cattle ranching, contributing directly to the expansion of ranching on the várzea. Since the várzea grasslands are treated as an open access commons, there is an incentive to overstock. Overgrazing, combined with seasonal burning of the grasslands and clearing of remaining várzea forests, would lead to the progressive degradation of várzea ecosystems, reducing the productive capacity of várzea fisheries. At the same time, the inability of state agencies to adequately monitor stocks and regulate fishing effort means that the state has little ability to prevent overfishing and the eventual exhaustion of fish stocks. Thus, floodplain fisheries would be caught in a two-way squeeze: degradation of habitat on one side and uncontrolled fishing pressure on the other.

There is, then, a strong likelihood that if fisheries policy follows the technocrat model, development of the fishery will have profoundly negative consequences for floodplain population and ecosystems. Rather than fueling a process of rural development, development of the fisheries is likely to result in the impoverishment of várzea smallholders, exhaustion of the fishery, and degradation of várzea ecosystems. The main point here, though, is that even if development of the fisheries is "successful," increasing total catch without exhausting stocks, the consequences for smallholders are likely to be the same. This is also true for the environmental degradation resulting from the expansion of ranching, encouraged both directly and indirectly by development of the fisheries.

Community Management Model

In the community management model, government development policy would direct access to the fishery to the rural smallholder population (Fig. 4b). Here communities and local fishers' organizations would be responsible for local-level management of floodplain fisheries within the context of a regional program for fisheries development. Organized as cooperatives or community associations, these local-level organizations would assume responsibility for managing lake fisheries and marketing local catch. Government fisheries-development policies would be oriented toward increasing the efficiency of local management. With formal control over local fisheries and effective management organizations, fishers would have an incentive to invest in the long-term management of local fisheries (MacKenzie, 1992).

Under this model, government development programs would be oriented toward increasing the productivity of smallholder production systems and strengthening local organizational capacity. First, following the logic of the smallholder economy described earlier, extension programs would encourage smallholders to invest fishing income in productive activities such as intensive agricultural and agroforestry systems, small animal production, and especially integrated systems involving terrestrial and aquatic species. Over time, as smallholders capitalize, growth in household income would come increasingly from development of productive activities rather than from fishing. Second, government extensionists would work with groups of smallholders to create collective organizations to manage fisheries and market output. Through these organizations, communities could also invest in local infrastructure, electricity, treated water, more reliable transportation, even improvements in the quality of local educational and health services—rural needs that municipal governments have proven unable to meet.

Within the community model, the high-productivity fishery would function as a kind of rural credit program, but one with important advantages. It would be far more efficient than conventional credit programs: It would require no costly government bureaucracy to administer, it would not require documents farmers rarely have, it would provide credit when farmers needed it, and, since it does not need to be repaid, it would not absorb most of the profits from sale of the harvest. Furthermore, the credit would get directly to the target population with no loss due to corruption and bureaucratic inefficiency. Finally, since the fishery is potentially a renewable resource, the program could be continued indefinitely with no burden to the state. Building on the logic of the smallholder economy, the community model provides a mechanism through which development of the fishery directly finances smallholder accumulation on the várzea.

In summary, in the community model, fisheries development, rather than taking the resource away from smallholders, focuses on increasing the efficiency with which they exploit local fisheries. In contrast to the technocrat model, in which smallholder income from fisheries declines, fisheries development following the community model could contribute to the capitalization of smallholder production. At the same time, increased income as fisheries are developed could be invested in complementary activities both intensifying and diversifying smallholder production systems. With secure control over lakes and collective organizations that assure equal access and punishment of infractors, fisheries could be managed sustainably to maintain or even increase harvests over time (Ostrum, 1990; MacKenzie, 1992). Furthermore, since fish are more productive than cattle, there would be an incentive to manage for fish rather than cattle, maintaining forest cover and limiting grazing on flooded grasslands. Overall, then, fisheries development in this trajectory could fuel a process of rural development which could lead to the growth of a stable and prosperous smallholder population and at the same time to the sustainable management of the várzea ecosystem which sustains the floodplain economy and society.

Barriers to Implementing the Community Management Model

Thus far we have concentrated on describing the community management model and on evaluating its potential to serve as the basis of a regional program for the sustainable development of regional fisheries. In this section we assess the feasibility of implementing such a model on a regional scale. In this regard two major issues need to be addressed which constitute major as yet unresolved barriers to the successful implementation of the lake reserve model: community organizational capacity and the land tenure and agrarian structure of the várzea.

Collective Management

While lake reserves appear superficially to be a contradiction of Hardin's (1968) thesis of the "tragedy of the commons," the reality is that the dynamic of self-interest which is responsible for the "tragedy" is a powerful force with which community organizations must come to terms if collective management is to work (Ostrom, 1990). In this regard, the major challenge to developing a functioning system of community lake reserves is not technical but organizational: how to create effective and resilient collective organizations capable of sustainably managing local fisheries. The problem of limited organizational capacity is critical. Unless functional community-based collective organizations can be developed, the community model is simply not an alternative. As a management strategy, it would be little different in practice from an open access fishery, but with the additional complication that the várzea population would control access to the resource, obstructing attempts to increase the efficiency and sustainability of fisheries management.

Throughout much of the Brazilian Amazon, existing rural communities are largely the result of the efforts of Catholic Church programs initiated in the 1960s. Organization of smallholders and fishers intensified with the development of the rural labor movement in the 1980s (Leroy, 1992; Schönenberg, 1994). While the success of these programs in creating physically coherent communities and training community leaders in basic organizational skills is unquestionably impressive, the resulting community-based organizations were designed to address the organizational needs of the Church and later the political objectives of rural labor movements. With the decline of these movements in the 1990s, organizational structures have been maintained by the Church in support of its religious activities. Although these community organizational structures have provided the basis for most of the successful efforts to develop community lake reserves (Lima, 1994; C.P.T., 1992b), they are, for the most part, quite fragile. Local successes tend to reflect the efforts of exceptionally capable individuals rather than that of enduring community organizations. Even where such organizations do exist, their efforts are often hamstrung by the fact that community leaders and institutions lack legal authority to impose decisions and punish transgressors.

With regard to community management of lake fisheries, the problem is compounded by the fact that there is no legal basis for lake reserves. In fact, to the extent that lake reserves exclude nonmembers, they are illegal, directly contradicting federal legislation regarding free access to inland waterways (Isaac et al., 1993). Furthermore, the measures most frequently employed by communities, restrictions on the use of gill nets and indirect measures to limit catch, have no legal basis. One result is that most community management systems are fairly crude, reflecting the limited ability of community organizations to control fishing activity. (For instance, if the community with the managed lake in McGrath et al.'s (1994) comparative study were to permit use of gill nets, fishing productivity would increase by 50–100% over present levels. But the community prefers to prohibit use of gill nets—despite the fact that fishing productivity is lowered considerably by this measure—because it does not have the capacity to regulate catch to insure that fishing is for subsistence and not sale.)

If lake reserves are to become a significant component of regional fisheries policy, these impediments to the development of functional community-based management systems must be overcome. Efforts are underway to address these deficiencies. In widely scattered locations along the Amazon floodplain, communities, fishers' unions, and supporting organizations are developing promising experiments in managing fisheries collectively (see Ruffino, this volume; Lima, 1994; C.P.T., 1992b; McGrath et al., 1993a). While IBAMA has shown increasing interest in and support for user-group participation in fisheries management, this support has not yet led to the concrete changes in fisheries policy that are necessary if these experiments are to succeed over the medium and long term. Policy changes alone will not be enough, however; what is needed is a much more systematic approach to training community leaders in the skills necessary for administering collective organizations for managing local fisheries.

Land Tenure and Agrarian Structure

The other major challenge to the implementation of lake reserves is the ambiguity of várzea land tenure systems and the prevailing agrarian structure of the várzea (Santos Vieira, 1992). This effort is complicated by the fact that tenure systems are undergoing profound and contradictory changes: on one hand, with a tendency toward privatization; on the other, with a tendency toward the formalization of common property relations. Furthermore, the concept of the lake reserve and the struggle for control of lake resources often draw on the contradictory principles of open access and local ownership.

In principle, the floodplain is the property of the state. Private property does not exist on the floodplain, although the state can concede long-term use rights to private individuals. In practice, however, much of the Amazonian várzea has been privately owned since the colonial period. Today, while few landowners have gone to the trouble of gaining formal recognition of use rights from the federal government, most have locally recognized deeds, and várzea properties are routinely bought and sold.

The ambiguity of prevailing tenure systems is exacerbated by the prevailing agrarian structure. While interior grasslands have traditionally been treated as a commons, with the expansion of cattle ranching on the várzea, these areas are also coming under private control. As a result of this process, significant portions of the lake systems on which many várzea communities depend are now located "within," or largely surrounded by, individual ranches. This can be a problem because ranchers and smallholders use the same criteria to justify control over lake resources. Thus, in many areas, várzea communities are exploiting resources which, by their own definition, are the property of neighboring ranchers. On the island of Ituqui in the municipality of Santarém, for example, 76% of the island is controlled by a small group of large landholders, while the six island communities control only the remaining 24% (Figure 5). Furthermore, several of the largest and most important lakes are wholly within the property boundaries of a single landowner. Were the rancher to prohibit access to lakes within his property, fishers would be cut off from an important part of the Ituqui lake system fisheries.

In general, the fact that communities depend on lakes that, by common acknowledgment, are within neighboring ranches has not been the problem that it is in regions such as eastern Marajó (Brabo, 1981). Although conflicts and disputes over access to

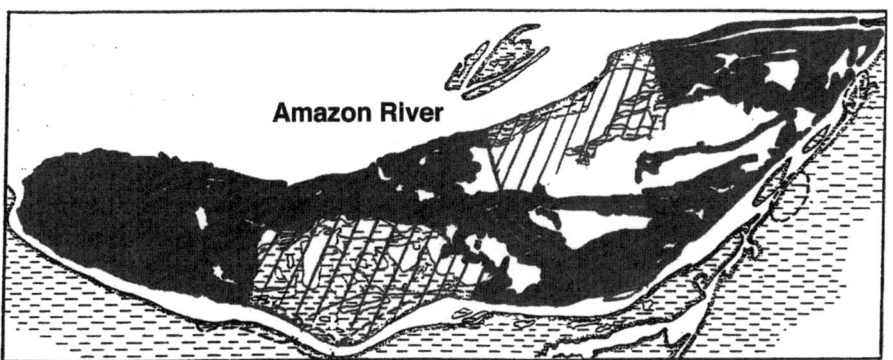

Figure 5. Community territory and lake access, Ituqui Island. Solid gray shading, rancher property; gray cross-hatching, community property; blank areas, lakes outside community property.

lake fisheries are frequent in the region as a whole, communities and local ranchers generally coexist peacefully and community fishers have more or less free access to lakes "within" ranches. In Ituqui, for example, community efforts to manage lakes have the passive support of local ranchers.

However, formalization of lake reserves or of community control of lake fisheries could easily disrupt traditional relations of access and use. Ranchers would likely regard lake reserves as a threat to the legitimacy of their claims of ownership. Even if it were argued that lake reserves are for managing lake fisheries and not floodplain property per se, the fact that for part of the year the lake covers the entire floodplain complicates considerably the distinction between lake and floodplain property as a whole. Large landowners are likely to respond to this threat by asserting control over lakes within their property, expelling local fishers, and cutting communities off from the lake fisheries on which they have traditionally depended. This trend is already well underway in the eastern half of Marajó Island (Brabo, 1981).

This conflict between communities and neighboring landowners highlights another complicated issue: What criteria are to be used in evaluating the claims of different types of landowners ("ranchers" and "communities") and other organized groups for access to or control over local fisheries? While várzea landowners, large and small, base their claims on the same concepts of property relations, these concepts are in conflict with the concept of "open access" which is the basis of national and regional fisheries policy. The state does not recognize private control over lake fisheries and, in theory at least, supports the right of fishers to fish in any body of water they can reach by boat. This places community fishers in the contradictory position of apparently supporting the principle of open access when fishing in lakes within large landholdings and asserting the principle of restricted access and local control when outside fishers attempt to enter lakes within community territory. This problem is compounded when one takes into account the lack of a definition of what constitutes a "community" as opposed to a group of property owners. When, for example, is a claim of jurisdiction over a lake a valid "community" claim, and when is it simply a resource grab by a group of land-owners?

Formalization of lake reserves as a category of land use or property will have profound effects on the existing tenure system and agrarian structure. If the history of land reform in Brazil is any indication, resistance to these changes will be fierce. However, unless these issues are resolved, it is unlikely that lake reserves can become an effective management strategy on a regional scale.

Conclusion

Amazonian fisheries are in a critical phase in their development. Two clear choices are emerging which are based on different interpretations of the current situation, have different assumptions and priorities, lead to different policies, and result in different outcomes for regional fisheries. Both models have serious unresolved questions regarding their efficacy under Amazonian conditions. While both alternatives are conceptually and theoretically coherent and reasonably compelling as alternatives, in both cases there are considerable barriers to effective implementation. In neither case does there exist the minimum institutional conditions necessary to implement the model with any degree of certainty of achieving medium- and long-term objectives.

The two are not equal contenders, however, if the objective is the sustainable development of the Amazon floodplain. Even if the basic institutional conditions for the technocrat model could be met, its track record elsewhere in the world is not encouraging, even for much simpler fisheries systems (McGoodwin, 1990; McEvoy, 1990; Swardson, 1994; Weber, 1993, 1994). Efficacy of this approach in the medium and long term is even more dubious in the context of Amazonian fisheries development. Furthermore, even if the technocrat model were to work in the narrow sense of maintaining a higher level of catch, the social and ecological consequences are likely to be problematic (McGoodwin, 1990; Neal, 1982; McCully, 1991). Fisheries development, especially of floodplain fisheries, cannot be treated as separate from the economic, social, and ecological context of the floodplain (Junk, 1989; Goulding, 1996; McGrath et al., 1993a).

While barriers to the effective implementation of the community model are considerable, it appears to be the most effective way of ensuring the sustainable development of not just Amazonian fisheries but also the Amazon várzea as a whole. Three aspects of the community model are especially relevant. First, the community model empowers local organizations of fishers, the user group with a vested interest in the sustainable management of the fishery. Second, it secures smallholder access to local fisheries the central element of the smallholder economy. And third, it provides a framework for developing lake fisheries as the lead sector within an overall strategy for the sustainable development of the várzea resource base.

But this does not mean simply turning responsibility for managing fisheries over to várzea communities and fishers' organizations, for the successful implementation of the community management model will depend on the nature and quality of government support at regional, state, and municipal levels. In this regard, what is needed is a broad-based government-sponsored program that aims to strengthen local institutional capacity to manage not just fisheries but also the sustainable development of the várzea resource base as a whole. Such a program should focus on three broad areas: fisheries management legislation, research on floodplain fisheries, and leadership development.

First, a thorough revision of Brazilian fisheries-management legislation is needed to permit the decentralization of management decision-making from national to basin/watershed, state, municipal, and local community levels, so that local user groups have the authority needed to effectively manage local fisheries. (IBAMA's fisheries sector has been involved in this process for some time; however, it does not appear that revisions will affect some of the fundamental tenets of fisheries policy such as open access, nor provide mechanisms for decentralization of management decision-making down to the level of local user groups.) This decentralization should be part of a basin-level management plan for Amazonian fisheries which distributes management authority to the appropriate regional level and provides a policy framework to guide management decision-making at regional, municipal, and local levels. Within this management framework, the state would play a central role, not just as the ultimate source of legal authority to enforce local management decisions, but also in evaluating conflicting claims to lake fisheries, monitoring the performance of community lake reserves, and providing technical assistance to local fisheries-management organizations.

Second, science also has a critical role to play. The proposal here is not to promote traditional management per se but, rather, to modernize várzea fisheries by building on traditional knowledge and utilization of lake fisheries. To this end, the success of the community model will depend on the scientific community's ability to systematize and extend local knowledge through research on the ecology of flooplain fisheries, the biology of commercially important species, and the improvement of technology for harvesting, processing, and storing fish products.

Finally, the single most important factor for the successful implementation of the model is the presence of local managers capable of adapting the community model to local social, economic, and environmental conditions. To this end, the state must invest in training a new generation of rural leaders skilled in administering collective organizations and managing common property resources.

If these three broad issues can be addressed effectively, it may be possible to surmount the barriers facing the community model and so realize the opportunity that development of Amazonian fisheries represents, in terms of increasing the quality of life of the smallholder population while retaining the biodiversity and functional integrity of várzea ecosystems.

Literature Cited

Anderson, L. 1986. The economics of fisheries management. Johns Hopkins University Press, Baltimore.

Barthem, R., M. Ribeiro & M. Petrere. 1991. Life strategies of some long-distance migratory catfishes in relation to hydroelectric dams in the Amazon basin. Biological Conservation 55: 339–345.

Bayley, P. 1995a. Understanding large river-floodplain ecosystems. Bioscience 45(3): 153–157.

————. 1995b. Sustainability in tropical inland fisheries: The manager's dilemma and a proposed solution. Pages 321–328 *in* M. Munasinghe & W. Shearer, eds., Defining and measuring sustainability: The biogeophysical foundations. The World Bank, Washington, DC.

——— & M. Petrere. 1989. Amazon fisheries: Assessment methods, current status and management options. Canadian Special Publications in Fisheries and Aquatic Science 106: 385–398.

Berkes, F., ed. 1989. Common property resources: Ecology and community-based sustainable development. Belhaven Press, London.

Brabo, M. J. C. 1981. Pescadores, geleiros, fazendeiros—Os conflitos da pesca em Cachoeira do Arari. Boletim do Museu Paraense Emílio Goeldi, Antropologia No. 77. Belém, Pará.

Bromley, D. & M. Cernea. 1989. The management of common property resources. World Bank Discussion Paper No. 57. The World Bank, Washington, DC.

Chernela, J. 1985. Indigenous fishing in the Neotropics: The Tukanoan Unanano of the blackwater Vaupés River basin in Brazil and Colombia. Interciência 12: 78–86.

Comissão Pastoral da Terra (C.P.T.). 1992a. Ribeirinhos: Uma estação de luta. Dossier 1992. Comissão Pastoral da Terra, Regional Amazonas e Roraima, Manaus.

———. 1992b. Os ribeirinhos: Preservação dos lagos, defesa do meio ambiente e a pesca comercial. Comissão Pastoral da Terra, Regional Amazonas e Roraima, Manaus. Environmental Conservation 16(4): 331–337.

Cordell, J. C. & M. A. McKean. 1986. Sea tenure in Bahia, Brazil. Pages 85–112 *in* Proceedings of the Conference on Common Property Resource Management. National Academy Press, Washington, DC.

Denevan, W. M. 1984. Ecological heterogeneity and horizontal zonation of agriculture in the Amazon floodplain. Pages 311–336 *in* M. Schmink & C. H. Wood, eds., Frontier expansion in Amazonia. University of Florida Press, Gainesville.

Fairlie, S., M. Hagler & B. O'Riordan. 1994. The politics of overfishing. The Ecologist 25(2/3): 46–73.

Feeny, D., F. Berkes, B. McCay & J. Acheson. 1990. The tragedy of the commons: Twenty-two years later. Human Ecology 18(1): 1–19.

Furtado, L. G. 1990. Características gerais e problemas da pesca Amazônica no Pará. Boletim do Museu Paraense Emílio Goeldi, Antropologia 6(1): 41–93.

Gentil, J. 1988. A juta na agricutura na várzea na área de Santarém—Médio Amazonas. Boletim do Museu Paraense Emílio Goeldi, Antropologia 79: 1–50.

Goulding, M. 1980. The fishes and the forest. University of California Press, Berkeley.

———. 1983. Amazonian fisheries. Pages 189–210 *in* E. Moran, ed., The dilemma of Amazonian development. Westview Press, Boulder, Colorado.

———. 1989. Amazon: The flooded forest. BBC Books, London.

———. 1993. Flooded forests of the Amazon. Scientific American 266(3): 114–120.

———. 1996. Pescarias amazônicas, proteção de habitats e fazendas nas várzeas: Uma visão ecológica e econômica. Consultant's report to the Pilot Program for the Conservation of the Amazon Rainforest. The World Bank, Brasília.

Hardin, G. 1968. The tragedy of the commons. Science 162: 1243–1248.

Hartmann, W. 1989. Conflitos de pesca em águas interiores da Amazônia e tentativas para sua solução. Pages 103–118 *in* Pesca artesanal: Tradição e modernidade. III. Encontro de ciências sociais e o mar. Programa de Pesquisa e Conservação de Áreas Úmidas no Brasil, São Paulo.

Hecht, S. 1993. The logic of livestock and deforestation in Amazonia. Bioscience 43: 687–695.

———, A. B. Anderson & P. May. 1988. The subsidy from nature: Shifting cultivation, successional palm forests, and rural development. Human Organization 47(1): 25–35.

Isaac, V., V. Rocha & S. Mota. 1993. Algumas considerações sobre a legislação da "piracema" e outras restrições da pesca do Médio Amazonas. Pages 187–211 in L. Furtado, A. Mello & W. Leitão, eds., Povos das águas: Realidade e perspectiva na Amazônia. Museu Paraense Emilio Goeldi, Belém.

Junk, W. 1984a. Ecology of the várzea of Amazonian whitewater rivers. Pages 215–244 *in* H. Sioli, ed., The Amazon: Limnology and landscape ecology of a mighty tropical river and its basin. W. Junk, Dordrecht.

———. 1984b. Ecology, fisheries and fish culture in Amazonia. Pages 443–475 *in* H. Sioli, ed., The Amazon: Limnology and landscape ecology of a mighty tropical river and its basin. W. Junk, Dordrecht.

———. 1989. The use of Amazonian floodplains under an ecological perspective. Interciência 14(6): 317–321.

———, P. Bayley & R. Sparks. 1989. The flood pulse concept in river-floodplain systems. Canadian Special Publications in Fisheries and Aquatic Science 106: 110–127.

Kurien, J. 1994. Resistance to multinationals in Indian waters. The Ecologist 25(2/3): 115–119.

Larkin, P. A. 1977. An epitaph for the concept of maximum sustained yield. Transactions of the American Fisheries Society 106(1): 1–11.

————. 1978. Fisheries management—An essay for ecologists. Annual Review of Ecology and Systematics 9: 57–73.

Leroy, J. 1992. Uma Chama na Amazônia. Vozes, Rio de Janeiro.

Lima, D. 1994. A implantação de uma unidade de conservação em área de várzea: A experiência de Mamirauá. Pages 403–412 in M. A. D'Incao & I Maciel da Oliveira, organizers, Amazônia e a crise da modernização. Museu Paraense Emilio Goeldi, Belém.

McCay, B. & J. Acheson. 1987. The question of the commons: The culture and ecology of communal resources. University of Arizona Press, Tucson.

McCully, P. 1991. FAO and fisheries development. Ecologist 21(2): 77–80.

McEvoy, A. 1990. The fishermen's problem. Stanford University Press, Palo Alto.

McGoodwin, R. 1990. Crisis in the world's fisheries. Stanford University Press, Palo Alto.

McGrath, D., F. de Castro, C. Futemma, B. Amaral & J. Calabria. 1993a. Fisheries and resource management on the Lower Amazon floodplain. Human Ecology 21(2): 167–195.

————, ————, ————, ———— & ————. 1993b. Manejo comunitário da pesca nos lagos de várzea do Baixo Amazonas. Pages 213–229 in L. Furtado, A. Mello & W. Leitão, Povos das águas: Realidade e perspectiva na Amazônia. Museu Paraense Emilio Goeldi, Belém.

————, ———— & ————. 1994. Reservas de lago e o manejo comunitário da pesca no baixo Amazonas: Uma avaliação preliminar. In M. A. Díncao & I. M. Silveira, eds., Amazônia e a crise da modernização. Museu Paraense Emílio Goeldi, Belém.

MacKenzie, W. 1992. An introduction to the economics of fisheries management. FAO Fisheries Technical Paper 226, FAO, Rome.

Mathews, D. 1994. Commons versus open access: The Canadian experience. The Ecologist 25(2/3): 86–96.

Mattos, M. & C. Uhl. 1994. Economic and ecological perspectives on ranching in the Eastern Amazon. World Development 22(2): 145–158.

Meschkat, A. 1960. Report to the Government of Brazil on the fisheries of the Amazon region. Technical Report 1305. FAO, Rome.

Neal, R. 1982. Dilemma of the small-scale fishermen. ICLARM Newsletter (July 1982): 7–9.

Nielsen, L. 1976. The evolution of fisheries management philosophy. Marine Fisheries Review (December 1976): 15–22.

Ostrum, E. 1990. Governing the commons. Cambridge University Press, Cambridge.

Piñedo-Vasquez, M., D. Zarin & P. Jipp. 1992. Community forest and lake reserves in the Peruvian Amazon: A local alternative for sustainable use of tropical forests. Pages 79-86 in D. Nepstad & S. Schwartzman, eds., Non-timber products from tropical forests. Advances in Economic Botany 9. The New York Botanical Garden, Bronx.

Santos Viera, R. 1992. Várzeas Amazônicas e a legislação ambiental Brasileira. INPA, Manaus.

Schönenberg, R. 1994. As formas institucionais e organizacionais, de articular interesses na área da pesca no Baixo Amazonas em particular, e na Amazônia em geral. Relatório Preliminar para Projeto IARA, IBAMA, Santarém, Pará.

Scott, A. 1979. Development of economic theory on fisheries regulation. Journal of the Fisheries Research Board of Canada 36: 725–741.

Smith, N. 1985. The impact of cultural and ecological change on Amazonian fisheries. Biological Conservation 32: 355–373.

Sparks, R. 1995. Need for ecosystem management of large rivers and their floodplains. Bioscience 45(3): 168–182.

Stocks, A. 1987. Resource management in an Amazon várzea lake ecosystem: The Cocamilla case. Pages 108–120 in B. McCay & J. Acheson, eds., The question of the commons: The culture and ecology of communal resources. University of Arizona Press, Tucson.

Swardson, A. 1994. Net losses: Fishing decimates oceans' "unlimited" bounty. Washington Post, 14 August.

Weber, P. 1993. Abandoned seas: Reversing the decline of the oceans. Worldwatch Paper 116. Worldwatch Institute, Washington, DC.

———. 1994. Net loss: Fish, jobs, and the marine environment. Worldwatch Paper 120. Worldwatch Institute, Washington, DC.

Welcomme, R. L. 1985. River Fisheries. FAO Technical Paper 262. FAO, Rome.

Artisanal Fisheries of Pirarucu at the Mamirauá Ecological Station

Helder Lima de Queiroz

The Pirarucu

The Osteoglossidae family is found in the tropical regions of both the Old World and New World, but it is represented by only two genera in the Neotropics: *Osteoglossum* and *Arapaima* (Sterba, 1973). There are two known species of *Osteoglossum*, known as *aruanãs*, or "water monkeys," because they jump out of the water to capture prey on the lower branches of trees near watercourses. They are predators, and the adults can be slightly over 1 m long and weigh approximately 3 kg (Aragão, 1984). Although not considered high-quality fish, they are frequently consumed by the population of the middle Solimões region, because they are abundant and low priced. The aruanãs are being studied at the Mamirauá Ecological Station (MES) but have not been included in this paper.

The genus *Arapaima* is monospecific and represented by the *pirarucu*, probably the largest-scaled fish found in the Amazon basin, and believed by some authors to be the largest-scaled fish in the world (Wootton, 1990). The pirarucu can attain a total length of 3 m and weight of 150 kg. *Arapaima gigas*, like the two *Osteoglossum* species, are important predators (Goulding, 1980) that feed on smaller fish species, although their diet can include other items such as mollusks, crustaceans, and insects.

The name *pirarucu* originates in the Tupi language (*pira* = fish, *urucu* = red, a reference to the red fruit of the *urucu*, *Bixa orellana*). In adult fish, the regions near the caudal, dorsal, and anal fins are red. This characteristic is stronger in the males during the reproduction period, when the dorsal region remains dark while the ventral region acquires a whitish coloration (Fontenele, 1948).

Pirarucus are traditionally consumed by the present Amazon population (Castelo, 1983), as they have been for at least 150 years. They have also been exported to other regions of the country and to other countries (Veríssimo, 1895). Pirarucus constitute an important share of all fish species unloaded in the various ports of the Amazon region (Petrere, 1978b; Smith, 1979; Flores, 1980).

As large predators, pirarucus, freshwater dolphins, and caimans occupy the highest trophic level in Amazon aquatic systems. They are found throughout the whole region, especially in the large rivers and tributaries of the Solimões–Amazonas and Orinoco watersheds. Popular belief has it that pirarucus prefer the várzeas of the sediment-rich rivers, but they can also be found in black- and clear-water sub-basins. Their geographic distribution and selective habitat, however, are still insufficiently known.

Like most fish species in the Amazon basin, the reproductive life of *Arapaima gigas* is closely linked to the annual high- and low-water cycles that characterize the rivers of the region (Lowe-McConnell, 1987). The peak of reproduction occurs immediately after the waters begin to rise and flood the forests along the banks of the rivers, making such areas available for mating. The pirarucus build their nests in recently flooded areas, on the bottom substrate, usually the clayey soil of the várzea. The eggs are laid on the nests ready for spawning. After a short incubation period (Fontenele, 1953), the larvae hatch and receive parental care for several months. The male seems to take on most of the responsibility for parenting.

The first year of life is marked by rapid growth, and the young can attain a length of 1 m and fresh weight of almost 10 kg (Menezes, 1951). Rapid growth continues until the beginning of the reproductive period; weights can double annually. Even after achieving reproductive maturity, pirarucus continue to increase in length and weight. The males are usually longer and thinner, the females shorter and thicker.

Pirarucus are mixed breathers. The fish use their highly vascularized air bladder as an accessory respiration organ, in addition to regular bronchial breathing (Sawaya, 1946). This evolutionary characteristic may be associated with the low levels of dissolved oxygen in Amazon watercourses. Pirarucus must use both breathing systems throughout their whole lives and, therefore, must come to the surface every 10–20 minutes to take in atmospheric oxygen (Lüling, 1964). Upon surfacing, the fish become easy targets for harpoon fishermen (Veríssimo, 1895).

Resource Exploitation

Pirarucu fisheries were described in the last century (Nunes-Pereira, 1893; Veríssimo, 1895). One hundred years later, the artisanal fishing methods used by the riverine communities are almost the same. Harpoons are the most frequently used fishing device, and the fish are still located when they break the surface to breath atmospheric oxygen. The continued use of this fishing method shows its adaptation to this artisanal mode of production. Other fishing gear traditionally used by riverine communities includes the *espinhéis*, a fishing line with one or more large hooks, baited with dead fish. At the MES, the few fishermen who still use this technique report using *traíras* and *acarás* as bait. The technique, however, was used in less than 1% of known catches; its infrequent use precludes any type of analysis. The technological changes in Amazon basin fisheries pointed out by other authors (Smith, 1979; Goulding, 1980; Lowe-McConnell, 1987) have also been felt in the pirarucu fisheries at the MES. Twenty years after nets began being widely used to catch pirarucu in the region, many local inhabitants believe that there has been a substantial reduction in the yield. There are no local historical records to back up this statement, but the bankruptcy of long-standing commercial enterprises that exported preserved pirarucu to other states or countries may be a useful indicator.

The local population claims that the reduction has been felt throughout the whole Amazon basin, although, again, there are few records to corroborate such claims. The lack of historical data is a problem. Nevertheless, in discussing other issues, some authors provide very valuable information. There is evidence that at the end of the 19th century, 11,500 tons of dry pirarucu were unloaded in the port of Belém during a nine-year period, or an average of 1300 tons/year (Veríssimo, 1895). By 1925, only 300 tons per year were being unloaded at the same port (Menezes, 1951). A marked decrease in the average size of the fish landed is mentioned in the literature, although no numbers accompany the allegations (Lowe-McConnell, 1987). According to a report by a fisheries specialist working in the Amazon basin in recent years, this decrease is quite real and has already had a negative impact on the stocks of some producing regions (Goulding, 1979; Lowe-McConnell, 1987; Pereira-Filho et al., 1991; R. Barthem, personal communication).

In the face of the alleged risk of overfishing to which the pirarucus are exposed and the acknowledged drop in production and fish size in the traditional producing centers, the Brazilian environmental authorities (Instituto Brasileiro do Meio Ambiente e dos Recursos Naturais Renováveis-IBAMA) issued regulations on two important aspects of pirarucu exploitation, namely, minimum fish size and prohibition of fishing during the reproduction period, which is called *defeso*.

In an administrative ordinance dated 20 December 1989, IBAMA forbade catching and selling pirarucu less than 150 cm long. This minimum length is not the minimum required for sustainable fisheries but, rather, the average size believed to represent the beginning of the reproductive life of the species. The following year, in an ordinance dated 4 March 1990, IBAMA forbade pirarucu fishing throughout the whole Amazon basin from 1 December through 31 March. In the Araguaia and Tocantins Rivers and their affluents, the defeso period stipulated for the pirarucu begins on 1 October and ends 31 March, during which time pirarucu fishing is forbidden.

The regulations gave rise to much complaining among fishermen and merchants, many of whom claimed that there was insufficient scientific knowledge to provide a sound basis for the decision. Nevertheless, the preliminary results of this study indicate that the sizes and periods set forth in the IBAMA regulations closely reflect the real situation.

The average size of pirarucu at the time of the first reproduction event is being studied at the MES and seems to be slightly less than 160 cm. A limit slightly lower than the real minimum size can be tolerated, if one takes into account the relatively high fecundity of the species.

In the middle Solimões region, the main egg-laying period and subsequent parental care phase coincide with the legal fishing prohibition period (defeso). The dates, however, do not seem to be as accurate for other regions of the Amazon. The fish lay their eggs immediately after the beginning of the floods, which vary substantially throughout the Solimões–Amazonas watershed.

Despite scientific or other justification, the above-mentioned administrative ordinances are indiscriminately disregarded, and the regulating agency does not have an adequate infrastructure to enforce the law and punish, when necessary, the fishermen and fish merchants who break the law. This is particularly true in view of the extensive area covered by the Amazon basin.

Pirarucu Fisheries at the Mamirauá Ecological Station

A comprehensive study of pirarucu fisheries, as well as of the biological, ecological, and behavioral aspects of the species, was started in June 1993 at the Mamirauá Ecological

Station. The purpose of the study is to draw up management regulations for the sustainable use of the species. There are approximately 60 communities that use the MES, four of which have been monitored since the beginning of the study. Another six communities have been monitored since June 1994. In each community, a local inhabitant was trained to record all pirarucu caught by the community members. The total length and weight of the fish are measured and recorded using a special measuring device and a suspended scale furnished by the project. The data collector also removes the stomach, some scales from the region behind the gills (above the lateral line), and the ovaries of female fish. Information is obtained about the lake where the catch was made, how long the fishing expedition lasted, the fishing gear used, and the number of fishermen involved in each expedition. The data collectors also record the weight and length of the salted fish produced, based on a subsample of the pirarucu landed, as well as the amount of salt used in processing the fish.

The first noteworthy feature of pirarucu fisheries is their marked seasonality, shown in Figure 1. Most fish are caught during a period of approximately four months, from the end of the low-water period to the beginning of the high-water period. As a rule, fishing takes place from September to December of each year. Surprisingly, the defeso does not influence the seasonality of the fisheries. When interviewed, the fishermen emphatically state that they continue to fish during the defeso period. Considering the high degree of trust between data collectors and fishermen, I do no believe that there is any systematic underreporting during the defeso.

Two important factors help establish the seasonality of pirarucu fisheries. First, the

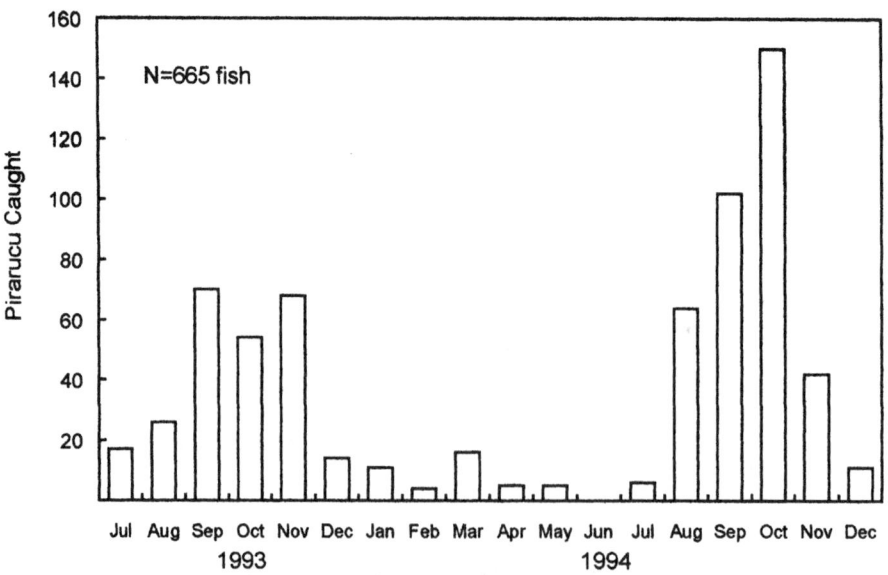

Figure 1. Monthly pirarucu catch in four MES communities in 1993 and 1994, with the peaks of production that characterize the fabrico and usually last three to four months. The December–May period corresponds to the period of IBAMA fishing prohibition.

peak fishing period is characterized by low water levels, when the species are trapped and concentrated in the lakes and canals, thus becoming easy prey for the fishermen. During the defeso period, pirarucu easily hide in the flooded forest, making both detection and capture with nets much more difficult (Smith, 1979). Second, the high-water and low-water regime establishes a rather rigid cycle for the subsistence activities of the riverine communities. Until the water begins to rise, the inhabitants are involved in logging (Albernaz, this volume). As soon as the timber rafts are sent to the mills, the first pieces of land are exposed by the rapidly falling water. The soil is prepared for short-cycle crops such as cassava. Only then can the riverine communities devote any time to pirarucu fishing. This period is called the *"fabrico do pirarucu"* and is necessarily limited by the need to harvest the crops before the water once again floods the farming areas. At that point, most várzea farmers go back to their subsistence farming activities. Ultimately, therefore, the fabrico do pirarucu is seasonal, because it basically depends on seasonal variation in water level.

The 1993 and 1994 fabrico periods will be analyzed in this chapter. Only the four communities sampled during those two years will be considered, except when otherwise indicated, so as to permit comparison. The communities are Vila Alencar (03°07'56.3"S, 64°48'41.4"W), Boca do Mamirauá (03°07'07.9"S, 64°47'34.1"W), Sítio São José (03°06'39.1"S, 64°47'07.2"W), and Jarauá (02°51'55.5"S, 64°55'33.4"W); all are communities within the Mamirauá Ecological Station.

The four communities differ in several aspects, particularly the size of the population, use of less traditional fishing technologies in their artisanal fishing efforts (Table I), and degree of dependence on extractivist activities, which is inversely proportional to their involvement in subsistence farming. It may be stated that all MES communities engage in both subsistence agriculture and extractivism, the only distinguishing feature being the share of each type of activity in the overall balance of their economies. The degree of dependence on extractivism determines, among other things, the relations of the community with merchants in neighboring towns, or even itinerant trading boats, called *"regatões."* These merchants operate on a patronage system, supplying foodstuffs in exchange for a promise of future supply of the products extracted from the várzea. Since riverine inhabitants are always in debt to their "patrons," extractivist activities are forever intensive.

Table I. Population characteristics and technology used in pirarucu fisheries in four communities

Community	Population	Fishermen	Craft[a]	Fishing gear[b]
Vila Alencar	96	15	4	Harpoon and espinhel
Boca do Mamirauá	59	10[c]	1	Harpoon
Sítio São José	52	10	2	Harpoon
Jarauá	102	25	16	Harpoon and gill net

Source: Lima-Ayres, 1994.
[a] Craft with center-mounted diesel engine (batelões) or with outboard motors (rabetas).
[b] Type of fishing gear used for catching pirarucu.
[c] The number of fishermen decreased from 10 in 1993 to 5 in 1994.

It is not yet possible to precisely determine the annual pirarucu catch at the MES. Also, it is impossible to tell to what extent the results obtained in these four communities can be extrapolated to the other 50 communities that use the station. Nevertheless, it is possible to estimate the annual production at about 63–115 tons of fresh pirarucu, on the basis of the production of the four communities in 1993 and 1994, which were, respectively, 3.1 tons and 7.7 tons of fresh fish (249 and 414 fish).

There are two important reservations that need to be stated, the first of which relates to withholding information. During the first fabrico period studied, the data collection structure had just been set up, and many fishermen were still concerned about the possibility of the research results leading to repressive measures against them, since the minimum fish size established in the law was being systematically disregarded. Only in 1994 did the fishermen really begin to trust researchers enough to show all pirarucu caught, including very small specimens.

Comparing the structure of the distribution of total lengths of the fish caught in 1993 with that of 1994 (Fig. 2), it may be seen that, during the first year, only 62.6% of fish were smaller than the size stipulated in the law. The following year the figure rose to 64.7%. It is believed that the almost 2% difference (close to 29 fish) is due to withheld information. Thus, the annual production of the four communities in 1993 would rise to 3.4 tons and the minimum production at the MES in 1993 would be approximately 67.5 tons of fresh pirarucu.

A second important reservation is possible overestimation. The floods of both 1993 and 1994 may be regarded as exceptionally high (probably the third and second highest floods, respectively, of the second half of this century), and in the second year of the

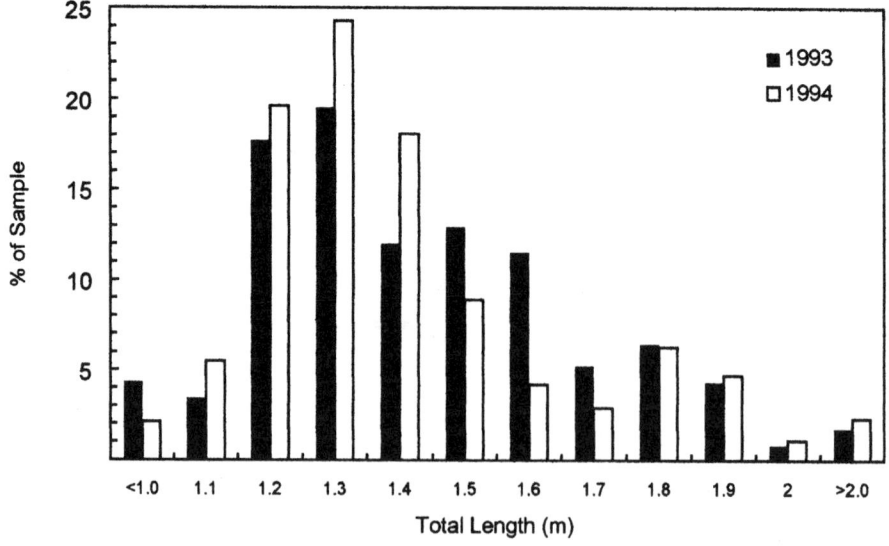

Figure 2. Distribution of the frequency of total length (in meters) of pirarucu caught in four MES communities during the fabricos of 1993 and 1994, of a subsample of 478 fish.

study the water level was 50 cm higher than the maximum level for the first year (13.8 m and 14.3 m, respectively). Unfortunately, there are no historical water-level data for the MES. All local fishermen assert that large floods result in big catches during the following low-water period. Thus, it is possible that the two fabrico periods shown in this study were exceptionally productive. Continuing work will help establish the actual relation between water level and pirarucu catches at the MES.

These two points notwithstanding, there is a clear disparity between the production of the first and second years, as shown in Table II. The disparity may be explained by the water-level dynamics for the two years rather than by the height of the water during the flood period. Such dynamics determine how long the fabrico will last, as explained above. In the first year, the end of the fabrico was unexpected because of the abrupt rise of the water level in the beginning of November (approximately three months). The forests were soon flooded, offering shelter to the fish. In the second year, the fabrico was extended until the end of December (approximately four months), since the rise in water level followed the pattern expected by the fishermen and took place in late December. It is useful, therefore, to consider how the catch is related to the unit of fishing effort (days worked), which determines productivity, as will be discussed later. Other aspects must be considered first.

The production of the four communities during the two years of the study is shown in Table II. Boca do Mamiraué and Sítio São José had a reasonable annual-catch increase. The drop in the catch at Vila Alencar was due to three main factors. First, Vila Alencar is the most farm-oriented of the four communities, and is increasingly less involved in extractivist activities. Second, a good part of this community moved away to terra firme, outside the MES area, during the 1994 flood, in order to protect itself and its crops from the flood. Last, the interviews demonstrate that Vila Alencar's 1993 catch was unduly large, due to almost accidental circumstances. The fast rise of the water in that year trapped a large number of pirarucu near the community, and the Vila Alencar fishermen joined the fishermen in a neighboring community in a fishing *ajuri* (communal enterprise). In two days, 25 fish of various sizes were caught. The phenomenal increase in the annual catch in Jarauá, as well as its significantly higher production when compared with the other three communities, in the two years under consideration, may be attributed to the fishing technology used by its inhabitants. The Jarauá fishermen use a combination of the traditional tool, the harpoon, with a more recently introduced fishing device, the *malhadeira* (gill net). The catch using gill nets is propor-

Table II. Annual pirarucu catch in four communities

Community	Annual catch (kg fresh fish)	
	1993	1994
Vila Alencar	766.6	471.7
Boca do Mamiraué	759.8	1015.2
Sítio São José	439.8	683.2
Jarauá	1214.3	5565.0
Total	3180.5	7735.1

tionally much larger than in the other communities. Furthermore, the large number of small craft in Jarauá means that the area fished is larger than that in the other communities. Both aspects will be discussed later.

Only about 30% of the fish caught during the two years exceeded the minimum fish size permitted by IBAMA. For both 1993 and 1994, there are records of fish more than 2 m long, albeit in very low frequencies (2% and 3.5%, respectively). Similarly, fish under 1 m in length were also caught (4% and 6%, respectively).

By the end of the 1994 fabrico, and considering only the four communities mentioned above, the mean size of all pirarucu fished at the MES since the beginning of the study was 138.4 cm (S.D. = 26.11 cm; var. coeff. = 0.17); during the second year, the mean size dropped to 130.6 cm (S.D. = 25.95 cm; var. coeff. = 0.19). It must be emphasized, however, that there is still a strong possibility that information was withheld during the first year, as regards 30 small-length fish.

The decrease in total length of the pirarucu caught, from one year to another, is not statistically significant, although it was consistent in all four communities studied, as shown in Table III.

It is interesting to note that the greater the distance from the community to the consumer and regional export market (Tefé), the greater the mean total length of the fish caught by the community ($r = 0.910$; $P = 0.013$), when all six communities studied in 1994 are taken into account. The distances were measured linearly, as the bird flies, with the assistance of a global positioning system.

The decrease in the average total length of the pirarucu caught may only reflect withholding of information in 1993, or it may be an artifact of the small sample size. But it may also be the first sign of rise of overfishing recorded for the MES populations. It is still early to make any statement, and the information on the fabrico of the years to come will clarify any doubts.

The pressure of the regional market tends to increase, since the demand for the pirarucu is growing in all consumer markets. This pressure is also felt in the size of the fish caught. Studying the pirarucu populations in the MES lakes, experimental fishing, as well as marking and recapture procedures, were carried out to determine the abundance and distribution of the species. This aspect of the study will be discussed elsewhere. It must be noted, however, that pirarucu density varied from 80 to 170 individuals per hectare of water surface in the preserved lakes of the station, whereas only 50 individuals per hectare of water surface were found in the fished lakes, after the

Table III. Mean total length, in centimeters, of pirarucu caught by four communities in 1993 and 1994

Community	1993	1994
Vila Alencar	144.1	129.2
Boca do Mamirauá	136.8	121.6
Sítio São José	135.1	134.7
Jarauá	137.8	136.8
Total	138.4	130.6

fabrico. In the preserved lakes, the pirarucu biomass varied from 1.5 to 3.7 tons per hectare of water surface and included fish of all sizes. In the fished lakes, the biomass dropped to 0.15 tons per hectare of water surface after the fabrico, and most of the individuals found were small. Only 22% of the fish remaining in the lakes after fishing weighed more than 20 kg, whereas in protected lakes, around 48% of the fish were above this weight.

Despite the overfishing problems reported (albeit undocumented) for other areas of the Amazon region, I believe that there is as yet no confirmation of the overfishing hypothesis in Mamirauá. The pirarucu populations at the MES seem to be rather healthy. Before reaching any premature conclusion it is necessary to build a consistent historical database on the existing pirarucus populations, as will be discussed later in this paper.

Fishing Expeditions

Another important feature of pirarucu fishing at the MES and, particularly, in the Jarauá community are the fishing expeditions. During the first month of the fabrico, the fishermen, either alone or in small family groups, combine the preparation of their farming areas near the community with pirarucu fishing in lakes and canals. The pirarucu caught are brought back to the community, in the evening, to be processed. As the fabrico progresses, neighboring areas are no longer productive and the fishermen form groups (based on either kinship or affinity) and travel to increasingly farther lakes, where they spend several days fishing and processing their catch.

In order to sample this type of fishery, 30 Jarauá expeditions were monitored during the two years of the study. The average length of these expeditions was 5.5 days, varying from one to ten days. The groups comprised between one and five individuals, with an average of 2.2. The average catch of each fishing expedition was 8.7 fish, with a production of 125.8 kg of salted fish.

The measurement of fishing effort adopted in the study was men/work days (Petrere, 1978a), whereas the catch measurement was kilograms of salted fish *manta*, which is the product sold in the market. Thus the productivity of the fishing expeditions was calculated. In the two years of the study, the average productivity was 16.02 kg/man/day. There was a small variation between the two years of the study. In 1993, the productivity was 14 kg/man/day, whereas in 1994 it rose to 16.5 kg/man/day. It is possible that the increased productivity was due to higher water level in 1994, as local fishermen believe.

The variation in average monthly productivity reveals the strategy adopted by the MES riverine farmers. During the first month of the fabrico—usually August or September, depending on the dynamics of the waters—the fishing expeditions are longer (an average of 9.5 days) and take place in lakes near the community. During this period the productivity did not exceed 0.5 fish/man/day, or 6 kg/man/day. During the second month of the fabrico, the expeditions are shorter and the areas fished are farther away from the community. The average length of the expeditions was 5.6 days, and production was 0.8 to 1.5 kg of fish/man/day. Since, as a rule, the total length (and fresh weight) of pirarucu from the more distant areas is greater, the productivity varies from

10.7 to 19.3 kg/man/day. In the last months of the fabrico, the expeditions are even shorter (an average of 4 days), the areas fished are increasingly remote, and the fish are even larger. Productivity varied from 17.2 to 29.4 kg/man/day. At the end of the fourth month of fabrico, the water begins to flood the forest, providing a refuge for the fish, and the crops must be immediately harvested. No more fishing expeditions are made, and catches abruptly decline.

The fishermen, therefore, go to richer and more productive areas only after nearby areas have been exhausted. In order to fish in more distant areas, the fishermen must use outboard motorboats, thus incurring fuel and repair expenditures. Although it is remarkable that the more productive areas are fished for shorter periods of time, there is an explanation, namely, the limited transport capacity of the small craft used. The boats cannot easily carry all the fishermen, their gear and belongings, the salt, and, on the way back, a load of fish. Since the more distant areas are more productive and the fish are much larger, the transportation capacity is soon filled.

This pattern is much more marked in the Jarauá community, where the resource area is larger and includes a great number of nearby and distant lakes. Also, the Jarauá fishermen own motorboats and, therefore, can travel farther away on their fishing expeditions. The fishing gear used by this community is another important feature of Jarauá fishing, as is explained below. In the other three communities studied, most fishermen do not have motorboats. Fishing expeditions are rarer and, as a rule, are made up of one or two fishermen in tiny canoes. The fishing areas are much smaller and, on average, so are the fish. Although it is less clearly perceived, members of the other communities also travel farther away as the fabrico period progresses. Thus, the productivity of these communities varied from 3.3 to 8.3 kg/man/day. It is quite possible that these values come closer to the pattern of the other MES user communities than to the high Jarauá value, since the latter is an important fishing center in the Mamirauá Ecological Station. It is believed that there are only two or three other communities with productivity as high as that of Jarauá.

Another technological factor that may contribute to high productivity in Jarauá is the fishing gear used. As a rule the other communities use harpoons to catch pirarucu. In Jarauá, 21% of fish were caught using only harpoons and 79% were caught with gill nets (8.6% with only gill net and 71.4% with harpoon and gill net together). In the other communities, gill nets are used only in approximately 10% of catches, and these communities also presented instances of use of hooks and *espinhéis* (another traditional fishing device).

Malhadeiras (or *caçoeiras*, as gill nets are also known) are nets used to entangle fish. The fish are caught perpendicularly to the net when they try to cross it. They are usually caught by the operculum (gill cover). Gill nets were introduced into the region in the early 1970s. Although all local fishermen have general knowledge about the more relevant biological aspects of the life of the pirarucu and refer to the introduction of the gill nets as the main cause of the alleged decline of production, these devices are increasingly being used. Whenever an opportunity arises, local fishermen buy and use industrially manufactured gill nets. Although flimsy, these gill nets are cheaper and ensure reasonable profits from fishing activities. Good descriptions of the above-mentioned fishing devices may be found in works by Petrere (1978b) and Smith (1979).

When used together with harpoons, gill nets really increase productivity. On the other hand, they are a source of concern for the survival of pirarucu populations. Harpoons accounted for 0.7 fish/man/day, whereas gill nets caught 1.2 fish/man/day. Transforming these numbers into productivity values, harpoons produce 19.8 kg/man/day since they can catch fish of all sizes, including large fish. Gill nets, by themselves, produce less—16.4 kg/man/day—because they catch mostly small fish, from 110 to 140 cm long. The distribution of pirarucu by total length caught by each fishing device is shown in Figure 3. Fishermen can estimate the length and weight of pirarucu before killing them with the harpoon, because this type of fishing is based on careful observation of the fish whenever it surfaces to breathe. Consequently, harpoon fishermen can select the fish they wish to catch, on the basis of its size.

When both harpoons and gill nets are used, there are more fishing opportunities and a better distribution of tasks among the members of the expedition. Hence, the two devices together permit a much higher productivity: 43.7 kg/man/day.

Another important argument for the use of the gill nets is the fact that a harpoon catches one fish at a time, whereas each placing of the gill net enables fishermen to catch up to 12 fish at once, even if some of them are small. It seems that the larger fish not only can avoid becoming ensnared in the nets but, indeed, can even disentangle themselves by tearing the net. Only very low-quality gill nets are used, because the material to make good nets is expensive. The smaller fish, which have not yet arrived at their first reproductive event, are weaker and less experienced. Consequently, they get caught in the gill net. The continuous use of gill nets seems to be detrimental to

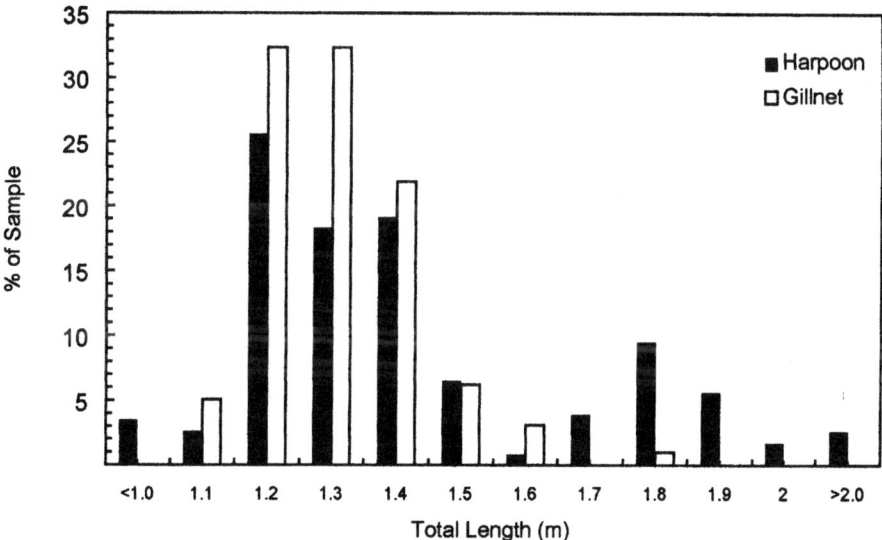

Figure 3. Distribution of the frequency of total lengths (in meters) of pirarucu caught in four MES communities using two different fishing devices, harpoons and gillnets, of a subsample of 300 fish.

localized subpopulations, but, because of the lack of information about the renewal of populations in each flood period, we do not know what the effect of this fishing device may be when used on a larger scale.

Fishermen in other parts of the Amazon region, such as Itacoatiara (Smith, 1979), also report the deleterious effects of gill nets on pirarucu populations mentioned by the Mamirauá fishermen. And yet there seems to be increasing use of these devices throughout the region.

Production of Salted Fish

The processing of pirarucu has been described by other authors (Botelho, 1956; Dias, 1983). The pirárucu of the MES are processed either in the communities or where the fish are caught. As a rule, the fishermen use only a small knife, although more experienced or skilled people substitute a larger knife. The scales, together with the skin, are first removed from behind the head, in the region of the gill cover, down to the tail. The fins are removed, leaving only the basal part of their rays. Beginning in dorsal region, a longitudinal cut is made along the backbone, from the end of the head to the tip of the tail. The flesh is separated laterally, on both sides, and the ribs and remaining fin rays are carefully loosened from the flesh. Finally, the ligaments linking the viscera to the peritoneum are cut in order to completely separate the bones, viscera and head from the flesh, which remains in a single, almost triangular piece.

The anterior part of this triangular piece corresponds to the back of the head, and the posterior part to the caudal peduncle. The central part of the piece corresponds to the ventral region, locally called "*ventrecha*" and much prized throughout the region. The sides of the piece correspond to the dorsal region, where the initial cut was made. The whole piece is called a "*manta*."

As a rule, the mantas are sold whole or divided into halves, or bands, by a longitudinal median cut. The manta from a large pirarucu may be so thick that it can be divided into layers near the sides and ventrecha. The resulting layers, or *coletes*, become separate pieces. A single fish may yield up to three large pieces.

After the fish is cut into pieces, salt is applied directly to the meat and then is reapplied a day later, when no more water oozes from the meat. At that point, the mantas are still fresh. After two or three days in the sun they will be dry and ready for market. At the MES, 1 kg of salt is used for every 12 to 16.7 kg of fresh pirarucu, or 7 to 10 kg of fresh meat. On the average, 1 kg of salt is used for every 14.2 kg of fish, or 1 kg of salt for 8.5 kg of fresh manta.

Of a sample of 194 fish monitored from the time they were caught to the production of the respective manta, it was observed that only 99 were sold dry. The average length of a fresh manta is close to 92% of the total length of the whole fish, whereas the average length of a dry manta is only 83% of the total length of the whole fish, or 90% of the fresh manta. The weight loss is even more striking. The average weight of fresh mantas is close to 44% of the total weight of the fish, while the weight of dry mantas is only 28% of the whole weight of the fish, or 63% of the fresh-manta weight.

One of the difficulties that enforcement personnel face in trying to prevent the fishing of pirarucu smaller than the minimum size established by law is that the products found are either fresh or dry mantas. The measurement of the mantas in this sample enabled the researcher to construct regression curves for the weight and total length of both fresh and dry mantas. On the basis of such regression curves, which are shown in Table IV, it was possible to ascertain that fresh mantas less than 126 cm long, or dry mantas less than 115 cm long, are illegal, because they were made from fish less than 150 cm long, below the minimum legal size.

The price of dry pirarucu mantas in the Tefé retail market has remained in the range of US$2.00–2.50/kg. The price paid to producers, however, has undergone fluctuations since the study was begun. An important factor in the price paid to the producer is the distance from Tefé. The price per kilogram of manta paid by traveling merchants is inversely proportional to the distance from Tefé, since the expenditures of boat travel are figured in. Thus, the communities farther away from Tefé sell their product for a lower price and pay much more for their supplies. The existence of motorboats in a community can again make a difference, since the boats enable fishermen to take their product directly to the consumer market and get better prices at the port.

During 1993 and in the first half of 1994, the price paid to producers was US$1.00–1.50/kg of dry manta in fishing communities near Tefé. In the more distant communities, the price was US$0.80–0.90/kg. After the abrupt changes in the Brazilian foreign exchange policy enacted by the economic authorities during the second half of 1994, the price of the dry manta paid to suppliers near Tefé ranged from US$1.90 to US$2.00, whereas that paid to more distant suppliers ranged from US$1.00 to US$1.50. Fresh, recently salted mantas command 80% of the price paid for dry mantas. Producers sell such mantas only when they urgently need money, or when no visits from trading boats are foreseen for a long time.

It is estimated that pirarucu fisheries at the Mamirauá Ecological Station generated from US$45,900.00 to US$81,000.00 in 1993, and from US$61,200.00 to US$108,000.00 in 1994. The productivity of the communities that use more traditional technology varied from US$6.00 to US$16.00/man/day, whereas that of more modern communities was close to US$33.00/man/day.

The Market

The biomass of pirarucu mantas sold each year in Tefé has not yet been determined. It is certain that all regional production finds an outlet at Tefé, with only a small part going to the nearby city of Alvarães. Fishermen and small merchants who run trading

Table IV. Regression formulas for the length and weight of fresh and dry pirarucu based on a sampling of 194 whole fish. G, weight of fresh fish; W, total weight of fish; F, length of fresh fish; L, total length of the fish; T, weight of dry fish; S, length of dry fish

G = 1.568 + 0.534 (W)	F = 8.913 + 0.782 (L)
T = 2.915 + 0.354 (W)	S = 24.071 + 0.602 (L)
T = 0.344 + 0.661 (G)	S = 10.087 + 0.810 (F)

boats sell to only a few large merchants in Tefé, who control regional sales and distribution to Manaus (the other large state market) or even to neighboring states. This oligopoly may explain the regional prices for pirarucu mantas. Although they pay US$1.50–2.00/kg to producers, these merchants are able to sell the product for up to US$6.00 in the large consumer markets of the region.

The effort to enforce prohibitions on the fishing of animals below the legal minimum size should focus on these few merchants. This action would be much simpler and much less expensive than trying to cover the many producers of pirarucu mantas, who are widely scattered throughout the region. Enforcement applied at the top of the economic pyramid would be transmitted to the base in a less traumatic and much more ethical manner, and would be more easily and effectively accepted. The local merchants themselves would be forced to convince fishermen to limit their catch to fish above the legal minumum size.

General Considerations on the Management of the Species at the MES

It is important to mention that the main objective of the Mamiraué Ecological Station is to protect the biodiversity of the area. Thus, when talking about the management of the pirarucu populations we are not referring to a model for optimizing fisheries, which would seek a maximum sustainable catch and ideal fishing effort. When finally defined, the management of pirarucu at the MES will aim at the maintenance of the demographic levels of the population and at assuring that it is as unaffected by fishing activities as possible. It is more important for the station to perpetuate the normal genetic stock and evolutionary processes of the species or resource than to promote the optimal utilization of such resource.

The conservation model tested at Mamiraué contemplates the permanence of the human populations that live in the area, as well as the continuation of their economic subsistence activities in a sustainable manner. It will be necessary, therefore, to establish zoning regulations that provide total protection to pirarucu populations, as well as fishing areas with well-defined rules of usage.

The information provided in this paper enables us to conclude that pirarucu fisheries at the MES are an essential activity within the local, traditional subsistence economy. The pirarucu, throughout the whole area of distribution of the species, is vitally important for maintaining the way of life of riverine communities, as well as for their sustenance. While I have not in this paper considered larger-scale commercial fisheries carried out by large craft from more developed fishing centers, regional priorities must include the perpetuation of this type of activity as well as local societies' traditional way of life. At the same time, managers must also implement conservation strategies for large, commercially valuable predatory species such as *Arapaima gigas*, which plays an extremely important role in maintaining biodiversity in the Amazon. Their disappearance would bring unpredictable consequences to the structure of the regional aquatic systems.

All the previously mentioned motives unequivocally point to the need to develop sustainable methods of managing the species. It is necessary to determine guidelines that will enable fishermen to continue exploiting the resource, even if at suboptimal levels, and ensure the maintenance of sound reproductive stock representing the whole range of variability of the species.

To this end, it is essential to determine the size of the stocks and their distribution. Knowing the available pirarucu biomass within the area of influence of the MES will be a necessary datum for monitoring the biomass fished—perhaps even much more important than the more stochastic indicators, such as the average size of pirarucu landed, found in such a small annual sample.

It is important to establish clearly how exploitation patterns are reflected in population parameters that might be, on one hand, sensitive enough to show changes in levels of risk and also, on the other hand, immediate enough to allow for an appropriate response before it is too late to prevent irreversible decline. In order to meet such needs, the study included inquiries into the reproduction (and population recruitment through birth and immigration), growth (and age determination), and population dynamics of the species, as well as its use of the habitat and·available food resources.

Studies on the abundance and structure of the populations in protected and unprotected areas of the MES may show the distinct effects of natural death and death resulting from fishing of the species in the region. Nevertheless, the most important result will be a historical database being built, with information on catch and fishing effort, together with variations in water level. So far, the database has data from only three years. I believe that five years of continuous data collection will be required before it is possible to determine whether the Mamirauá populations are seriously threatened.

The available data are being used to analyze cohorts and the dynamics of pirarucu biomass (Hilborn & Walters, 1992); the analyses will become increasingly accurate as the number of annual fabricos analyzed increases. A preliminary definition of age classes was obtained, and fecundity tables and life tables were drawn up for 1993 and 1994. The results of these analyses will be published elsewhere, since they are beyond the scope of this paper.

Until all data are in, some measures must be implemented in order to prevent any risk to unidentified stocks. For the time being, the regulations established by IBAMA seem to be quite appropriate for the area. The exceptional conditions offered by the Mamirauá Project permit the actual implementation of the measures, thanks to the continuous presence of researchers and extension workers in the area and the environmental-education activities being carried out. Virtually all communities within the MES focal area are involved in implementing the measures through their participation in the Mamirauá Project.

Longer-term management measures may be defined in 1996, based on the data from the first three years of catch monitoring. As a rule, fisheries management strategies, depending on their specific objectives, involve limiting either production or fishing effort (Caughley & Sinclair, 1994), and actions vary from restricting or limiting the fishing gear or craft to placing minimum size limits on the fish caught to establishing fishing stations and total catch ceilings. The uniqueness of the Mamirauá Ecological Station will probably require a mixed system including various types of action, implemented at various scales.

Acknowledgments

The work is being carried out with the invaluable assistance of a large number of riverine community members who live in the Mamirauá Ecological Station. These are almost 20 people who collect data, act as guides, catch and mark the fish, and engage in many other activities. I am and will always be extremely grateful for their help. The MES research team has been extremely important in discussions on all the points of this research, including the drawing up of the manuscript. The whole study and ideas expounded here are my own responsibility, although I must also thank M. Ayres, R. Barthem, P. Henderson, and B. Robertson for their help in the original design of the research. Support for the work is provided by ODA, WWF, WCS, CI, SEMACT-AM, IBAMA, CNPq-PTU, MPEG, INPA, UFPA, and many other Brazilian and foreign institutions.

Literature Cited

Aragão, L. P. 1984. Contribuição ao estudo da biologia do Aruanã, *Osteoglossum bicirrhosum* Vandelli, 1829, no Lago Janauacá—Estado do Amazonas, Brasil. 1. Desenvolvimento e alimentação larval. Boletim Cearense de Agronomia 15(1/2): 7–17.

Botelho, A. T. 1956. Preparação e Salga do Pirarucu. SPVEA, Belém. 35p.

Castelo, F. P. 1983. O pirarucu do Amazonas, alimento indígena de alto valor protéico. Manaus.

Caughley, G. & A. R. E. Sinclair. 1994. Wildlife ecology and management. Blackwell Scientific, Oxford.

Dias, A. F. 1983. Salga e secagem do pirarucu *Arapaima gigas* (Cuvier 1829) com aplicação de coletores solares. Tese de Mestrado, INPA/FUA, Manaus.

Flores, H. G. 1980. Desarrollo sexual del paiche (*Arapaima gigas*) en las zonas reservadas del estado (Rios Pacaya y Samiria) 1971–1975. Informe Instituto del Mar del Peru 67.

Fontenele, O. 1948. Contribuição para o conhecimento da biologia do pirarucu, *Arapaima gigas* (Cuvier), em cativeiro (Actinopterygii, Osteoglossidae). Revista Brasileira de Biologia 8(4): 445–459.

———. 1953. Hábitos de desova do pirarucu, *Arapaima gigas* (Cuvier) (Pisces: Isospondyli, Arapaimidae), e a evolução de sua larva. DENOCS Publication 153, ser. I-C.

Goulding, M. 1979. A pesca no Rio Madeira. INPA/CNPq, Manaus.

———. 1980. The fishes and the forest. Exploitations in Amazonian natural history. University of California Press, Berkeley.

Hilborn, R. & C. J. Walters. 1992. Quantitative fisheries stock assessment. Choice, dynamics and uncertainty. Chapman & Hall, London.

Lima-Ayres, D. 1994. Análise do crescimento demográfico da EEM: 1990–1994. Projeto Mamirauá: Relatório Semestral 5. Unpublished report.

Lowe-McConnell, R. H. 1987. Ecological studies in tropical fish communities. Cambridge University Press, Cambridge.

Lüling, K. 1964. Wiss Ergebnisse der Amazonas-Ucayali Expedition. Dr. K. Lüling 1959/1960. Zur Biologie und Okologie von *Arapaima gigas* (Pisces, Osteoglossidae). Zeitschrift Morphologie Ökologische Trere 54: 436–530.

Menezes, R. S. 1951. Notas biológicas e econômicas sobre o pirarucu. Servicia de Informacão Agricultura Série Estudos Técnicos 3. Rio de Janeiro.

Nunes-Pereira, M. 1893. O pirarucu. L. C. Alves & Cia. Rio de Janeiro.

Pereira-Filho, M., S. F. Guimarães, A. Storti-Filho & E. W. Graef. 1991. Psicultura na Amazônia Brasileira: Entraves ao seu desenvolvimento. *In* A. L. Val, R. Figliuolo & E. Feldberg, eds., Bases científicas para estratégias de preservação e desenvolvimento da Amazônia: Fatos e perspectivas, vol. 1. INPA, Manaus.

Petrere, M., Jr. 1978a. Pesca e esforço de pesca no estado do Amazonas. I—Esforço e captura por unidade de esforço. Acta Amazonica 8(3): 439–454.

————. 1978b. Pesca e esforço de pesca no estado do Amazonas. II—Locais, aparelhos de captura e estatísticas de desembarque. Acta Amazonica 8(3): Suppl. 2.

Sawaya, P. 1946. Sobre a biologia de alguns peixes de respiraçao aerea *Lepidosiren paradoxa* (Fitz.) e *Arapaima gigas* (Cuv.). Boletin da Faculdade de Ciencias e Letras (Zoologia), Universidade de São Paulo 11: 255–278.

Smith, N. J. H. 1979. A pesca no Rio Amazonas. CNPq/INPA, Manaus.

Sterba, G. 1973. Freshwater fishes of the world. T. F. H. Publications, London.

Veríssimo, J. 1895. A pesca na Amazônia. Monographias Brasileiras III. Livraria Clássica Alves, Rio de Janeiro.

Wootton, R. J. 1990. Ecology of teleost fishes. Fish and Fisheries Series 1. Chapman & Hall, London.

Fisheries Development in the Lower Amazon River

Mauro Luis Ruffino

Introduction

The Amazon basin covers more than 6,000,000 km², of which 3,900,000 km² are located in Brazil. According to Bayley and Petrere (1989), there are approximately 180,360 km² of várzea in the basin, along the Solimões and Amazon Rivers.

Várzea plays a central role in the regional economy. Its fertile soils and high concentration of fish and aquatic vertebrates support one of the largest population densities in the basin (McGrath et al., 1993a; Meggers, 1971). Várzea lakes and igapós are extremely important for aquatic life, as a rich source of nutrients and a habitat for macrophytes whose fruit, leaves, and seeds provide food for the fish (Goulding, 1980; Rai & Hill, 1980). Consequently, adequate management is necessary in order to prevent changes in the fertility and diversity of this biota (Pereira Filho, 1991).

The decline of traditional extractivist activities, such as hunting and gathering, and the collapse of várzea agriculture have made riverine populations increasingly dependent on fishing, both as a source of income and for subsistence (Furtado, 1988). As a result of these changes and the development of new technologies for capture, transport, and storage, as well as the population growth in the lower Amazon, there has been an increase of fishing activities, which has led to conflicts between commercial and subsistence fishermen over the use of the fisheries (McGrath et al., 1993a).

Since the management of fisheries is a dynamic and continuous process, the lines of action must follow the strategic guidelines of a program and produce the feedback required by any ongoing process. Thus, for the purpose of diminishing conflicts in the lower Amazon region and providing input to adapt fishing regulations to the actual situation, the Instituto Brasileiro do Meio Ambiente e dos Recursos Naturais Renováveis (IBAMA) created the IARA Project (IBAMA, 1994).

The present state of knowledge about fishing activities in the lower Amazon is presented, some lines of action being developed for the purpose of adapting existing regulations are discussed, and new management instruments that may help establish fishing zones are shown.

Material and Methods

Surveys were carried out in the cities and fishing communities of the Lago Grande de Monte Alegre (Monte Alegre municipality), Lago Jauari and Lago Dos Botos (Alenquer and Óbidos municipalities) and the Maicá and Tapará regions (Santarém municipality),

101

using an interdisciplinary approach that integrates direct action, research, training, environmental education, and monitoring subprojects/activities. Among the methodologies used in the study were the statistical community census (Isaac et al., 1998a), rapid rural survey (Mitlewski, in press), length-based methods (Pauly & Morgan, 1987), fishing statistics (Isaac & Ruffino, 1998), and participative management (Bonato & Oliveira, 1994).

Results and Discussion

Fishing Fleet

To date, the IARA project has registered 423 craft in the municipalities of Prainha (2.4%), Alenquer (9.5%), Monte Alegre (17.5%), Óbidos (10.9%), and Santarém (59.7%) (Isaac & Ruffino, 1996). The average age of boats is 11 years; average length, 10.2 m; and average storage capacity, 1.5 tons (Table I). All craft are made of wood and propelled by sails and/or, in the case of canoes, paddles. Some canoes have an outboard motor. Center-mounted motors are found on regular-line boats (the local means of transportation), on boats used for fishing, cargo, and tender, and on boats used by middlemen.

Commercial Fishing

Very little information is available for the period prior to 1991, when the IARA project began collecting fish-catch data at three markets and one processing plant in Santarém. In September 1993, data collection activities were expanded to include the municipalities of Alenquer, Monte Alegre, Óbidos, and Parintins; collection in the Almeirim, Prainha, and Oriximiná municipalities began in October 1994.

In the city of Santarém, 3700 tons and 4412 tons of fish were unloaded in 1992 and 1993, respectively (Isaac & Ruffino, 1996). Catches included more than 70 fish species, although only 10 species account for close to 80% of the total catch. Unloading at the Frigorífico Edifrigo fish-processing plant accounted for 50% of total unloaded tonnage and consisted mainly of catfish such as dourada (*Brachyplatystoma flavicans*), filhote (*Brachyplatystoma filamentosum*), piramutaba (*Brachyplatystoma vaillantii*), surubim

Table I. Number of craft registered by the IARA/IBAMA Project in the lower Amazon

Type	Year (avg.)	Mean length (m)	Mean power (Hp)	Mean TAB (ton)	Mean storage capacity (kg)	Number
Canoe	86	5.6	—	—	—	21
Motor Canoe	81	7.7	8.5	1.4	400.0	13
Fishing Boats	84	10.6	15.2	5.2	2475.3	299
Middlemen Boats	86	12.7	33.4	5.9	3011.9	21
Tender Boats	84	11.9	17.2	6.9	1064.8	41
Regular Line Boats	84	10.9	19.7	5.6	924.3	19
Large regular line boats	86	15.0	103.0	13.0	7000.0	1

(*Pseudoplatystoma* spp.), and mapará (*Hypophythalmus* spp.) (Ruffino & Isaac, 1994). The remaining fish were unloaded at three markets and consisted mainly of caracidae and sciaenidae, such as curimatá (*Prochilodus nigricans*), jaraqui (*Semaprochilodus* spp.), pescada (*Plagioscion* spp.), aracu (*Schizodon fasciatus*), tambaqui (*Colossoma macropomum*), and pacu (*Mylossoma* spp., *Myleus* spp.). Monthly unloading in Santarém varied from 100 to 500 tons, depending on the season, and the average catch per trip (the average trip being 3–4 days long) was 380 kg. The mean price paid to the fishermen varied from US$0.15 to US$1.50/kg, the average price being US$0.50/kg, depending on the fish species and the season of the year. The most popular fishing method is the gill net; during the peak season (July to October), however, the variety and combination of fishing techniques increases substantially. Fishing in the lakes occurs throughout the whole year, while river fishing is seasonal and linked to the hydrological cycle and the species' life cycle.

Table II shows fish unloadings during a one-year period in the cities of Alenquer, Monte Alegre, Óbidos, and Santarém. It must be emphasized that the Óbidos numbers are underestimated, since data collection was interrupted for four months.

Commercialization

The inadequate fishing norms and standards in the region, in addition to the lack of health inspection in the markets, have a great impact on the commercialization of fish, and in most cases consumers buy an inferior-quality product.

Commercialization usually takes place in the fish markets, where fish are transported in baskets and not packed in ice, so that they attain a temperature of 20°C. The fish not sold during the morning are either packed in ice to be marketed in the afternoon or are pickled in brine or salted at the market itself for later commercialization. In some municipalities, fish are sold in the open air, on the river banks, over wooden planks, and without ice; here the temperature of the fish may reach 28°C, depending on the time of day (Ruffino & Carvalho, 1995).

The fish-processing plants recently began financing fishing operations in the lower Amazon. Selling the fish is a streamlined operation that attracts fishermen (Barthem, pers. comm.). Nevertheless, the price paid by the fish-processing plants is lower than that obtained at the fish market.

Only five of the six fish-processing plants in the lower Amazon (McGrath et al., 1993a) are still operating: two in Óbidos and three in Santarém. Only Frigorífico Ed-

Table II. Fish unloadings at four cities in the lower
Amazon, from September 1993 to August 1994

City	Fish (kg)
Alenquer	525,067
Monte Alegre	352,562
Óbidos	449,425
Santarém	4,386,150
Total	5,713,204

ifrigo operates the year round—the others function only during the peak catfish season, approximately from July to November. Hygienic conditions are satisfactory at the processing plants. Fish are classified, washed in chlorinated water, beheaded and cleaned, filleted, and frozen. The consumer market for fish fillets is the southeastern-southern region of the country, where quality requirements have created a commercial niche for the product. Unfortunately, the backbone, bowels, and heads are not commercialized (though the frozen heads infrequently are sold as lobster bait in the northeastern region of Brazil).

In most cases, fishermen who work with the processing plants receive loans to purchase fishing material and/or boats; they also get the ice from the plants, either free of charge or to be paid later. The lack of adequate state assistance to the artisanal fishing sector has led some fish-processing plant owners to think about creating fishing cooperatives, to be finance by the Constitutional Financing Fund for the Northern Region (Schönenberg, 1994).

There are four main actors involved in commercial fishing activities: fishermen, bankers, ice makers, and processing plants, all of whom tend to work in partnership. The bankers sell fish in the local markets and the ice makers own the ice boats (boats with ice and fish storage boxes or chests). According to McGrath et al. (1993a), the relationship between ice makers and fishermen may follow various patterns with regard to the financing of fishing equipment and the prices paid per kilogram of fish.

Subsistence Fishing

Preliminary data from the IARA project indicate an average per capita consumption of fish of 454 g/day in the riverain communities of Lago Grande de Monte Alegre (Cerdeira, IARA project, pers. comm.), a rather high number when compared with the daily consumption estimated by Shrimpton et al. (1979) for Manaus (155 g), Smith (1979) for Itacoatiara (194 g), and Amoroso (1981) for Manaus (102 g).

Subsistence fishing seems to be very important both from an socioeconomic standpoint and with respect to fisheries management. The IARA project unloading data show that boats from Macapá, Breves, Belém, Abaetetuba, Santarém, and Monte Alegre caught close to 1094 tons of fish at Lago Grande de Monte Alegre, from September 1993 to August 1994 (Table III).

If we consider Cerdeira's estimated value (IARA project, pers. comm.) and a population of 8275 inhabitants living in 31 lake communities (Isaac et al., 1998b), fish consumption by the Lago Grande de Monte Alegre riverain population would be approximately 1371 tons per year, i.e., 25.3% higher than the commercial catch in the lake unloaded at the six cities. Such information matches the data presented by Merona (1985), who estimated that subsistence fishing exceeded commercial fishing in the Rio Tocantins basin.

It is impossible to effectively manage fish stocks without considering both subsistence and commercial fishing, as well as fishing pressure. The management model must also take into account the socioeconomic variables, which may or may not reduce social conflicts.

Table III. Fish catches at Lago Grande de Monte
Alegre from September 1993 to August 1994,
unloaded at six lower Amazon cities

City	Fish (kg)
Abaetetuba	739,771
Belém	86,971
Breves	29,162
Macapá	33,677
Monte Alegre	39,834
Santarém	164,643
Total	1,094,058

Statistical Community Census

The Statistical Community Census (CEC) (Isaac et al., 1998a) was developed for the purpose of obtaining general information about the fishing communities in the region—with regard to infrastructure, organization, economic and cultural activities, fishing conflicts and agreements, and fishing activities (number of commercial and subsistence fishermen, number and types of boats)—as well as fisheries management suggestions.

So far, 89 communities have been surveyed using this methodology, 31 of which are located at Lago Norte de Monte Alegre (Monte Alegre municipality), 48 at Lago dos Botos and Jauari (Alenquer and Óbidos municipality) and 10 in Maicá and Tapará regions (Santarém municipality), with an overall population of 24,524 (Table IV). Of the 5184 people who declared that they were fishermen, 39.6% were commercial fishermen and 60.4% were subsistence fishermen.

There were 4349 boats in the 89 communities surveyed, as follows: hulls (41.5%), canoes (41.9%), motorized canoes (5.6%), fishing boats (3.3%), and tender boats (3.1%) (Table V). As a rule, regions with the more intense commercial fishing have a larger number of powerful boats.

Biological Studies

Ferreira et al. (1996) identified 129 fish species that are captured and sold in the Santarém markets. The largest family was the Pimelodidae, which accounted for 22 species, followed by the Cichlidae (20 species) and Serrasalmidae (15 species). The data show that many more species are consumed in the region at the present time than is suggested by the common names (68) used by the population, which are used to identify more than one species.

Isaac et al. (in press) studied the reproductive cycle of eight commercial species in the lower Amazon and classified the aracu (*Schizodon fasciatus*), curimatá (*Prochilodus nigricans*), and tambaqui (*Colossoma macropomum*) among those species having a *seasonal* strategy. These are migrant fish species, almost perfectly adapted to river-level fluctu-

Table IV. Number of communities, inhabitants, and commercial and subsistence fishermen in the IARA project census

Region	Maicá	Taparã	Lago Grande de Monte Alegre	Lago Jauari and Lago Dos Botos	Total
Communities	7	3	31	48	89
Inhabitants	1862	775	8273	13,614	24,524
Commercial fishermen	46	153	626	1230	2055
Subsistence fishermen	318	46	1236	1529	3129

ations and annual rainfall patterns, with high fertility rates and small eggs that require no parental care. The tucunaré (*Cichla monoculus*) was classified as a species with a *balanced* strategy: sedentary, local distribution, long and probably parceled-out spawning period, low fertility, large eggs, and requiring parental care. The observations of catfish species, such as dourada (*Brachyplatystoma flavicans*), filhote (*Brachyplatystoma filamentosum*), and surubim-tigre (*Pseudoplatystoma tigrinum*) indicate the existence of special spawning grounds, probably located beyond the reach of the regional fishing fleet, since no mature fish were detected in the lower Amazon.

According to Ruffino and Isaac (1994), tambaqui are fished throughout the year, and most of the fish caught are juveniles. The average length of fish sold in the Santarém markets is 40.25 cm, which is equivalent to the average length at the first sexual maturation ($L_{50} = 55$ cm). Another recent study (Isaac & Ruffino, 1994) revealed that the fishing recruitment of the tambaqui occurs between the first and second year of life, when fish length is 23–35 cm. An analysis of the yield per recruitment shows that tambaquis are overexploited in the lower Amazon: Only an increase in the first-catch length can increase the yield per recruitment and ensure maximum sustainable exploitation.

Some of these findings are already being used by IBAMA in the Fisheries Man-

Table V. Number and type of craft by region

Type of craft	Maicá	Taparã	Lago Grande de Monte Alegre	Lago Jauari and Lago Dos Botos	Total
Boat	164	109	685	1003	1961
Canoe	90	83	609	1043	1825
Motor canoe	3	4	60	180	247
Fishing boats	4	0	71	70	145
Tender boats	2	28	3	103	136
Middlemen boats	0	0	2	1	3
Regular line boats	1	0	0	6	7
Freight boat	3	0	2	20	25
Total	267	224	1432	2426	4349

agement Program for the Amazon basin, particularly with regard to the appropriateness of the fishing statutes and their adaptation to the characteristics of the species, fishing techniques, reproduction period, and so forth. A good example of the existing discrepancies that cannot be justified by physiographic reasons are the administrative ordinances pertaining to minimum catch length. In the case of the jaraqui (*Semaprochilodus insignis*), for instance, the minimum catch length was 25 cm in the states of Pará and Amapá, but only 20 cm in the Amazon basin. The same is true for the pacu, a name that covers several species with different minimum lengths (30–40 cm). The most abundant species of pacu in the Santarém markets belongs to the *Mylossoma* genus, whose total length never reaches 30 cm.

Because of the great diversity of fish species in the Amazon basin (both migratory and sedentary species), stock management should vary by species. It is important to classify such stocks, and a basic problem with migrating species is precisely the so-called stocks. There is no information available on, nor are there means of establishing, stock lengths or size of exploited populations. Nevertheless, the problem of stock unity is very important for management strategies, particularly when the resources are exploited by several countries, as is the case with some Siluriforms such as dourada, filhote, surubins, piramutaba, and others.

Existing Management Measures

Government Regulations

There are many different government regulations in effect in Brazil. Nevertheless, all fishing activities must obey Decree-Law 221/67 and Law 7679/88, which are valid for the whole country. On the basis of these statutes, various national, state, and regional norms and administrative ordinances have been issued, with a view to regulating specific cases such as the minimum length at first capture, permissible fishing methods, acceptable mesh size and placement of nets, prohibition of fishing during spawning migrations, and issuance of fishing licenses.

Fishing in Brazilian inland watercourses has been ignored by government agencies, despite its considerable socioeconomic importance (Fischer et al., 1992). On the basis of the lower Amazon fisheries, Isaac et al. (1993) discuss the various decree-laws and administrative ordinances that regulate minimum catch length, types of fishing methods, minimum mesh size, and protected areas, and conclude that both the theoretical basis of the statutes and their practical application are plagued with problems.

Actually, administrative ordinances are specific, aiming at solving isolated problems in particular locations. In areas not contemplated by specific regulations, fishing is controlled by Administrative Ordinance 466, which establishes fishing norms for all continental waters in Brazil. This administrative ordinance is flawed because it considers all continental waters as a single environment and does not take into account the characteristics of the environment where the fishing takes place, not even in terms of the main Brazilian ecosystems. Administrative Ordinance 466 was issued in 1972, with the apparent purpose of meeting the needs of the southeastern and southern regions of the country.

Isaac et al. (1993) suggest that the period in which fishing is forbidden should be defined regionally, since the various states in a region have adopted slightly different regulations and thus are favoring black market activities, which are harmful to regular fishermen.

Nevertheless, with a view to correcting distortions resulting from the lack of a fisheries policy, IBAMA has established the basis of an integrated Fisheries Management Program. The program would ensure sustainability by taking into account both the impact on the resources and the cultural, ecological, economic, and social conditioning factors (Fischer et al., 1992).

Community Management

A growing number of riverain communities, concerned with their livelihood, are getting organized to draw up fishing agreements. Such agreements spell out informal fishing rules whose purpose is to decrease fishing activities in várzea lakes and thus to ensure the productivity of the environment. The rules usually involve one or more fishing restrictions, which may be seasonal or for an indeterminate period of time. Examples of these rules are prohibiting ice boats (motor boats with ice chests for fish storage), commercial fishing, certain fishing gear, fishing in given spots, and fishing certain species, as well as restricting the number of fishing days and storage capacity of boats.

These rules were apparently developed for the purpose of protecting the fishing rights of community members and are totally unrelated to any environmental considerations. Most people, in fact, think that fishing resources are infinitely large and unending, which is a great fallacy.

McGrath et al. (1993b) discuss the knowledge, experience, and initiative of the lake reserves by riverain communities in the lower Amazon as a local alternative fisheries management mechanism and show their fragility within the existing fishing policy context. The authors, however, acknowledge that IBAMA is really interested, for the first time, in involving riverain communities in the management of the fishing resources of local lakes.

McGrath et al. (1994) made a comparative study of the fishing production of two communities participating in some type of fishing agreement. The data showed bigger catches and higher incomes per fishing time in managed lakes. Furthermore, the study demonstrates that community management is a feasible alternative for the conservation of sedentary fish species and suggests a possible protection for short-distance migratory species (McGrath et al., 1994).

Nevertheless, despite the ecological feasibility of lake management by the community and the increasing number of such management schemes in the lower Amazon, few communities have been successful in this endeavor, mostly due to difficulties in setting up an internal control system. Such a system would require the concerted efforts of commercial fishermen from other regions (itinerant) and local fishermen (sedentary), as well as an efficient internal organization of communities.

Fishermen Awareness and Organization

After studying the institutional and organizational schemes used by fishermen in the lower Amazon, Schönenberg (1994) concluded that in order to make fisheries management proposals it is necessary to know the target groups and their forms of social organizations, that is, their spokespersons. The same author reports on the heterogeneity of the numerous organizational structures of the fishing sector in the lower Amazon and the randomness of their relations with municipal, state, and federal institutions. At the community level, traditional forms of organization do not explicitly represent fishermen's interests, although the interests of fishing communities are implicit in their social, commercial, and spiritual interests, which are much sounder and dominant forms (Schönenberg, 1994).

According to Schönenberg (1994), one-fourth of the population in the state of Pará depends on fishing for survival, or approximately 1.2 million people, 200,000 of whom are active fishermen. Of these active fishermen, 20–30% belong to "Fishermen Colonies" (*Colônia de Pescadores*). This means that 40,000 to 60,000 fishermen are formally registered and organized as members of their category, although being registered does not necessarily mean an active participation in the colony, nor that the colony itself is active.

In order to ensure the success of any future legal regulation of fisheries by IBAMA, it would be necessary to know the intra- and interinstitutional relations of the organizations of the various target groups, which are characterized by socioeconomic and ecological-cultural heterogeneity, as well as to obtain the participation of such organizations.

To that end, the IARA project is taking the first steps, through the Communication and Environmental Education Subproject, to work with the Fishermen Colonies in Santarém, Monte Alegre, Alenquer, Óbidos, Prainha, Almeirim, and Oriximiná. The IARA project is also mediating conflicts between riverain communities, providing information about existing legislation, and sharing its macroscopic overview of fisheries problems in the region (Bonatto & Oliveira, 1994). Another result of these contacts has been an increase in Fishermen Colony membership, which strengthens these organizations. The number of fully paid-up members rose from 8 to 1300 in Alenquer; from 300 to 1500 in Monte Alegre; and from 1100 to 2000 in Santarém.

Furtado (1993) provides a synthetic view of fishing conflicts in the lower Amazon and reports on the proposals made by the fishermen in meetings and documents and, particularly, on the ideas presented to the Federation of Fishermen of the State of Pará, showing that fishermen are becoming organized and creating mechanisms for participating in the management of fisheries resources.

Conclusion

The overall problem of fisheries management in the Amazon region is the conflict resulting from the use of the resources by different groups of users. Thus, in order to achieve sustainability, fisheries management should take into account the efficient use of the resource, the conservation of ecological/environmental diversity, and the distribution of the benefits arising from the appropriation of the resource.

Fishing, as practiced by riverain communities (sedentary fishermen), does not have a formal (legal) mechanism at present to place the management of fisheries resources in the hands of these communities. A strong community organization could ensure and/ or help legalize the forms of management being discussed.

Considering the extension of the Amazon River basin, only measures accepted and defended by the local population have the slightest chance of being effective. Thus, it is necessary for fishermen and fisheries resource users, as a whole, to harmonize their interests and be represented in the decision-making process of fisheries-management agencies. To this end, it is necessary to develop mechanisms for strengthening all formal and informal organizations that are linked, in one way or another, to the fisheries sector.

Acknowledgments

The author thanks IBAMA, GOPA/GTZ, and PTU/CNPq for the financial support that made it possible to obtain the data; and the data collectors and keyboard operator of the IARA/IBAMA project. Special thanks are due to Victoria Isaac and two anonymous referees for their comments and contributions to this manuscript.

Literature Cited

Amoroso, M. C. M. 1981. Alimentação em um bairro pobre de Manaus, Amazonas. Acta Amazônica (Suppl.) 11(3).

Bayley, P. B. & M. Petrere Jr. 1989. Amazon fisheries: Assessment methods, current status, and management options. Pages 385–398 *in* D. P. Dodge, ed., Proceedings of the International Large River Symposium. Canadian Special Publications in Fisheries and Aquatic Science 106.

Bonatto, M. P. O. & P. R. S. Oliveira. 1994. O papel da comunicação e educação ambiental no manejo da pesca na Amazônia. Pages 574–590 *in* Anais do I Encontro Brasileiro de Ciências Ambientais, vol. 2. UFRJ, Rio de Janeiro.

Castro, F. & D. G. McGrath. 1994. O manejo comunitario de lagos no Baixo Amazonas: Efeitos ecologicos e potancial para o gerenciamento pesqueiro. Manuscript.

Ferreira, E. J. G., J. Zuanon & G. M. dos Santos. 1996. List of commercial fish species from Santarém, State of Pará, Brazil. NAGA, the ICLARM Quarterly 19(3): 41–44.

Fischer, C. F. A., A. L. das G. A. Chagas & L. D. C. Dornelles. 1992. Pesca de águas interiores. Brasília. Série Estudos Pesca 2. IBAMA, Coleção Meio Ambiente.

Furtado, L. G. 1988. Os caboclos pescadores do baixo rio Amazonas e o processo de mudança social e econômica. *In* Ciências sociais e o mar no Brasil. II. Programa de Pesquisa e Conservação de Áreas Úmidas no Brasil, São Paulo.

———. 1993. "Reservas pesqueiras," uma alternativa de subsistência e de preservação ambiental: Reflexões a partir de uma proposta de pescadores do médio Amazonas. Pages 243–276 *in* L. Furtado, W. Leitão & A. F. Mello, eds., Povos das águas—Realidade e perspectivas na Amazônia. MCT/CNPq/MPEG/, Belém.

Goulding, M. 1980. The fishes and the forest: Explorations in Amazonian natural history. University of California Press, Berkeley.

IBAMA (Instituto Brasileiro do Meio Ambiente e dos Recursos Naturais Renováveis). 1994. Relatório técnico preliminar sobre ordenamento pesqueiro para a bacia Amazônica, Brasília, 20–24 September 1993. Mimeograph.

Isaac, V. J. & M. L. Ruffino. 1998. Informe estatístico do desembarque pesqueiro na cidade de Santarém, PA: 1992–1993. Série Estudos Pesca. IBAMA, Coleção Meio Ambiente. (In press.)

——— & ———. 1996. Population dynamic of tambaqui, *Colossoma macropomum* Cuvier 1818, in the Lower Amazon, Brazil. Fisheries Management & Ecology 3(4): 315–333.

———, B. Mitlewski & P. R. S. Oliveira. 1998a. Censo estatístico comunitário (CEC): Metodologia para primeiros contatos, levantamentos participativos e multidisciplinares e incentivo à cooperação junto aos grupos alvos de projetos de desenvolvimento. Série Estudos Pesca. IBAMA, Coleção Meio Ambiente. (In press.)

————, ————, M. L. Ruffino & P. R. S. Oliveira. 1998b. Lago Grande de Monte Alegre: uma análise de suas comunidades. Série Estudos Pesca. IBAMA, Coleção Meio Ambiente. (In press.)

————, V. L. C. Rocha & S. Mota. 1993. Considerações sobre a legislação da "piracema" e outras restrições da pesca da região do Médio Amazonas. Pages 187–211 in L. Furtado, W. Leitão & A. F. Mello, eds., Povos das águas—Realidade e perspectivas na Amazônia. MCT/CNPq/MPEG/, Belém.

————, ———— & ————. In press. Ciclo reprodutivo de algumas espécies de peixes de valor comercial do Baixo Amazonas. Série Estudos Pesca. IBAMA, Coleção Meio Ambiente.

McGrath, D. G., F. Castro, C. Futemma, B. D. Amaral & J. Calabria. 1993a. Fisheries and evolution of resource management on the lower Amazon floodplain. Human Ecology 21(2): 167–195.

————, ————, ————, ———— & ————. 1993b. Manejo comunitário da pesca nos lagos de várzea do Baixo Amazonas. Pages 213–229 in L. Furtado, W. Leitão & A. F. Mello, eds., Povos das águas— Realidade e perspectivas na Amazônia. MCT/CNPq/MPEG/, Belém.

————, ———— & ————. 1994. Reservas de lago e o manejo comunitario da pesca no Baixo Amazonas: Uma avaliacão preliminar. In M. A. Dincao & I. M. Silveira, eds., Amazonia e a crise da modernizacão. Museu Paraense Emilio Goeldo, Belém, Pará.

Meggers, B. 1971. Amazon: Man and culture in a counterfeit paradise. Aldine Press, Chicago.

Merona, B. 1985. Les peuplements des poissons et la peche de la bas Tocantins (Amazonie bresilenne) avant la fermeture du barrage de Tucurui. Verhandlungen Internazionale Vereinigung Limnologischen 22: 2698–2703.

Pereira-Filho, M., S. F. Guimaraes, A. Storti-Filho & E. W. Graef. 1991. Psicultura na Amazonia Brasileira: Entraves ao seu desenvolvimento. Pages 373–380 in A. L. Val, R. Figliuolo & E. Feldberg, eds., Bases cientificas para estrategias de preservação e desenvolvimento da Amazonia: Fatos e perspectivas. Vol. 1. INPA, Manaus.

Rai, H. & G. Hill. 1980. Classification of central Amazon lakes on the basis of their microbiological and physicochemical characteristics. Hydrobiology 72: 85–99.

Ruffino, M. L. & N. L. de A. Carvalho. 1995. Aspectos da conservação, armazenamento e comercialização do pescado no Baixo Amazonas. Pages 99–107 in Anais da 2° Conferência Regional da AIM na América Latina: Saúde e Atenção à Saúde na Região Amazônica. AIM/Fundação Esperança.

———— & V. J. Isaac. 1994. The fisheries of the lower Amazon: Questions of management and development. Acta Biologica Venezuelica 15(3): 37–46.

Schonenberg, R. 1994. As formas institucionais e organizacionais, de articular interesses na area da pesca no Baixo Amazona em Particular, e na Amazonia em geral. Relatorio Preliminar para Projeto IARA. IBAMA, Santarém, Pará.

Shrimpton, R., R. Giugliano & N. M. Rodrigues. 1979. Consumo de alimentos e alguns nutrientes em Manaus, 1973–1976. Acta Amazônica 9: 117–141.

Smith, N. J. H. 1979. A pesca no rio Amazonas. CNPq/INPA, Belém.

Section 2:
Forests and Forestry

Introduction

Charles M. Peters

Amazonian floodplains are a difficult place to grow. Newly deposited seeds must germinate and grow before they are covered by several meters of water, and any seedlings that do manage to survive the first year must then tolerate for the rest of their lives an annual cycle of submersion, sediment deposition, and immersion. Flooding may last for up to 300 days in some years, with water levels exceeding 16 meters along the major rivers. Sediment loads in western Amazonia are commonly 1–1.5 m per flood cycle.

The rigors of the floodplain habitat give rise to forests that are notably different from those found on upland sites. In terms of floristics, várzea forests usually contain fewer species than upland forests. It is not uncommon, for example, to find more than 200 different tree species in a single hectare of upland Amazonian forest. A similar inventory in várzea forest would record fewer than half this many. Coupled with this reduced diversity is an increase in the density of individual species. Tree populations in upland forests frequently contain only one or two individuals per hectare. Some várzea species, most notably palms, form populations containing several hundred individuals per hectare. There are also pronounced structural differences between the two forest types. Várzea forests typically have a lower and more patchy canopy, a higher percentage of large-diameter trees, and an open, less vigorous understory.

These ecological characteristics greatly simplify the expoitation of forest resources. The low diversity and high density of tree populations in the várzea mean that a fruit collector, rubber tapper, or logger needn't look as long or hard to find a particular resource as would be necessary in upland forest; moreover, once the resource is located, usually a larger quantity of material is available for harvest. Add to this the fact that the forest is accessible by boat during much of the year, and it is easy to understand why the collection of plant products from várzea has been an integral part of Amazonian subsistence for several millennia.

More recent forms of forest exploitation in Amazonia have also focused on the várzea. Today, however, the great majority of the plant resources collected from the floodplain are destined for the market or the sawmill. Over the past fifty years, the volume and value of forest products extracted from the várzea has increased dramatically, especially in the case of timber and palm products (i.e., palm heart and fruits). As is documented in the five papers that follow, the rapid spread of commercial resource extraction in várzea forests has had a major impact on both the people and the plants of the Amazonian floodplain.

Anderson et al. open the section with an analysis of virola (*Virola surinamensis*) logging in the Amazon floodplain. Virola, one of the most valuable timber species in várzea forest, is heavily exploited for both sawtimber and plywood. The authors estimate that 350 small sawmills and two plywood factories in the Amazon estuary processed over 700,000 m³ of virola logs in 1989 alone, providing employment for thousands of local people and generating gross revenues of approximately US$50 million. Although the species responds well to silvicultural treatment and has great potential for sustainable production in managed forests or plantations, the natural populations of virola are suffering rapid depletion by intensive, uncontrolled logging and inadequate reforestation. The authors suggest several policy alternatives to avoid the imminent demise of this valuable resource, yet they caution that nothing will be done without the public awareness and political will that are thus far lacking. The net result, they predict, will be that the lucrative wood-processing industries will eventually move out of the Amazon estuary in search of more abundant sources of timber, thus repeating the boom-and-bust cycles that have historically characterized the exploitation of many other forest resources.

Albernaz and Ayres continue the discussion of logging in the várzea by describing the dynamics of timber extraction in the Mamirauá Ecological Station on the middle Solimões River. Timber is an important resource in this region, and an average of 20,000 m³ of sawtimber and poles were harvested during 1993 and 1994. Although a total of 38 species are exploited commercially, the majority of the material harvested at Mamirauá was produced by three species: *Hura crepitans*, *Xylopia* cf. *calophyllum*, and *Virola* sp. The authors document patterns of decreasing harvest diameter and gradual species substitution that are classic symptoms of resource overexploitation. For example, they observed that the maximum diameter of the *Hura* and *Virola* trees harvested in 1993 was 10–15 cm larger than that recorded in 1994. As old, large-diameter trees become increasingly rare and hard to find, local loggers are switching to more abundant, albeit inferior timbers. The gradual shift from *Cedrelinga* to *Calycophyllum*, and the current *Ceiba* to *Hura* to *Couroupita* transition, are good illustrations of this trend. Albernaz and Ayres offer particularly useful insights into the problem of wastage during várzea logging. Due to unpredictable flooding, loggers are frequently forced to create excessively tall stumps or to leave felled trees on the site. Based on preliminary surveys, they estimate that one tree is left in the forest for every two removed.

A final glimpse of várzea logging is provided by Barreras and Uhl, who offer a detailed description of the wood industry in the Amazon estuary. Their analysis moves stepwise through each stage of the production process including extraction, transport, and final processing in either small or medium-sized sawmills or plywood factories. Due primarily to the low costs of extraction and the availablity of fluvial transport, the economics of timber extraction in the estuary are very attractive, and the region currently suppports more than 1000 wood-based industries. These industries employ almost 24,000 people, consume more than 1,000,000 m³ of wood per year, and generate annual revenues in excess of $93 million. Using estimates of current harvest volumes, projected demand, and the areal extent of várzea forest, the authors found that the wood resources in many areas are being overexploited and that concerted management efforts will be needed to maintain the local wood industry. They emphasize, however, that forest management is only one component of the sustainability equation and that new forestry legislation, improved monitoring and support of the forestry sector, and clarification of tenurial arrangements are also sorely needed.

Palms are undoubtedly the most conspicuous and well-known plants in várzea forests. They also produce some of the most valuable nontimber products found in these habitats. Henderson and Kahn provide a useful overview of the ecology and use of nine genera of important várzea palms. Building their narrative around the original observations of such famous Amazonian explorers as Martius, Bates, Spruce, and Wallace, the authors take the reader on a journey upriver from Belém and point out the interesting palms and palm products encountered along the way. The journey starts with the açaí (*Euterpe oleracea*), valued for palm heart and edible fruits; allows brief visits with *Raphia, Mauritia, Manicaria, Socratea, Iriartea, Astrocaryum,* and *Phytelephas*; and ends with a discussion of *Oenocarpus bataua,* a widespread species whose seed kernels contain a high-quality oil. Although várzea palms are extensively and, in many cases, destructively harvested, the authors report that most species appear to be maintaining themselves in the forest.

The final contribution in this section is a report by Hiraoka on the use and management of the mirití palm (*Mauritia flexuosa*). This dioecious palm is one of the most abundant and characteristic species in várzea forest and also one of the most widely used plant resources in Amazonia. The species is selectively protected in young swiddens, and seedlings are occasionally planted in house gardens to provide a handy source of leaf fibers. Interestingly, ribeirinhos commonly removed the male trees form natural stands to increase the percentage of fruit-producing trees. In spite of these management practices, the local abundance of *Mauritia* palms has declined notably in recent years, largely in response to the extensive cultivation and management of another important várzea palm, açaí (*Euterpe oleracea*). *Mauritia* and *Euterpre* share essentially the same habitat, yet there is a large commercial demand for *Euterpe* fruit in urban markets whereas *Mauritia* is primarily a subsistence resource in the estuary. The interplay between these two palms is a fascinating example of comparative economics and its effect on resource management and the distribution and abundance of native species.

We can draw several general conclusions from the five papers in this section. The first and most obvious is that várzea forests contain a host of valuable resources and that the exploitation of the resources can generate millions of dollars in revenues and a large number of jobs for local people. The second conclusion is that exploitation of many of these resources, especially timber, is being conducted in a nonsustainable manner that is gradually depleting the original resource base. The third is that forestry industries will be forced to look elsewhere for raw material once the local supply of resources has been exhausted. The unfortunate result is that the jobs, the local infrastructure, and the annual revenues will accompany the industries when they move out of the várzea. The fourth conclusion is a bit more hopeful: As is already practiced with many palm species, forest resources in the várzea can be managed, if there is sufficient motivation to do so. This course of action would help preserve many of the economic, social, and ecological benefits currently derived from the Amazonian floodplain, and would be a major step forward in trying to smooth the inevitable booms and busts of várzea forestry. We are, in essence, faced with two alternatives for dealing with plant resources in the várzea: We can use them wisely or we can use them up.

Logging of *Virola surinamensis* in the Amazon Floodplain: Impacts and Alternatives

Anthony B. Anderson, Igor Mousasticoshvily Jr., and Domingos S. Macedo

Introduction

Logging is the fastest growing land use in Brazil's Amazon region, and, with the possible exception of mining, the most lucrative. Since the 1980s, the spread of this land use has been fueled by construction of a network of roads in the region and by declining stocks of native forest resources in southern Brazil, which induced wood-processing industries to relocate to Amazonia. As major sources of timber in Southeast Asia dry up by the early twenty-first century, the Amazon region is likely to assume paramount importance in the international tropical timber trade.

Despite its growing importance, logging is a relatively recent phenomenon in much of the Amazon region. Prior to the 1980s, it was largely confined to bottomlands adjacent to major rivers, which facilitated transport and permitted extraction of the bulk of Amazonian timber during 300 years (Rankin, 1985). As a result, floodplain logging has a historical dimension that is largely missing from the uplands, and this dimension provides a basis for discerning the future of logging in the region as a whole.

This paper focuses on virola [*Virola surinamensis* (Rol.) Warb.], the most economically important wood in the Amazonian floodplain and, until 1989, the second most important wood export (after mahogany) in the Brazilian Amazon. A source of raw material for hundreds of sawmills in the Amazon estuary, virola is a crucial element in the regional economy. In addition, it shows excellent potential for sustained management in plantations and in natural stands. Despite its potential for sustained use, however, current harvesting practices and associated policies are leading to the rapid exhaustion of this resource.

Virola's economic importance, unsustainable exploitation, and potential for sustained use make it an appropriate case for understanding the problems and opportunities associated with wood extraction in the Amazon region. Summarizing the results of a three-year study (Anderson et al., 1994; Macedo & Anderson, 1993), this paper begins with an analysis of current logging practices and their environmental impacts. Then we look at the dimensions and economic efficiency of the diverse industrial sectors that utilize virola. Next, we investigate the effectiveness of current policies designed to regulate supply and demand of virola. Finally, based on our findings, we propose alternative policies that could lead to sustained use of this resource. Although these alternatives are technically feasible, their adoption requires a degree of public awareness

119

and political will that is currently lacking. The case of virola—one of the most promising cases for sustainable use of a tropical timber species—thus provides a sobering insight into the future of logging in the Amazon region.

Logging

The Regional Context

Logging of virola is concentrated in and around Marajó, a 49,606 km² island in the Amazon River estuary (Fig. 1). This region—and especially the western portions of Marajó and adjacent areas—is covered by a complex web of watercourses, and much of its area is subject to periodic floods generated by tides. Historically, the estuarine region has served primarily as a source of forest and riverine products, while agriculture has played a secondary role, usually one of subsistence. This dependence on wild products has led to repeated boom-and-bust cycles that have become more pronounced over the past century due to the increasing penetration of international markets into the Amazon region. In the late nineteenth century, the estuary provided an important source of fuelwood for steamships that traveled the Amazon River and its tributaries, and also of natural rubber for the world market. By the early twentieth century, however, the decline of steamships reduced demand for fuelwood, and the market for local

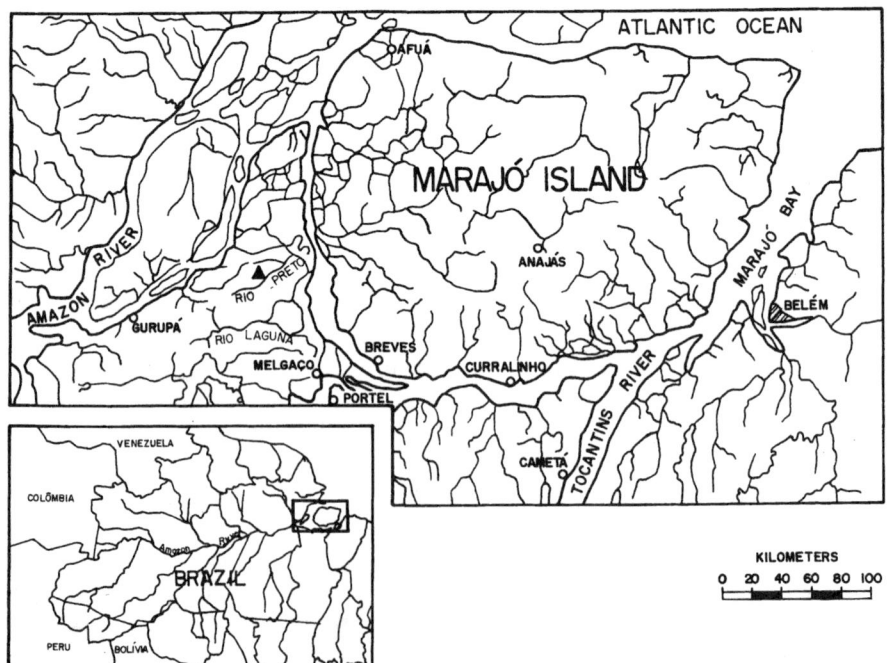

Figure 1. Map of the Amazon estuary. Triangle represents study site on the Preto River.

sources of rubber also dwindled with the initiation of plantation production in Southeast Asia. Rising international prices produced a brief boom in rubber production during World War II, which ended abruptly with the restoration of plantation sources in the late 1940s. Large-scale extraction of timber for export began in the 1950s and now—together with the harvesting of palm heart from the açaí palm (*Euterpe oleracea* Mart.)—constitutes the major market-oriented activity in the Amazon estuary.

The Rise of Virola

The estuary's current logging boom traces its origins to 1956, when Georgia Pacific established a plywood plant in the town of Portel (Fig. 1). Today six plywood manufacturers operate in the region, although most have diversified their raw material sources and only two continue to rely primarily on virola. Through the 1970s, local plywood manufacturers concentrated almost exclusively on virola due to its unique combination of ecological, physical, and mechanical characteristics: The species is abundant in bottomlands, and its easily peeled wood is low in density (0.45 g/cm^3), light in color, and highly porous—ideal qualities for production of plywood veneer. Plywood manufacturers favor large-girthed trees (>45 cm diam.), which are more appropriate for peeling. Because such large-girthed trees are relatively rare, extraction of virola expanded beyond the estuarine region during the 1970s and by 1987 had reached the vicinity of Benjamin Constant on the Colombian frontier—more than 3000 km away. After nearly 40 years of exploitation, natural stands of virola no longer furnish the needs of most local plywood manufacturers, which have substituted it with inferior floodplain species such as samaúma (*Ceiba pentandra* Gaertn.) and ventosa (*Hernandia* spp.).

Virola wood is also highly prized by sawmills. Light in color and lacking well-defined growth rings, its wood is ideal for production of moldings used in house construction. The first sawmill specializing in virola was established during the 1960s, near the town of Breves (Fig. 1), and today hundreds of sawmills operate in the Amazon estuary, focusing most of their production on virola. In contrast to plywood manufacturers, sawmills utilize wood obtained from small-girthed trees (<30 cm diam.), which greatly increases the intensity of forest exploitation. As a result, extraction of virola for sawnwood is still largely concentrated in the estuarine region, although growing scarcity is leading sawmills to seek more distant sources.

Ecological Impacts

Logging of virola follows the distribution of the species, which proliferates in inundated forests from the mouth of the Amazon River to its headwaters, including principal tributaries such as the Madeira and Purus Rivers (Rodrigues, 1972). In the Amazon estuary, logging of virola is exclusively manual and takes place primarily during the rainy season (January–June), when high floodwaters facilitate removal of logs. During the dry season (July–December), when wood extraction is much more difficult, loggers generally dedicate themselves to other activities such as fishing, rubber tapping, harvesting fruits of the açaí palm, and clearing plots for subsistence agriculture. Cutting trees and sectioning logs is carried out manually (Fig. 2A), and the logs are removed by

Figure 2. Logging of virola in the Amazon estuary. **A.** Sectioning a virola bole into logs, Mocoões River. **B.** Manual removal of a virola log from the forest, Mocoões River. **C.** Transport of a virola log along a logging canal, Preto River. **D.** Transport of virola logs in a raftlike platform (*jangada*), Preto River. (Photographs by I. Mousasticoshvily Jr.)

dragging them over the ground (Fig. 2B) or floating them along natural or man-made canals (Fig. 2C). Upon arriving at major watercourses, the logs are cabled together along their lengths to form an extensive, raftlike platform (Fig. 2D). This platform is generally sold to an agent or middleman and is subsequently floated to a wood-processing plant.

The intensity of virola extraction varies according to the type of forest. Virola occurs in seasonally inundated floodplain forest (*mata de várzea*) and permanently inundated swamp forest (*mata de igapó*; cf. Prance, 1980). In the Amazon estuary, these two vegetation types cover an estimated area of 25,000 km² (Calzavara, 1972). While floodplain forest is far more extensive, it generally contains a much lower concentration of virola than does swamp forest (Fig. 3). As a result, logging of virola tends to be extensive in floodplain forest and intensive in swamp forest. Two case studies reveal the ecological impacts of these contrasting types of logging.

In a floodplain forest site on the Mocoões River in central Marajó (Fig. 1), Uhl (1990) found that extensive logging for plywood manufacturers concentrates on large trees (avg. 52 cm dbh), removes a low volume per unit area (ca. 5 m³/ha), and damages only 10% of the remaining trees. Forest regeneration after such a low level of disturbance is probably rapid, especially on relatively fertile floodplain soils. Furthermore, as virola grows well on disturbed sites, the highly selective logging carried out on the Mocoões River may actually enhance growth of the remnant stand. Yet the long-term effects on

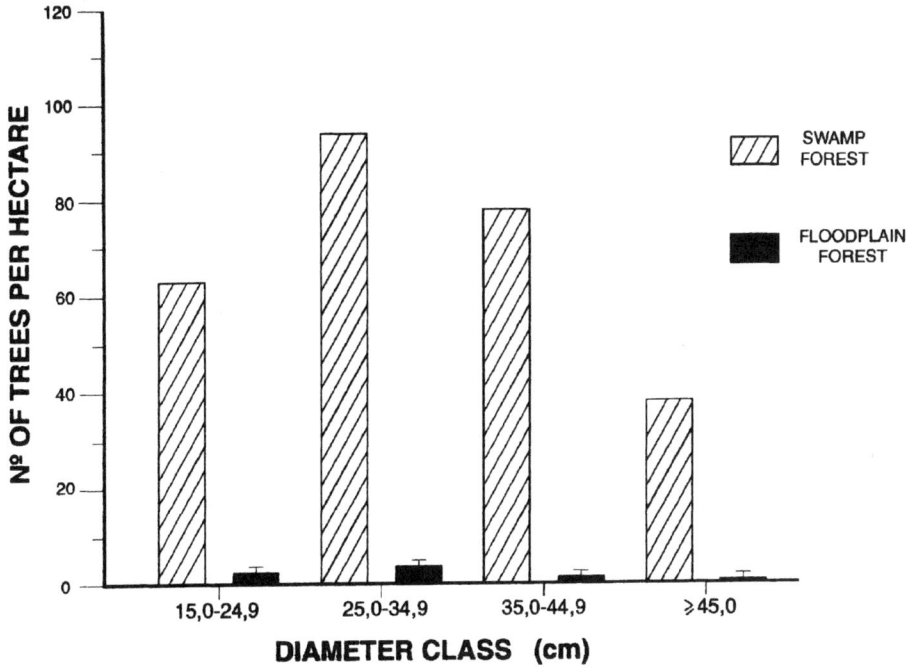

Figure 3. Abundance of virola according to diameter class in two inundated forest types in the Amazon estuary. (Data from the floodplain forest sites courtesy of H. Knowles.)

the stand are less certain, as selective logging targets genetically superior trees and consequently leaves behind less desirable germplasm (Kageyama, 1981).

In contrast, in a virola-dominated swamp forest on the Preto River 50 km west of Marajó Island (Fig. 1), we found that logging targets relatively small trees (avg. 35 cm dbh, min. 26 cm dbh) and consequently removes a high volume per unit area (145 m³ ha)—representing 90% of the original virola stand and 56% of the entire swamp forest (Macedo & Anderson, 1993). After a five-year period, we also found that such intensive logging results in the complete elimination of virola seedlings from the understory. This elimination appears to be due to the presence of a thick surface layer of organic matter, which impedes root penetration by seedlings and causes them to desiccate during the dry season. As a result, intensive logging leads to the eventual demise of virola in swamp forest—an ecosystem that, prior to logging, is frequently dominated by this species.

Industrial Use

Demand

In the Amazon estuary, the main types of industries that utilize virola include approximately 350 small sawmills (annual production <1000 m³), 24 large sawmills (annual production >1000 m³), and 2 plywood manufacturers. Based on analysis of responses to a questionnaire applied to a sample of these industries (Mousasticoshvily, 1991), we estimate that in 1989 they provided 5045 direct jobs and indirect employment for thousands, and generated roughly US$50 million in gross receipts. Again using the questionnaire data, we constructed a flowchart illustrating the movement of virola wood during 1989 (Fig. 4). The flowchart shows that wood-processing industries in the estuarine region purchased approximately 722,000 m³ of virola logs. Large sawmills were far and away the largest consumers (435,000 m³), followed by small sawmills (242,000 m³) and plywood manufacturers (45,000 m³). The share of large sawmills was, in fact, greater, since they purchased all of the sawnwood produced by small sawmills. As a result, we estimate that 86% of virola wood passed directly or indirectly through 24 large sawmills located in the Amazon estuary. This finding has important policy implications, which we discuss below.

Processing and Marketing

According to our estimates (Fig. 4), the 722,000 m³ in virola logs purchased in 1989 generated roughly 287,000 m³ in final products—a yield of about 40%, which is ex-

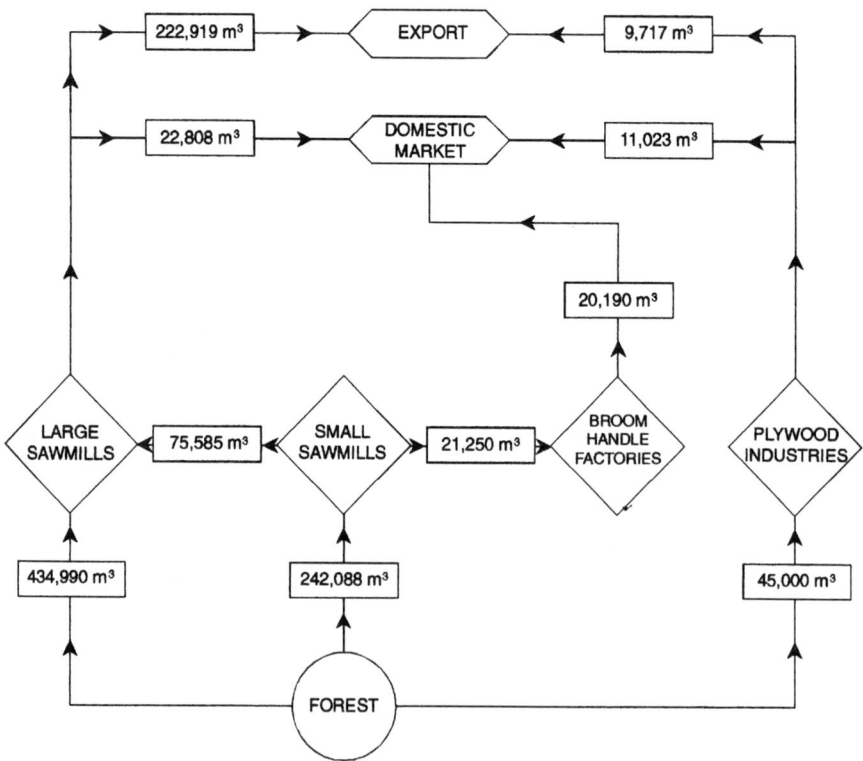

Figure 4. The movement of virola wood through industries based in the Amazon estuary region during 1989. (From Mousasticoshvily, 1991.)

tremely low by modern wood-processing standards. Most (91%) of the sawnwood produced was exported, primarily to the United States; in contrast, the plywood produced
from virola (including a mix of other species) was split fairly evenly between domestic
(52%) and international (48%) markets. Interestingly, we found that our 1989 estimate
of virola sawnwood exported by local industries (222,919 m³) was 2.3 times the official
government figure (93,860 m³; cf. Fig. 5), which suggests that considerable quantities
of virola wood are marketed informally. The policy implications of this finding are
discussed below.

Performance

A comparative analysis reveals the performance of the three major types of industries
that process virola (Table I). In environmental terms, as plywood manufacturers consume a lower volume of virola logs and convert a higher proportion into finished
product, they exert less pressure on the stands of virola than do large or small sawmills.
In financial terms, the profit margin of plywood industries is much higher than that of
large or small sawmills. And in social terms, while plywood manufacturers employ fewer
people than do small sawmills, they provide higher salaries and job security than both
large and small sawmills, and they pay far higher taxes.

Current Policies and Patterns of Resource Use

Logging

The office of Brazil's environmental agency (IBAMA) responsible for the Amazon
estuary employs only one forest guard and lacks necessary infrastructure and financial
resources for effective monitoring of the virola market. To investigate that effectiveness,
we analyzed two legal parameters designed to regulate logging of virola.

First, IBAMA stipulates a minimum diameter limit of 29 cm (or 90 cm circum-
ference) for marketed virola logs. In a random sample of logs present on the Preto River
during two-day periods in the rainy and dry seasons, we found that a substantial proportion (22%) was below this limit (Macedo & Anderson, 1993). Most illegal logs are
extracted during the dry season: We found that 82% of the logs sampled in this period
were below the legal limit.

Second, the Brazilian government also defines a limit based on tree size: 45 cm
dbh for logging operations in native forest (Brazilian Forestry Code, Law 4771/65). To
illustrate the effectiveness of both tree- and log-size limits in regulating extraction of
virola, we calculated the volume of uncut, residual, and extracted wood in a hypothetical, intensive logging operation at our swamp forest study site near the Preto River
(Table II). Of the total volume of virola wood extracted per hectare (145 m³), only
33% would be below the legal minimum log size, while fully 70% would be derived
from diameter classes below the minimum tree size. Although it greatly diminishes the
impacts of logging, the minimum tree size is virtually impossible to enforce. On the
other hand, our field observations suggest that strict enforcement of the log-size limit

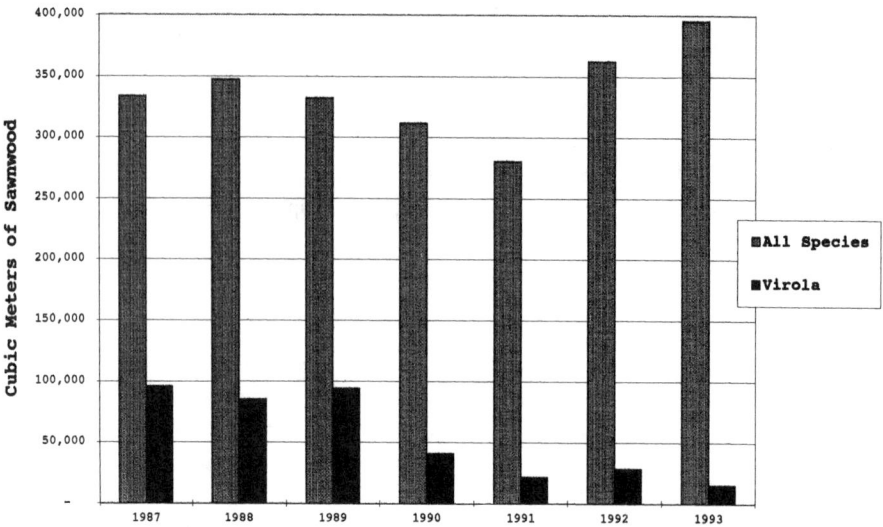

Figure 5. Sawnwood exports from Pará state, 1987–1993. (From an unpublished report by the Pará State Association of Wood Industries for Export.)

would probably induce loggers to extract fewer legal-sized logs per tree rather than to cut fewer trees. In short, the tree-size limit would reduce the impacts of current extraction but is impossible to enforce, while the log-size limit is enforceable but would probably not reduce ecological impacts. We conclude that current policies to regulate the logging of virola are not only poorly enforced but ill conceived.

Reforestation

The Brazilian Forestry Code establishes levels of reforestation according to the amount of wood consumed. For each cubic meter utilized by wood-processing industries, six seedlings of forest species must be planted (Regulation 441 of 9 August 1989). Planting costs are borne by the industries, which can pay IBAMA a tax to support reforestation by third parties or, with IBAMA approval, implement their own reforestation projects. Payment of the tax or direct involvement in reforestation is required in order to obtain official permits for transporting logs, which determine the amount of wood that industries can legally purchase.

Ten reforestation projects involving virola are currently underway in Brazil, encompassing a total area of approximately 4800 ha (Kanashiro & Yared, 1991). We examined in detail two projects in the Amazon estuary, carried out by the major plywood manufacturers utilizing virola (cf. Anderson et al., 1994): an 80-ha plantation established by EIDAI near the town of Portel; and a 2666-ha plantation established by TREVO along the Amazon River, downstream from the town of Gurupá (Fig. 1). For comparative purposes, we utilized data from a 2450-ha plantation of virola in floodplain forest in Suriname (Schulz & Rodriguez, 1966).

Table I. Comparison of small sawmills, large sawmills, and plywood industries

	Small sawmills (n = 46)	Large sawmills (n = 22)	Plywood industries (n = 2)
A. Raw material			
Demand for raw material (m³ of logs)	242,088	434,447	45,000
Efficiency of industrial use of raw material (percent m³ utilized per m³ of logs)	40	43.1	46.1
B. Labor			
Mean monthly salary (US$)	55.44	97.35	344.90
Number of employees per sector	1936	1434	1675
Stability of employment (percent of jobs maintained throughout year)	14	70	100
C. Financial aspects			
Value of investment (US$ per factory)	1,500	1,600,000	3,000,000
Operational costs (US$ per m³)	79.92	97.09	58.30
Final price (US$ per m³)	84.11	124.70	371.30
Net profit (percent of final price)	5.0	22.1	84.3
D. Governmental issues			
Legality (percent of licensed firms per sector)	50	100	100
Contribution in taxes (US$ per year)	0	355,000	9,000,000

Source: Mousasticoshvily, 1991.

The plantation of EIDAI (Table IIIA) was established on a cleared upland site that was periodically weeded. In plantings established 4–12 years previously, we found that 37% of the trees were dead and an additional 19% were defective in form, probably due to virola's poor adaptation to upland soils. Furthermore, under the high light conditions characteristic of this plantation, virola trees tend to branch and bifurcate at low heights, thus reducing the volume of trunk available for log extraction. As a result, after 10 years the plantation had attained a mean total height of only 4.2 m and an exploitable volume of 56.1 m³/ha.

The plantation of TREVO (Table IIIB) was established in the understory of a floodplain forest previously logged for virola. Other than occasional thinning of understory vegetation, no silvicultural treatments were carried out in the forest. We found that the form of the trees was, in general, superior to that of the plantation established by EIDAI in full sun. Yet, due to the low levels of light in the untreated forest understory, yields of the TREVO plantation were negligible: after 10 years the plantation had attained a mean total height of 2.2 m and an exploitable volume of <10 m³/ha.

The plantation in Suriname (Table IIIC) was also established in the understory of a floodplain forest that had been logged for virola. But in contrast to the previous case,

Table II. Volumes of uncut wood, residual wood, illegal logs, and legal logs as a function of tree diameter in a hypothetical logging operation in mature swamp forest in Rio Preto

Tree diameter class (cm)	Cubic meters per hectare				
	Uncut wood	Residual wood	Illegal logs	Legal logs	Total volume
15.0–19.9	5.5	0	0	0	5.5
20.0–24.9	12.5	0	0	0	12.5
25.0–29.9	4.3	7.2	17.1	0	28.5
30.0–34.9	0	10.2	14.2	8.3	32.7
35.0–39.9	0	18.5	12.9	20.5	52.0
40.0–44.9	0	13.3	4.1	23.6	41.1
45.0–49.9	0	14.8	0	24.5	39.3
50.0–54.9	0	5.2	0	8.4	13.6
55.0–59.9	0	2.7	0	6.8	9.5
≥60.0	0	0.5	0	4.2	4.7
Total	22.3	72.4	48.3	96.3	239.4

this forest was selectively thinned to increase luminosity to 65–80% of full sun, and was subsequently weeded twice per year to reduce competition. This combination of treatments resulted in a relatively high yield: After 10 years the plantation had attained a mean total height of 16 m and an exploitable volume of 208 m³/ha.

The yields of the Suriname plantation are relatively high for a native species not yet subjected to genetic improvement. Based on the results obtained after 12 years (Table IIIC), Schulz and Rodrigues (1966) estimated that this plantation would yield 440 m³/ha under a 40-year rotation—equivalent to a mean annual productivity of 11 m³/ha. By comparison, plantations of improved strains of *Pinus caribea* on cleared sites in Suriname attained a mean annual productivity of 17 m³/ha. Through simple selection of seeds from high-quality trees, we believe that virola could attain similar yields.

The experiences summarized above indicate the best strategy for obtaining productive plantations of virola. This species grows best on moist, bottomland soils under intermediate light levels maintained by periodic silvicultural interventions. In the Amazon estuary, such conditions are found in extensive areas of floodplain forest currently logged for virola, where establishment of understory plantations could guarantee future supplies. Yet this strategy has not been adopted by the industries that are most dependent on virola. Despite 12–20 years of experience in reforestation, the two plywood manufacturers investigated in our study continue to maintain plantations under inadequate ecological conditions and without appropriate silvicultural treatments. This situation appears to reflect the general absence of forest policy enforcement noted above. Due to non-enforcement, the taxes for reforestation actually paid are well below the levels specified by law, which weakens incentives for direct investment in reforestation projects. Together with weak economic incentives, the risks of long-term investment in a historically inflationary economy and insufficient technical orientation probably explain the disappointing results of reforestation projects involving virola in the Amazon estuary.

Table III. Responses of virola in plantations in full sun and in shade without silvicultural treatment and with silvicultural treatment

Locale	Age (yr)	Trees (n/ha)	Mean height (m)	Maximum height (m)	DBH (cm)	Basal area (m²/ha)	Volume (m³/ha)
			A. Plantations in full sun				
Portel, Pará	4	1899	4.5	6.2	7.0	7.0	23.2
	6	1910	6.7	9.1	8.2	10.6	46.1
	7	1019	6.6	8.7	11.1	9.9	40.6
	8	1019	6.0	7.1	11.6	11.0	40.5
	9	1324	5.3	7.5	10.5	11.4	39.9
	10	1798	4.2	7.4	10.7	18.2	56.1
	11	997	9.0	12.1	14.2	16.4	93.9
	12	477	9.5	12.5	16.3	10.4	66.4
			B. Plantations in shade without silvicultural treatment				
Macapá, Amapá	*3*	*986*	*0.6*	*2.0*	—	—	—
	8	1134	1.4	6.0	2.0	—	—
	10	902	2.2	8.9	2.2	—	—
	15	1245	3.6	18.0	5.3	4.9	19.7
	20	839	7.8	21.0	13.6	21.7	143.9
			C. Plantations in shade with silvicultural treatment				
Suriname	8	550	15.0	16.0	19.0	16.0	135.0
	9	550	15.0	17.0	22.0	21.0	175.0
	10	550	16.0	18.0	24.0	25.0	208.0
	11	450	18.0	19.0	26.0	23.0	179.0
	12	450	19.0	19.0	28.0	26.0	210.0

Sources: A, Anderson et al., 1994; B, Macedo, unpubl. data; C, Schultz & Rodriguez, 1966.

Forest Management

Virola has several qualities that would facilitate management of natural stands. The species is widely distributed in the floodplain of the Amazon River and its major tributaries (Rodrigues, 1972), where it regenerates profusely. In the Amazon estuary, the fruiting period of virola (January–March) coincides with peak floods, a coincidence that would ensure dissemination of seeds over extensive areas. In addition to water, birds such as toucans (*Ramphastos sulfuratus* Lesson and *R. swainsonii* Gould) serve as important dispersal agents (Howe & Schupp, 1985). The seeds of virola exhibit up to 90% germination in nurseries, and while its seedlings do not tolerate the deep shade of the forest understory, they do grow well in small gaps. Virola's strong apical dominance permits development of a straight, cylindrical stem, with minimal branching and a small crown—ideal characteristics for a timber species. In contrast with mahogany, virola appears to be free of serious phytosanitary problems. Its natural abundance, which has promoted exploitation thus far, could also facilitate management. Virola is common in floodplain forest and achieves extremely high densities and volumes in swamp forest. In this latter ecosystem, selective logging of large-girthed trees (>45 cm dbh) could probably be sustained if carried out under sufficiently long rotations (>10 years).

Despite this potential, however, there are currently no efforts to manage natural stands of virola on a commercial scale. As in the case of plantations, managing natural stands represents another lost opportunity to increase the supply of virola for local industries.

Current Prospects

Although it is impossible to determine precisely how long current stocks of virola will last, evidence suggests that after nearly 40 years of exploitation, and with growing demand by hundreds of sawmills in the estuary, most of the marketable sources of this species are already exhausted, and the collapse of the virola market is imminent. The abrupt decline in export of sawnwood since 1989 (Fig. 5), the difficulties in obtaining logs reported by all industrial sectors, the substitution of lower-quality timber species, and the lack of effective measures to augment supplies are clear signs of the demise of the virola market, which threatens to undermine locally based industries dependent on this resource, increasing the estuarine region's already high rates of unemployment and urban migration.

Alternative Policies and Patterns of Resource Use

Changing the above scenario requires alternative policies and resource-use patterns to decrease present demand and increase future supplies of virola. In this final section, we explore possible strategies for achieving each of these goals.

Demand

The combined demand of plywood manufacturers and large and small sawmills has placed unprecedented pressure on natural stands of virola. Furthermore, our study shows that existing plywood manufacturers absorb lower quantities of virola wood—and make slightly more efficient use of that wood—than do either large or small sawmills. Consequently, we would recommend a series of policies that reduce the overall flow of virola to wood-processing industries and, in addition, favor those industries that make more efficient use of this resource.

To reduce overall demand, we would first recommend concentrating monitoring efforts on large sawmills. As our study shows, the large sawmills in the Amazon estuary consume 60% of virola logs (by volume) and purchase all of the sawnwood produced by small sawmills, thus absorbing more than 80% of the total flow of virola wood through regional industries. More effective monitoring of large sawmills would be relatively easy, as these mills are few in number (24) and are mostly concentrated in or near major towns and cities. Effective monitoring would also enable government authorities to obtain more accurate measures of wood consumption and production—which are essential for regulating the virola market and exacting realistic taxation. The latter, in turn, could help support stepped-up monitoring by IBAMA and induce industries to engage more seriously in reforestation.

As a second and complementary measure, we would recommend limiting the purchase of virola timber by sawmills to a six-month period during the rainy season (January–June). This limitation would discourage extraction of small-girthed logs,.a practice that occurs primarily during the dry season and that greatly increases the impact of logging on natural stands. As above, this measure could be carried out by monitoring large sawmills and would be far more effective and easier to implement than the log-size limitation currently utilized.

Third, our study indicates the need for polices to induce more efficient raw-material use by industries involved with virola. Specifically, fiscal incentives could be provided to wood-processing industries investing in technologies that (1) increase final product value and (2) reduce raw-material waste. This measure could have critical implications not only for virola but also for all forest resources currently utilized by wood-processing industries in the Amazon region.

Supply

We would also recommend measures to increase supplies of virola. The most important measure is to impose realistic taxation on virola consumption, which would provide a stronger fiscal incentive for reforestation. As discussed above, we believe this policy could be effectively implemented by concentrating monitoring efforts on large sawmills.

To promote the development and dissemination of appropriate management techniques for reforestation, we would recommend establishment of a research program on virola. This program could involve joint collaboration between regional research institutions and local industries engaged in reforestation projects, and it could be supported by revenues derived from realistic taxation of those industries.

And finally, we would recommend promoting alternative markets for virola timber produced from sustainable sources. Such markets offer higher prices, which could discourage the currently wasteful use of virola. This measure would require labeling to distinguish timber derived from sustainable sources. Industries and not-for-profit groups have begun efforts to establish such labeling in Brazil.

Prospects for Change

The above recommendations are technically feasible and, if implemented, would promote dramatic transformations in the current exploitation of virola by regional industries in the Amazon. Indeed, similar recommendations could be developed to govern the use of most native timber resources in the Amazon region.

Despite the need for policy change, however, three factors are likely to impede their implementation in the near term. First, there is little awareness among local inhabitants in the Amazon estuary of the impending demise of the virola market, and even less of the possible social and economic impacts of that demise. Yet broad public awareness of these issues is an essential prelude to adopting appropriate solutions. Second, while representatives of wood-processing industries are acutely aware of virola's shrinking supply, they generally view substitution of other species as the most appropriate solution. This solution, however, does not take into account the lower quality of virola's substitutes, which could undermine final product competitiveness, especially in international markets.

Finally, implementation of the above recommendations would require substantial changes in the present structure and operation of government agencies involved in regulating the virola market, especially IBAMA. In addition to increased human and financial resources for monitoring local industries, IBAMA's regional offices need greater autonomy to administer funds and hire qualified personnel, both in the estuary and elsewhere in the Amazon region. As mentioned above, realistic taxing of wood-processing industries would probably cover most or all of these additional costs. The need for increased autonomy, however, runs counter to a long tradition of centralized administration that is a hallmark of federal agencies in Brazil. This tradition derives, in part, from efforts to stem corruption and control government spending, and abandoning it would require a degree of political will that virola's demise is unlikely to muster.

Given these considerations, the most probable scenario is that supplies of virola will continue to diminish and regional industries will increasingly substitute it with inferior species. This substitution, in turn, will lead to increased pressures on a wide range of forest resources, thus exacerbating the ecological impacts of current logging practices in the region. And as those resources are, in turn, depleted, wood-processing industries in the Amazon estuary—which employ thousands of workers and in 1989 generated US$50 million in revenues from virola alone—will eventually relocate to new areas where forest resources are more abundant. This scenario is a recurring historic pattern in Brazil, and it is destined to continue in the absence of policies that take into account the full value of those resources. Without such policies, the boom-and-bust cycles that have characterized Amazonia's past are likely to continue in the future.

Acknowledgments

This study is part of an interdisciplinary research project on logging in the Amazon estuary, supported by the Tropical Forestry Program of the World Wildlife Fund. Additional financial support was provided by the Brazilian Research Council (CNPq) and the Ford Foundation. We gratefully acknowledge the institutional and logistic support provided by the Emílio Goeldi Museum of Pará in Belém, Brazil.

Literature Cited

Anderson, A. B., I. Mousasticoshvily & D. S. Macedo. 1994. Impactos ecológicos e sócio-econômicos da extração seletiva de virola no estuário amazônico. World Wildlife Fund, Washington, DC.

Calzavara, B. B. G. 1972. As possibilidades do açaizeiro no estuário amazônico. Boletim da Faculdade de Ciências Agrárias do Pará 5: 1–103.

Howe, H. F. & E. E. Schupp. 1985. Early consequences of seed dispersal for a neotropical tree (*Virola surinamensis*). Ecology 66: 781–791.

Kageyama, P. Y. 1981. Endogamia em espécies florestais. Instituto de Pesquisas e Estudos Florestais (IPEF), Série Técnica 2(8): 1–40.

Kanashiro, M. & J. A. G. Yared. 1991. Experiências com plantios florestais na bacia amazônica. Pages 117–152 *in* O desafio das florestas tropicais. Universidade Federal do Pará, Belém.

Macedo, D. S. & A. B. Anderson. 1993. Early ecological changes associated with logging in an Amazon floodplain. Biotropica 25(2): 151–163.

Mousasticoshvily, I. 1991. Comercialização e ndustrialização da virola no estuário amazônico. Tese de Mestrado, Universidade Federal do Paraná, Curitiba.

Prance, G. T. 1980. A terminologia dos tipos florestais da Amazônia sujeitas a inundação. Acta Amazônica 10(3): 341–349.

Rankin, J. M. 1985. Forestry in the Brazilian Amazon. Pages 369–392 *in* G. T. Prance & T. E. Lovejoy, eds., Amazonia. Pergamon Press, Oxford.

Rodrigues, W. A. 1972. A ucuúba da várzea e suas aplicações. Acta Amazônica 2(2): 29–47.

Schulz, J. P. & L. P. Rodriguez. 1966. Forest plantations in Surinam. Revista Forestal Venezuelana 9(14): 5–36.

Uhl, C. 1990. Wood as an economic catalyst to ecological change in Amazonia. Pages 7–27 *in* Economic catalysts to ecological change. University of Florida, Gainesville.

Selective Logging along the Middle Solimões River

Ana Luisa K. M. Albernaz and J. Márcio Ayres

Introduction

Selective logging of floodplain forests along the middle Solimões River has been going on for more than 40 years. This practice is common in tropical forests, where there are many species per unit area and timber trees usually occur at low densities. A number of commercially important species from the Brazilian Amazon, such as mahogany, samaúma (*Ceiba pentandra*), and macacaúba (*Platymiscium ulei*), are endangered today because of the huge demands placed on their populations.

As timber stocks diminish in Southeast Asia, wood volume from Amazonia has shown a significant increase in export markets. In 1984, the region produced 43.6% of Brazil's total log-wood volume; a decade earlier, this figure was 14.3% (Browder, 1988). The state of Pará is responsible for most of today's output, but over the next few years, the state of Amazonas should increase its share of the timber market (Higuchi et al., 1994). Although floodplain forests occupy some 2–5% of Amazonia's total area, approximately 75% of the wood that enters the market is from these forests (Brazil Ministry of the Interior, 1982). In spite of the fact that these are the region's most heavily exploited areas, forestry management has been very poorly studied in the floodplain forests (Jansen & Alencar, 1991; Higuchi et al., 1994).

Along the middle Solimões, selective logging takes place only in flooded forests. There are extensive floodplains and upland swamps (*igapós*) in this central portion of the Amazon region. Because the uplands are located far from the Precambrian shields, it is difficult to obtain the crushed rock used in road construction, so an extensive road system has yet to be established. The more remote uplands being inaccessible, selective logging takes place only in the seasonally flooded forests and is especially intense along the middle Solimões during the high-water season. This paper focuses on timber exploitation in a very restricted area (the eastern sector of the Mamirauá Ecological Station, or MES) during the flood seasons of 1993 and 1994, when water levels were exceptionally high. Timber cutting is, however, an activity that is rapidly expanding to include all flooded forests along the larger rivers of the Amazon Basin.

Study Area and Survey Methods

This study was carried out in what is known as the focal area of the Mamirauá Ecological Station (Fig. 1), which is bounded by the Solimões and Japurá Rivers and by the Aranapú *paraná* (a natural canal linking two rivers). There are some 260,000 ha of

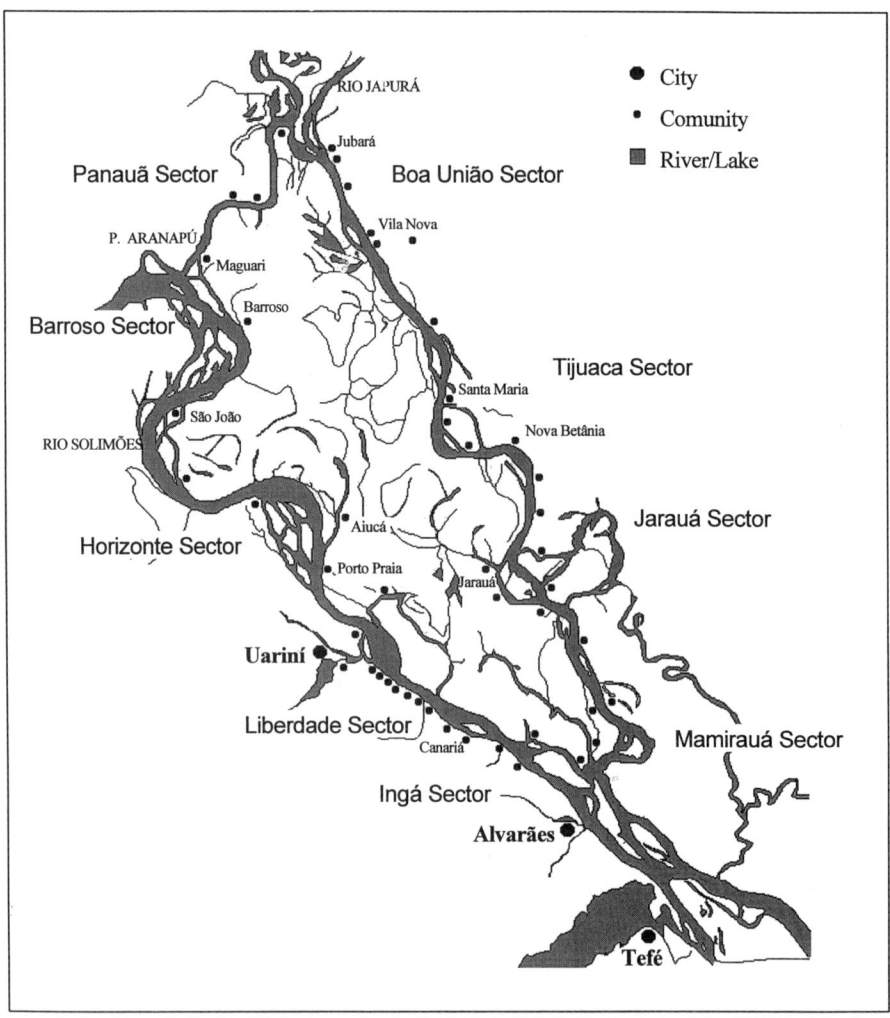

Figure 1. Map of the focal area of the Mamirauá Ecological Station (sectors and communities) and neighboring cities.

floodplain in this sector of the MES, where three municipalities (Tefé, Alvarães, and Uarini) and around 60 riverine communities find space for fishing, logging, and farming.

The data on logging sites and labor involved were obtained mainly from local informants; almost all of the communities were visited, beginning in April, during the tree-felling phase, in 1993 and 1994. The visits took place at regular intervals in order to keep tabs on the operation. When the logs were "ready" (i.e., tied into rafts near the banks of lakes and paranãs), the species were identified and the length and diameter of the logs measured. A description of how the rafts are set up is found in Higuchi et al., 1994.

Trees were identified by their common names (Table I). This list was compared to that of Ayres (1993) for the same study area to provide scientific names for these species, wherever possible. The loggers (when found) were questioned as to prices, use categories, ways of marketing and buyers, reasons for being involved in logging activities, selection of species for felling, harvest period, and tools used. Although it was possible to obtain this information for the timber rafts of individual loggers, it proved impossible in the cases of buyers who set up their own rafts using logs from different suppliers. For this reason, many tree trunks were identified and measured as they floated freely in the lakes; these were marked with spray paint to avoid measuring the same log twice. Other tree trunks were measured in the rafts made up of logs from various sources. Estimated mean raft size was based on both kinds of rafts, so the estimated number of people involved in the operation is not precise. For buyers' rafts, an attempt was made to obtain information on the destination of the wood and the marketing process (i.e., number of middlemen involved between loggers and the sawmill or factory).

Neighboring communities in Mamirauá are grouped into "sectors," or basic units of resource-use control (Reis, 1993; Ayres et al., 1994). The nine sectors are as follows: Mamirauá, Jarauá, Tijuaca, Boa União do Médio Japurá, Aranapú, Barroso, Horizonte, Liberdade, and Ingá. There was some variation in how some data were collected in particular sectors. The very large amount of wood of just one species, envira-vassourinha (*Xylopia* cf. *calophyllum*), harvested in 1994, made it possible to measure only a sample of trunks from those sectors where that species was exploited. The first three rafts found in each such sector were measured; the others were merely counted. The sample included 38.8% of the total number of trunks removed. The Barroso sector had the smallest percentage of trunks measured of the total removed (120 of 612, or 19.6%). In the Liberdade and Ingá sectors, the sample included all of the trunks removed (16 and 32, respectively). On rafts where no measurements were taken, trunks were counted and mean volume for the sector was used to calculate total volume for the category.

Economic Aspects

Marketing

Wood was marketed through various channels and in various ways. It proved impossible to quantify the precise amounts of wood that found their way to the market in each particular way, because marketing data were obtained mostly by indirect methods. The following modes of marketing were distinguished: through *patrões* (patrons; singular *patrão*), *aviamento* (advance of materials), leasing, and cash sales. The first two forms are the most common. Part of the wood harvested is also used by the loggers for construction of their own homes. Agreements with the sawmills return about one-third of the processed wood to the logger, while two-thirds is kept by the mill.

The patrões of today include itinerant traders, middlemen, and merchants; they own no property and conduct their business from boats. They supply goods to the communities all year long, and buy produce in season, be it fish, wood, cassava flour, or bananas. The producers are forever in debt to these people because the price of

Table I. List of species (common and Latin names) harvested in 1993 and 1994, plus number of trees and use categories

Common name	Scientific name	Number of trees 1993	Number of trees 1994	Category[a]
Andiroba	Carapa guianensis	10	17	H
Arapari	Macrolobium acaciaefolium	7	4	H
Araparirana	Macrolobium bifolium	0	14	S
Assacu	Hura crepitans	2104	1425	S
Biribarana	indet.	—[b]	311	F
Boieira	Apeiba sp.	140	15	F
Bolacheira	Apeiba sp.	299	262	F
Caucho branco	Styrax sp.	3	2	F,L
Caxinguba	Ficus maxima	0	1	F
C.-de-macaco	Couroupita sp.	0	2	S
Castanharana	Eschweilera sp.	0	1	S
Cedro	Cedrelinga odorata	51	120	H
Copaíba	Copaifera sp.	67	55	S
Embaúba	Cecropia latiloba	1	6	F
Envira-amargosa	Annonaceae	0	1	F
Envira-fofa	Annonaceae	17	0	F
Envira-do-igapó	Annonaceae	0	2	F
Envira-da-restinga	Annonaceae	0	1	F
Envira-preta	Annonaceae	0	1	F
Envira-vassourinha	Xylopia cf. calophyllum	528	1750	L,F
Envira-sangue	indet.	166	32	L
Envira-surucucu	Annonaceae	6	1	L
Faveira	Vatairea guianensis	0	1	H
Icezeiro	Luehea cymulosa	54	45	L,F
Itaúba	Mezilaurus sp.	0	3	H
Itaubarana	Acosmium sp.	1	0	H
Jacareúba	Calophyllum brasiliense	281	174	H
litó	Guarea sp.	0	20	H
Louro-abacate	Aniba sp.	1	2	H
Louro-amarelo	Nectandra sp.	9	11	H
Louro-jacaré	Lauraceae	4	24	H,L
Louro-inamuí	Ocotea cymbarum	622	555	H
Lourinho	Lauraceae	0	1	L
Louro-preto	Lauraceae	111	58	H,L
Macacaricuia	Couroupita sp.	168	413	S
Macacaúba	Platymiscium ulei	10	17	H
Maparajuba	Sapotaceae	0	1	H
Muiratinga	Maquira coriacea	278	269	S
Mulateiro	Calycophyllum spruceanum	148	442	H
Munguba	Pseudobombax munguba	35	54	F
Mungubarana	Pachira sp.	0	7	S
Murupita	Sapium hippomane	345	298	F,L
Mututi	Paramachaerium ormosioides opoormosioidesormosioides	12	22	S
Paricarana	Pithecellobium corymbosum	25	46	S,F
Piranheira	Piranhea trifoliata	23	36	H
Samaúma	Ceiba pentandra	538	400	S
Sardinheira	Homalium racemosum	3	0	F
Seringa-barriguda	Hevea spruceana	11	4	F
Seringarana	Micrandra siphonioides	11	2	F
Sucuúba	Himatanthus tarapotensis	2	0	S
Tacacazeiro	Sterculia elata	5	12	S
Tachi-branco	Pterocarpus sp.	0	2	F
Virola	Virola sp.	790	953	S,F,L
Total species		38	49	
Total trees		6897	7895	

[a] S, softwood; H, hardwood; F, floats; L, fuelwood.
[b] This species was probably included as boieira and bolacheira.

merchandise is high and that of produce is low. Most patrões operate in a defined area—that is, they have a fixed clientele. Some of the local people have more than one patrão at any one time.

Aviamento is a system of advancing food and other necessities to extractors in anticipation of their collecting a specified product—in this case, wood. Materials advanced to loggers include axes or chain saws (occasionally also a chain-saw operator), fuel, and cables and hammers for constructing the log rafts. The cost of all these materials is deducted from the value of the produce.

Local inhabitants sometimes allow outsiders to work at sites near the community in exchange for part of the product extracted or for some community benefit. In 1993, for example, land was leased in exchange for part of the wood extracted from the São João sector, whereas in Fortaleza de São José the exchange involved a chain saw. In both instances, the agreement was finally not honored, and the communities received nothing for their timber.

At the end of the high-water season, buyers from other regions, especially Manaus and Itacoatiara, visit the area in search of timber. Since they have no local ties, they pay cash and usually offer the highest prices. In 1993, these buyers were mainly after samaúma, and they took only small amounts of other species. In 1994, they added virola and muiratinga to the preferred species list.

Use Categories, Volume, and the Consumer Market

Wood harvested from the floodplain of the MES falls into four use categories: (1) hardwood, sold to sawmills (for furniture, trim, lumber for walls and floors); (2) softwood, for general construction and plywood factories; (3) fuelwood; and (4) floats, used to make the rafts that keep nonbuoyant wood from sinking. Floats have no commercial value.

Total wood volume harvested at the MES in 1993 and 1994 was about 20,000 m^3 (Table II). Softwood accounted for most of this volume and made up a larger percentage in 1993 than in 1994. While softwood volume fell in 1994, hardwood volume increased slightly and fuelwood showed the greatest increase.

All of the hardwood was sold to the local market in 1993; one sawmill directly

Table II. Trees removed for commercial use in 1993 and 1994, by volume, as percent of total removed, and as number of individuals

	1993			1994		
	m^3	%	N	m^3	%	N
Softwood	15,398.8	76.1	3603	11,956.6	58.5	3238
Hardwood	2421.6	12	1280	3223.4	15.8	1475
Floats	—	—	1209	—	—	1327
Fuelwood	2415	11.9	805	5250	25.7	1849
Total	20,235.4	100	6897	20,430	100	7889

purchased more than 50% of the MES production for this year. In 1994, another sawmill, built in the nearby city of Tefé, bought most of the hardwood removed from the MES and sold part of the production as lumber to Manaus. The softwood went mainly to Manaus and Itacoatiara, with a small portion going to Belém; for this reason, softwood sales generally involved more middlemen. The two sawmills, plus a merchant from Tefé, were the final buyers in a chain of middlemen. They bought from small buyers and sold to factories. All of the fuelwood was sent to Tefé, where 95% was purchased by the city's brick factory.

A small sawmill in the municipality of Alvarães bought 5% of the hardwood in 1993. In 1994, records show that no wood was purchased for the sawmill in Alvarães. Buyers from this small city, some of whom were patrões in a number of communities, ordered mostly softwood, and served as middlemen for sales to Manaus/Itacoatiara. The volume of wood they handled, however, was much smaller than that purchased by the Tefé wood merchants cited above.

The small sawmill at Uarini did not buy any wood from the MES in 1993 or 1994. Local wood extractors, who harvest timber to use in rebuilding their own homes, prefer to deal with the Uarini mill because it is the only one that returns 50% of the processed wood to the logger.

Prices

The price of a cubic meter of wood varied with the year, category, species, and buyer (Table III). In 1993, prices depended basically on two factors: category of use (softwood, hardwood, or fuelwood) and log diameter (first-grade wood, with diameter >50 cm; second-grade wood, with diameter <50 cm). In all categories of use, a cubic meter of first-grade wood was worth twice as much as a cubic meter of second-grade, regardless of category of use, while hardwood brought higher prices than softwood. In 1994, three

Table III. Prices (US$) of hardwood and softwood paid at MES by local buyers (from Tefé and Alvarães)

		1993		1994	
	Species in greatest demand	Grade 1[a]	Grade 2[b]	Grade 1[a]	Grade 2[b]
Hardwood	Ocotea cymbarum Calophyllum brasiliensis Calycophyllum spruceanum	5.76	2.88	8.45	4.23
Softwood	Virola sp. Ceiba pentandra Maquira coriaceae	1.3–3.8	0.75–1.3	3.3–10.3	1.75–5.15
	Couroupita sp.			2.4–6	1.2–3
	Hura crepitans			1.45–5	0.75–2.5

[a] Diameter >50 cm.
[b] Diameter <50 cm.

softwood species (sumaúma, virola, and muiratinga) commanded the highest prices, others (e.g., macacaricuia, tacacazeiro, and paricarana) brought intermediate prices, and assacú was sold for the lowest price in this category.

The price of softwood varied more widely than that of hardwood, probably because the former is not sold locally and marketing involves more intermediate steps. Those who sold softwood directly to Manaus or Itacoatiara obtained the highest prices: US$9.60 in 1993 and US$13.02 in 1994.

Fuelwood sales are based on a different unit of measurement: each tree is cut into 0.8 m logs, and the cubic-meter total is estimated by the number of $1 \times 1 \times 1$ m boxes filled by these logs. Prices were not given for 1993, but they were certainly lower than in 1994. In 1994, a cubic meter of fuelwood cost US$0.85, considered to be a fair price in relation to that of other types of wood, so harvesting for this use increased.

These price discrepancies between softwood species, plus higher prices and greater demand for fuelwood in 1994, were responsible for major changes in harvesting intensity from 1993 to 1994. For example, the quantity of envira-vassourinha trees (a highly valued fuelwood species) increased by more than a factor of three, from 1993 to 1994, while that of the devalued species, assacú, fell by almost 700 trees (Table III). The higher prices for virola encouraged loggers to fell trees >35 cm in diameter, which is the commercial limit established by buyers. In 1993, for example, virola was among the five most heavily exploited species in the softwood and float categories, and it was harvested for fuelwood as well. In 1994, it was no longer logged for fuelwood and its use as a float diminished.

Biological Aspects

Number of Trees Harvested

Both 1993 and 1994 were years of exceptionally high floods, which permitted easy access to and removal of timber from areas that are not reached by high water in most years. According to local residents, the amount of timber removed during these years was approximately twice that of years with "normal" flooding. In 1993, 6897 harvested tree trunks were identified and measured; in 1994, this figure rose to 7895 (Table II). The larger number of trees harvested in 1994 was possibly the result of improved sampling rather than an actual increase in timber removed. The area was better known in 1994, and sites could be reached that were not sampled in 1993. Also, since this was the second year of sampling, the loggers were less reluctant to give out information because they no longer feared negative repercussions from the study. The largest number of harvested trees belonged to the softwood category for both years (Table II). In 1993, both hardwood and floats accounted for more felled trees than fuelwood, but in 1994 the opposite was true. The number of trees in the float category was similar to that of hardwoods for both years.

Exploited Species and Selectivity

The logging survey included 38 morphospecies in 1993 and 49 in 1994. However, over 50% of the timber removed in each category came from only two or three species.

There were more morphospecies in the hardwood category in 1994 than in 1993 (Table I).

During both years, about 85% of the commercial timber removed came from just nine tree species: assacú (*Hura crepitans*: Euphorbiaceae), envira-vassourinha (*Xylopia* cf. *calophyllum*: Annonaceae), samaúma (*Ceiba pentandra*: Bombacaceae), virola (*Virola* sp.: Myristicaceae), muiratinga (*Maquira coriaceae*: Moraceae), macacaricuia (*Couroupita* sp.: Lecythidaceae), louro-inamuí (*Ocotea cymbarum*: Lauraceae), jacareúba (*Calophyllum brasiliense*: Guttiferae), and mulateiro (*Calycophyllum spruceanum*: Rubiaceae). The most sought after species were basically the same in both years, the greatest difference being the decrease in demand for assacú (*Hura crepitans*: Euphorbiaceae) and the increase in demand for envira-vassourinha (*Xylopia* cf. *calophyllum*: Annonaceae). For softwood species, there was an increase in the number of macacaricuia trees harvested, while for hardwoods, the harvest of mulateiro increased while that of jacareúba decreased (Table I).

Of the four designated categories of use, the float category contains the greatest number of exploited species, but these are also highly selected. Many of these species float only if they are cut before their roots become immersed in the floodwaters, which is the main reason cited for their restricted use.

Market prices are the main factor in selecting tree species for commercial exploitation. Besides the market, loggers also take into consideration the need and availability of floats when harvesting hardwoods. Cedro is buoyant and jacareúba and louro-inamuí also usually float. Mulateiro and piranheira, on the other hand, will always sink and require several large floats, which means a greater effort is expended per harvested tree.

Harvesting Diameters of the Ten Most Heavily Exploited Species

The law establishes a minimum diameter of 45 cm for timber tree removal. Although loggers are well aware of the law, this rule was not observed for the species harvested at the MES. For most of the commonly harvested species, both minimum and mean diameter of trees logged in 1994 were greater than they were in 1993 (Table IV).

Smaller minimum diameters in 1994 were found only in the cases of louro-inamuí, mulateiro, and envira-vassourinha. The latter two species had many more trees removed in 1994 than in 1993. This increase in demand may have been responsible for the lack of diameter selectivity. The other species, louro-inamuí, has been exploited for a much longer time (probably more than 35 years), and trees with large diameters are rare or are only found in remote areas. This is a common species in rafts headed for the sawmill, so it seems that the search for marketable wood is more important than the search for large trees.

Average diameters were larger in 1994 than in 1993 for all species except assacú, samaúma, macacaricuia, and envira-vassourinha. These species were the most intensely exploited within their respective use categories.

Maximum diameters of assacú, virola, samaúma, macacaricuia, and jacareúba were larger in 1993 than in 1994, which seems to indicate that large trees are becoming rarer.

Table IV. Minimum, mean, and maximum diameters of the 10 most heavily exploited species, in 1993 and 1994

| | Diameter (cm) | | | | | | Number measured | |
| | Minimum | | Mean | | Maximum | | | |
	1993	1994	1993	1994	1993	1994	1993	1994
Hura crepitans	20	35	76.0	74.8	185	175	2103	1420
Virola sp.	10	15	37.4	39.3	85	70	787	945
Ceiba pentandra	35	35	91.3	85.4	210	190	537	395
Couroupita sp.	35	40	64.8	63.0	130	115	169	410
Maquira coriacea	35	35	54.9	60.2	150	175	279	269
Calophyllum brasiliense	30	40	59.3	60.9	135	125	281	173
Ocotea cymbarum	25	20	46.4	49.0	80	80	622	550
Calycophyllum spruceanum	40	35	62.3	63.8	120	130	148	441
Xylopia cf. calophyllum	15	10	31.2	29.8	50	50	528	680
Sapium hippomane	10	15	30.6	33.5	55	55	341	297

Maximum diameters for mulateiro and muiratinga were larger in 1994 than in 1993, which may indicate that exploitation of these species is increasing.

Density of Felled Trees

During this two-year study, an estimate of 0.057 trees/ha, on average, were removed from the 260,109-ha area. Most of the commercial species, however, grow primarily on the *restingas* (strips of flooded forest on levees that border streams and rivers in Amazonia) that make up only about 30% of the total area (Ayres, 1993). So, within the area occupied by these forests, 0.19 trees/ha were felled in the two-year period. Preliminary forest surveys estimated the density of trees cut in the restingas as 0.77 trees/ha (Albernaz, 1994).

Estimated annual yields for several restingas showed some variation. Direct estimates of 0.06–0.24 trees/ha were obtained through mapping (using tape measure and compass) in areas harvested by loggers. This variation appears to be the result not of differing degrees of exploitation but of varying timber potential between restingas due to locally high species densities, a common phenomenon among many tree species of floodplain forests and other tropical forest habitats (Pinedo-Vasquez, pers. obs.).

Losses

Since logs are transported by water in the MES, the flood regime exerts a strong influence on final wood output. Because of the unpredictability of the river's cycle, planning is difficult and some of the felled trees are lost annually. Losses due to water-level fluctuation come about in three ways: (1) Trees were felled at sites not reached by rising water levels; (2) log removal was delayed; and (3) trees were felled above maximum water level. The first two cases take a heavier toll because they involve the loss of entire logs. The maximum water level in any given year is unpredictable. Some loggers begin work before the dry season has ended, and if the water level does not reach the site, the logs are left stranded. If the site is flooded the following year, hardwood logs can still be used, but softwood logs rot more quickly and are worthless if left until the next high-water season.

Delays in removing the logs may be caused by the logger himself or by the buyer. The logger usually keeps cutting trees as long as the waters continue to rise. When maximum water level is reached, he begins the work of skidding (taking logs from the felling site to a nearby body of water for shipping), floating and setting up the rafts in lakes and paranãs (Higuchi et al., 1994). The wood must be removed as soon as flood-waters begin to recede. The unpredictability of the cycle makes it hard for the logger to ascertain exactly how much time he will have for skidding, so he may cut down more trees than he can skid out during the slack. This type of loss has increased as more workers use chain saws, a practice that shortens the time it takes to fell trees and increases the number of trees felled. Delays may sometimes be caused by other factors such as accidents or sickness. When the logger causes the delay, the felled trunks remain in the forest and are removed only if the next year's floodwaters reach the site and if the wood has not rotted during this time.

Local residents tell stories of loggers who have left enormous piles of logs in the forest. In 1993, for example, a man cut down 150 mulateiro trees, of which he removed only 30. In 1994, a patrão financed the felling of 900 trees, of which fewer than 50 were skidded out. Although these data are only estimates, they serve to illustrate the scale of this type of waste. It is hard to put numbers on the total wood volume lost every year. Preliminary surveys of commercial-tree density at MES showed that for every two trees removed from the forest, one is left to rot (Albernaz, 1994). Hardwood species may lie for several years on the forest floor without rotting, so the number of trunks recorded in this category is cumulative.

For most of the wood production in the area, the buyer's agreement is signed when felling begins. Whether a patrão finances the operation or aviamento is involved, there is always a debt to settle, at least partially, which guarantees that the logger will remain true to a particular buyer. Sometimes, however, the buyer may be delayed in picking up the wood for a number of reasons—the motorboat breaks down or he has other business commitments, or he may even yield to indifference, because even if the buyer does not go after the wood, the debt remains. In cases where the delay is caused by the buyer, the rafts are usually set up at the edge of some lake ready for delivery. Sometimes a belated surge in flood waters allows the wood to be shipped out a few months later. When this does not happen, the raft must be reassembled with new floats and only the hardwood is usable. The buyer pays for the wood (or reduces the debt) only when he picks it up, measures it, and calculates the volume. When the buyer is late and the wood is stranded, it is the logger who takes the loss. Wood lost in this way is included in the totals of harvested wood; the loss percentage could not be assessed.

Sometimes, to avoid the risk of the wood being stranded, the logger chooses to fell only those trees that are already standing in water. In this case, the base of the trunk, representing 2–3 linear meters per tree, is not usable.

Use of Chain Saws

In 1993, measurements were taken on 9 rafts containing wood that had been cut by Tefé residents using a chain saw. At that time, only one person had this equipment. In 1994, 10 rafts from Tefé were measured, and 4 residents felled trees with their own chain saws. For each raft, in 1993, an average of 26 trees had been cut by ax and 139 by chain saw; in 1994, this proportion was 35:120. Felling by chain saw helps the logger to set up larger

rafts, but it also increases the risk of loss, as discussed above. The greatest danger of chain saw use, however, is not the increase in wood volume felled per season; rather, it is related to the elimination of site and season restrictions that once put a brake on the harvesting of hardwoods. Part of this wood is processed on site as sawn lumber and boards and is removed as a finished product. Control over this type of exploitation is impossible, because felling and removal take place year-round and the processed wood is transported in boat holds.

The small quantity of piranheira (*Piranhea trifoliata*) recorded in the sample is due in part to this practice. This heavy wood is not buoyant and requires several large, good-quality floats (such as assacú) or many low-quality floats. Most sawmills do not work with this species because it quickly wears down the saws. Lumber made from piranheira was seen in the holds of several rafting tugboats, which means that wood of this species is removed as a finished product, even when it is cut during the peak of the harvest season.

Social and Economic Aspects

The Work Cycle and the People Involved in Timber Harvesting

The economically most important output of the residents of the MES floodplain falls into three basic categories: cassava meal (*farinha*), fish, and wood. Cassava is harvested for preparation mainly in February and March. From March on, for as long as the waters keep rising, it is the time for cutting trees. During the two-year study period, wood hauling always began in late May or early June, but 80% of the wood was removed during mid-July to mid-August (1993) or to late August (1994). Commercial fishing then becomes the main activity and so remains until it is again time to prepare the cassava meal. Sometime in September, a few days are dedicated to preparing the fields and planting cassava (Fig. 2).

This is the general pattern of the cycle, but some communities or individuals may spend more time working at one or the other of these activities. Those who have more experience or are better equipped for harvesting trees sometimes begin logging in January, when the fish catch diminishes. On the other hand, those who are more inclined

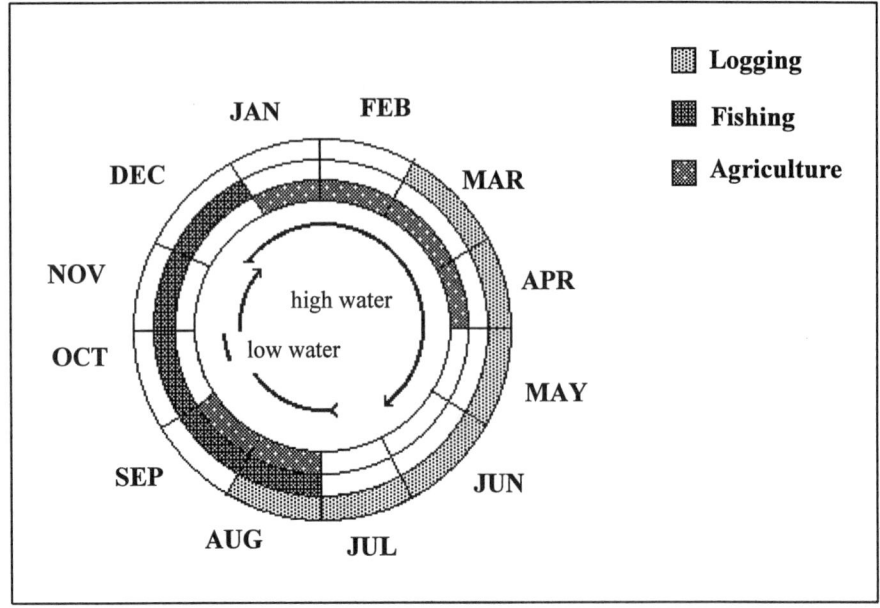

Figure 2. Timetable of residents' economic activities at the Mamiraüá Ecological Station.

to farm resort to logging only when the flood waters invade the restingas. This does not happen every year, but when it does, most of the agricultural production is lost.

Water levels therefore affect not only how many people are involved in timber harvesting but also how long this activity will go on. Heavy floods favor the removal of larger volumes of wood because they permit access to underexploited areas and restrict other activities (fishing and farming).

In 1993, MES inhabitants were expecting a heavy flood season, which in fact occurred. By mid-March, the field crops had already been lost and people began preparations for the tree-felling season. In 1994, a second season of high water levels was not expected, so even in late April, only those communities or individuals who traditionally harvest timber intended to do so. Many loggers started felling trees in late May and early June, showing how the duration of the activity depends not only on water levels but also on the people's expectations concerning flood intensity.

Harvest of softwoods and hardwoods by the local people was usually done in pairs, whereas residents of Tefé worked in teams of 2–4. Fuelwood was gathered individually. Output per worker during the 1993 season averaged 34.75 trees for those using chain saws (ca. 40 workers); 80.5 trees for those gathering fuelwood (ca. 10 workers) and 18.66 trees for those felling softwoods or hardwoods using an ax (ca. 250 workers). In 1994, mean output per worker was 35.17 trees for those using chain saws (ca. 50 workers), 74.96 for fuelwood (ca. 30 workers) and 14.53 for the remainder (ca. 300 workers). As mentioned above, these estimates are based on the number of rafts measured annually and therefore are not very accurate, but they are probably not too far removed from actual figures. Those who used chain saws showed similar results for both years, but there was a drop in output for the other loggers in 1994 because they began at a later date. On the other hand, the total number of people working in the timber industry in this area was estimated to be around 300 in 1993 and 380 in 1994, for a user population of ca. 5000, excluding the municipal seats (Lima-Ayres, 1994a). Each logger had an average income of around US$183.98 in 1993 and US$398.07 in 1994.

The local people usually travel back and forth to the cutting site every day. In general, chain saw users (both local residents and those from Tefé) set up floating campsites during the harvest. The Tefé workers receive provisions regularly from their backers. The local people usually return to their own communities on Sunday. So during harvest time, involvement is practically full-time, with short breaks to hunt and fish for food.

Patterns of Exploitation in the Focal Area

Timber harvesting is not uniform over the whole area; some sectors exploit a greater volume of wood. In 1993, around 40% of wood volume was removed from the Jarauá sector. This is the largest sector and is the only one that set aside land for use by the towns of Alvarães, Tefé, and Uarini. In 1993, the Horizonte and Tijuaca sectors had the second and third largest wood output, respectively. In 1994, the Horizonte sector produced the greatest volume of wood, followed by Tijuaca and Jarauá, in that order (Fig. 3). Horizonte is the second largest sector, with five communities participating in timber exploitation; production in three of these (Aiucá, Porto Braga, and São João) was well above that of the others. In Tijuaca, two of the six communities (Nova Betânia and Santa Maria do Cururú) are more deeply involved in timber harvesting.

The smallest volumes of wood were extracted in the Ingá and Boa União do Médio Japurá sectors. These settlements are located on the banks of the Solimões and Japurá Rivers, respectively, and enjoy direct access to the uplands. During the timber harvest, other activities are available in the uplands, such as Brazil-nut harvesting and hunting. Many people raise cattle in the Ingá sector.

It should be pointed out that most of the wood from the Horizonte and Tijuaca sectors was removed by local residents, whereas in Jarauá, timber extractors from Tefé were responsible for much of the timber removal.

Area of Community Influence

Several logging sites were visited in 11 communities, and the shortest distance between site (lake or paranã) and community was calculated using a global positioning system. Jarauá residents harvested timber up to 13.4 km away from their community (mean = 6.8; $N = 6$), farther away than any other community. Residents of Nova Colômbia traveled up to 10.6 km (mean = 5.9; $N = 6$), and those from Aiucá traveled up to 10.4 km. Distances were also noted for Barroso (8.9 km), São Francisco do Cururú (5.9 km), Porto Braga (4.7 km), Maiana (4.6 km), Sitio Maguari (4.5 km), São João (3.8 km), Santa Maria do Cururú (3.5 km), and Nova Betânia (2.9 km).

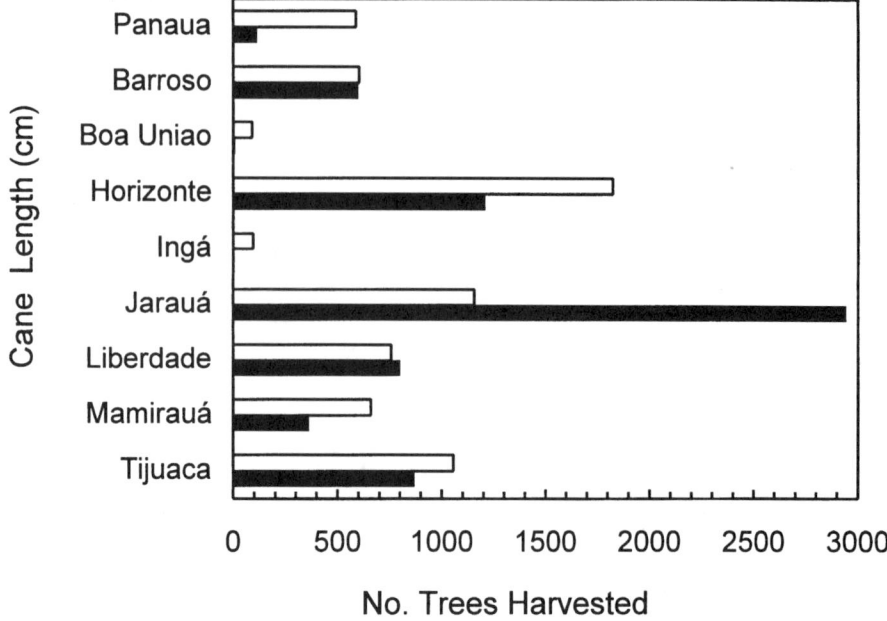

Figure 3. Number of trees harvested in 1993 and 1994, by sector.

In Jarauá and Aiucá communities, the greatest distances were covered by owners of boats powered by the locally well-adapted outboard engines known as "*rabeta*" as well as other small motor-powered craft. Residents of these two communities, as well as those from Barroso and Maguari, worked on several restingas; those from Porto Braga, Nova Betânia, and São Francisco do Cururú worked together on one restinga.

The Management Plan

The main objective of the management plan that is being drawn up for the MES is to provide for intensive-use zones (settlements and commercial centers), controlled-use zones (for maintenance), and preservation zones with no significant human activities permitted (see Ayres et al., this volume). Due to the characteristics of this area, these zones should form a mosaic; they should not be concentric, as has been recommended for other conservation units.

Fishing was the focus of this plan initially, and use categories were defined for many of the lakes (Lima-Ayres, 1994b). Forestry studies began later, and in 1993, timber was removed from the restingas that line the banks of several hatchling ponds. This made clear the need to emphasize the fact that many important timber trees furnish food for fish species. The idea that nurseries should be totally preserved was readily accepted and became a legislative resolution. In 1994, the only nursery sites where timber was exploited were those that are still in dispute today by some sectors or communities.

Besides the preservation issue, discussions on timber use have stressed the problems of loss and minimum cutting diameters. Although these issues are understood when they are discussed, they are difficult to put into practice. Most sectors established minimum diameters by species, but the 1994 harvest showed that these limits were not observed. A satisfactory solution for avoiding or reducing waste has yet to be found.

Discussion

The absence of logging operations in the uplands of the middle Solimões River and the increased demand for laminated wood in world markets have turned floodplain forests into the focal point of this activity throughout the Brazilian Amazon region. In order to use this resource in a sustainable manner, we must expand considerably our knowledge of the impacts of logging on local flora and fauna. The data we do have on certain aspects of these operations—such as increase in number of species utilized every year and changes that occur in use intensity—are cause for concern, because they show that this activity is not sustainable.

Data from 1993 and 1994 show that the extraction of macacaricuia (*Couroupita* sp.), paricarana (*Pithecellobium corymbosum*) and arapariana (*Macrolobium bifolium*)—all of them softwood species—is on the rise; only samaúma (*Ceiba pentandra*) exhibited a long-term decline. Exploitation of most softwoods, however, began rather recently. In 1984, at the outset of softwood exploitation, samaúma trees made up 15% of the total number of trees harvested in all categories, but 10 years later this figure had fallen to only 5%. In 1984, the most important softwoods were samaúma (almost one-half of the harvested timber in this category) and macacaricuia. At that time, assacú (*Hura crepitans*) and virola (*Virola* sp.) were used only to build houseboats (Ayres, 1994).

For hardwoods, the volume of mulateiro (*Calycophyllum spruceanum*) extracted has increased from one year to the next. In 1994, this species was the most important by volume for the category. For about two decades, the main species harvested had been cedro (*Cedrelinga odorata*: Meliaceae) and macacaúba (*Platymiscium ulei*: Papilionaceae). Today, only a small number of cedro trees are harvested, and these are often used directly by the logger. Macacaúba has practically disappeared from the market. The status of these species is critical because the few trees found in forest surveys were either twisted or hollow. In 1984, small amounts of cedro and macacaúba could still be found, but louro-inamuí (representing over one-half of the wood destined for sawmills), jacareúba, and smaller amounts of mulateiro were more important in terms of number of trees harvested (Ayres, 1994).

Although there are still stocks of these commercial trees with small diameters in the restinga forests of the middle Solimões, the problem of negative genetic selection through logging activities should not be forgotten. It is always the tallest, straightest, healthiest trees that are harvested, thus causing reproductive specimens to become rarer and also promoting negative selection in relation to wood quality. Although negative selection may be the most serious problem detected in today's harvesting methods, the increased diversification of species use may have an even greater impact on the environment.

A unique aspect of floodplain timber exploitation is the use of floats. The impact of this should not be ignored. Float species have no direct commercial value, so the few studies done on utilization of timber resources from the floodplains have not taken them into consideration (Higuchi et al., 1994). However, some of these species are important for the fauna, such as seringueira-barriguda (*Hevea spruceana*), envira-vassourinha (*Xylopia* cf. *calophyllum*), bolacheira (*Apeiba* sp.), and assacú—all of which furnish important food resources for fish and tree-dwelling fruit eaters of the region (Ayres, 1994). For the forest as a whole, extracting float species represents a loss that is equal to or even greater than that of hardwoods. The number of trees removed for floats is probably greater than that indicated by the sample, because some trees are not utilized when the rafts are being set up and therefore are not counted. Although these species have no direct commercial value for the logger, timber transporters informed us that those sawmills that process wood sell the floats for fuelwood. Management plans for floodplain forests should also consider float species as a resource.

From an economic viewpoint, timber exploitation is perhaps the most important activity on the floodplains of the middle Solimões. Pressure from outside interests is therefore intense, making it difficult to put into practice measures aimed at sustainable development. Furthermore, the inefficiency of environmental protection agencies contributes to the deterioration of this important floodplain-forest resource. The Mamirauá Ecological Station is the only conservation unit in Brazil that lies entirely within the várzea floodplain, and it plays a crucial role in establishing new models for sustained use of timber resources. This area also serves as a seed source for floodplain areas where forests have deteriorated because of overexploitation during the past few decades. This role of Mamirauá is of utmost importance for the entire Amazon region.

Acknowledgments

Arnaldo D. Carvalho helped with the fieldwork. This study was financed by the Conselho Nacional de Desenvolvimento Científico e Tecnológico–MCT, Overseas Development Administration–UK, World Wide Fund for Nature–UK, Wildlife Conservation Society–NY, Fundação Botânica Margaret Mee, I.E. Aqualung, and Fundo Nacional do Meio Ambiente, as part of the Projeto Mamirauá.

Literature Cited

Albernaz, A. L. K. M. 1994. Relatório semestral de atividades, programa de Sistemas Terrestres. *In* Relatório Semestral #4, Projeto Mamirauá, Outubro 1993– Março 1994. Manuscript.

Ayres, J. M. 1993. As matas de várzea do Mamirauá. CNPq/SCM, Brasília, DF.

———. 1994. The conservation status of the white-uakari. Primate Conservation 7: 22–26.

———, **E. A. F. Moura & D. Lima-Ayres.** 1994. Estação Ecológica Mamirauá: O desafio de preservar a várzea na Amazônia. *In* Trópico em Movimento—Alternativas contra a pobreza e a destruição ambiental no Trópico Úmido (H. B. Franco, org.), UFPa, POEMA, Belém, PA.

Projeto Radam Brasil. 1982. Levantamento de recursos naturais. Vol. 26. Ministério do Interior, Rio de Janeiro.

Browder, J. 1988. Public policy and deforestation in the Brazilian Amazon. *In* R. Repetto & M. Gillis, organizers, Public policies and the misuse of forest resources. Cambridge Univ. Press, New York; World Resources Institute, Washington, DC.

Higuchi, N., A. C. Hummel, J. V. de Freitas, J. R. Malinovski & B. J. Stokes. 1994. Exploração florestal nas várzeas do estado do Amazonas: Seleção de árvores, derrubada e transporte. VIII Seminário de Atualização sobre Sistemas de Colheita de Madeira e Transporte Florestal, Curitiba, 8–13 de Maio. Program abstract.

Jansen, M. R. A. & J. da C. Alencar. 1991. Contribuição à reposição florestal no estado do Amazonas. *In* A. L. Val, R. Figliuolo & E. Feldberg, eds., Bases científicas para estratégias de preservação e desenvolvimento da Amazônia: Fatos e perspectivas. Vol. 1. INPA, Manaus.

Lima-Ayres, D. 1994a. Relatório semestral de atividades, programa de sócio-economia e participação comunitária. *In* Relatório Semestral #5, Projeto Mamirauá, Abril–Outubro 1994. Manuscrito.

———. 1994b. A implantação de uma Unidade de Conservação em área de várzea: A experiência de Mamirauá. *In* M. A. D'Incao & I. M. da Silveira, organizers, Amazônia e a Crise da Modernização. Museu Paraense Emilio Goeldi, Belém.

Reis, M. B. 1993. Relatório de atividades semestral, sub-programa de Participação Comunitária. Pages 319–333 *in* Relatório Semestral no. 3. Projeto Mamirauá, Outubro 1992 a Março 1993. Manuscript.

The Economic and Social Significance of Logging Operations on the Floodplains of the Amazon Estuary and Prospects for Ecological Sustainability

Ana Cristina Barros and Christopher Uhl

Introduction

Timber has been exploited in Amazonia for over 300 years. From the sixteenth to the nineteenth centuries, wood was harvested from stands near the rivers and exported as logs to large European cities. Wood was of secondary importance in those days and was near the bottom of the list of exports from the Amazon region (Santos, 1980; Silva, 1987; Gentil, 1988).

In the latter part of the twentieth century, logging intensified in Amazonia. In 1950, there were only 25 sawmills in operation; 20 years later the number had risen to 287, plus 4 plywood and laminated-wood factories. Only a few species were utilized, and these were sold preferentially to foreign markets; virola (*Virola surinamensis*), a species from the floodplain forests, was the main timber tree all over the Amazon region. In 1972, this species alone represented almost one-half of the total log volume consumed by the wood industry (Bruce, 1976).

As the first roads were opened up and paved in the 1970s, the timber industry increased its activities in the upland forests of eastern Amazonia. Large laminated-wood factories and sawmills also increased in number in the Amazon estuary. In the 1980s there was a boom of small, family-run sawmills, based on estuarine floodplain-forest exploitation. Here, logging is not mechanized, the rivers are used to transport logs, and sawmills vary in size from small family businesses to those with more than 500 employees run by international groups.

The aim of this paper is to assess timber exploitation in floodplain forests of the Amazon estuary, including tree harvesting, log hauling, and industrial wood processing. Every phase of the operation is described and the social, economic, and environmental impacts of these activities are examined, based on data from each enterprise (number of jobs created, income from wood sales, volume of wood harvested). Finally, on the basis of historical factors and present-day operations of the timber industry, we analyze the future of logging activities in the region and perspectives for regional development of sustainable forest utilization.

Methods

Our data were gathered through personal interviews, conducted over a period of one year (1990–1991) in the Amazon estuary. This investigation was part of a project aimed at characterizing logging activities in the river-transport region of eastern Amazonia—that is, the estuary and the lower Amazon River (Barros & Uhl, 1995). The estuary includes the Marajó archipelago, at the mouth of the Amazon River, and the lower Tocantins River. Here each of the seats of 19 municipalities was visited for three or four days (Fig. 1). Much longer periods were spent in the more important cities such as Breves (Marajó) and Cametá (lower Tocantins) due to the abundance of data available and the number of sawmills in the area. In Breves, because of the many sawmills in the area and the dense network of rivers and *furos* (natural channels that link two rivers, or a river and a lake), we also spent 15 days traveling by boat to the more remote corners of the municipality.

Approximately 400 interviews were conducted during the trips. In the beginning

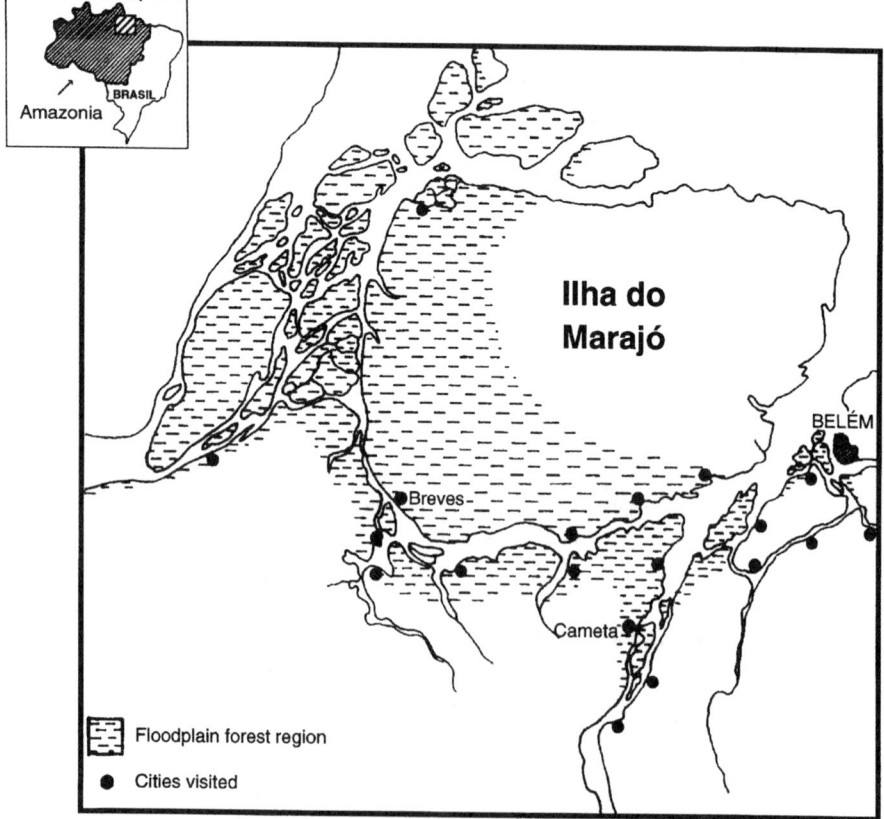

Figure 1. Location of the study area in Pará and of Amazonia within Brazil.

we interviewed people who have a comprehensive view of the region, including political leaders, rural extension agents, and experienced businessmen. They were questioned as to the location of sawmills, how long they had been in the area, and what
type of saw they used. These interviewees were also asked to give their views on regional
prospects for natural resource utilization. Data on sawmill site and history were checked
with at least one other source. Sawmill sites were plotted on municipal maps (created
by IBGE and updated in 1988) and these enterprises were classified according to type
of equipment, as follows: (1) small mills using circular saws, (2) medium-sized mills
using band saws, and (3) large plywood and laminated-wood factories.

A second group of interviews using a formal questionnaire targeted people who
had timber-exploitation experience in one of the following categories: harvesting, transportation, industrial processing, and trade. The questionnaire for loggers included items
related to the number of people involved in tree felling, equipment used, output, and
harvesting costs. The river boatmen furnished information on operational costs such as
fuel, labor, and upkeep on the tugboats that tow the rafts. Finally, the questionnaires
given to the sawmills dealt with mill installation (year, source of capital, owner's previous and current line of work), number of employees, period of operation, sawnwood
production, raw materials used (number of logs, logging agent, harvest site, type of log
transportation, species used, prices), and commercialization (trade route, selling price).

Results and Discussion

Timber Harvesting in the Floodplain Forests of the Estuary

Since colonial times, timber exploitation of floodplain forests has been mainly a hand-
powered operation. In the 1990s, axes were still used to fell trees in most cases (81%).
In only 19% of the sawmills visited were logs split by chain saw, and even in these cases
the trees had first been cut down manually ($n = 63$).

Power-driven machinery was not used to haul the logs from the forest. In 90% of
the 63 sawmills interviewed, lumberjacks skidded the logs by hand. During the high-
water season the logs were floated out, but as water levels fell, crude tracks had to be
built to skid the logs for distances of up to 2 km. In only 10% of the cases did we
observe the use of modern log-skidding techniques and equipment such as winches
(2%) and buffaloes (8%).

Floodplain logging teams usually included three men (3.3 men; $n = 19$) whose
daily output was 4.85 m³ of log wood (Table I). Harvesting was done by teams from
the sawmills in only 7 of the 81 mills interviewed (6 small mills and 1 medium-sized
mill). In the remainder of the sample, self-employed loggers handled tree removal (37%)
or they worked for the landowner or for whomever had financed the operation and
handled log-wood sales.

The sawmills that purchased the logs also financed harvesting by paying the middlemen in advance; these, in turn, gave credit to the loggers in the form of supplies or
cash (83% of the mills interviewed financed harvesting, $n = 53$).

The main logging expenses were workmen's wages and provisions, which ac-

Table I. Annual production and costs of a logging
team in the floodplain forests of the Amazon estuary

Annual production	
Log-wood volume removed (m³)	873
Production per person (m³)	265
Annual cost (US$)[a]	
Labor	$3338
Purchase of standing trees	$2532
Equipment[b]	$1.90
Total cost	$5872
Cost/m³	$6.73

[a] From Barros & Uhl, 1995.
[b] Trees were felled by ax; logs were skidded out of the
forest by hand.

counted for 57% of the total cost. Each team member received US$3.93/day ($n = 28$) plus $1.69 for food, 180 work days per year (dollar figures throughout this paper are in U.S. dollars). The remaining expenses (43%) went for the purchase of preharvest trees costing 30–35% of the value of the logs ($9/m³ of extracted wood), or $2.90/m³ of tree bole. So each cubic meter harvested from floodplain forests cost $6.70 (Table I). This figure is about one-half that of the cost of a cubic meter harvested from upland forests of the lower Amazon and lower Tocantins, where the use of trucks for log transportation and chain saws increases logging costs to $14.50/m³ (Barros & Uhl, 1995).

Transporting Logs from Forest to Sawmill

The logs extracted from the floodplain forests were either towed, 5–10 at a time, by hand-powered dugout canoes to nearby sawmills, or they were gathered together into large rafts that were towed by motorboats. The rafts destined for small sawmills were composed of up to 50 m³ of wood, whereas those that supplied medium-sized to large sawmills were much larger.

The larger rafts contained an average wood volume of 960 m³ ($n = 4$), or the equivalent of 1000 logs, towed by a 10–20-ton boat. These small craft managed to tow large rafts because, when loaded, they traveled only with the tide. In the Amazon estuary, the tide changes direction every six hours, so the tugboats stop and wait for the next tidal change before continuing on their way.

Due to the huge volume of wood transported and low navigation costs, towed rafts are the cheapest means of transportation in Amazonia. Although the rafts run the risk of breaking up during heavy tides or sudden storms, we estimated the cost of transporting 1 m³ of logs by raft over a 100 km stretch from the logging site to the sawmill to be approximately $0.90 (Table II).

Table II. Annual costs of using rafts (load capacity 960 m³) to transport 12,600 m³ of logs a distance of 100 km (logging site to sawmills) in the Amazon estuary

Item	Cost (US$)
Depreciation and upkeep[a]	5752
Labor[b]	3035
Cost of capital[c]	890
Fuel[d]	1676
Total cost for 12,600 m³	11,353
Total cost/m³	0.90

Source: Adapted from Barros & Uhl, 1995.

[a] New raft tugboats cost $29,668 (n = 6 interviews at boatyards) and annual depreciation based on a 20-year period with 10% residual value was $1335. Annual cost of maintaining these boats was estimated at $670; the material used to tie the logs into rafts was renewed yearly at a cost of $3747.

[b] Tugboat crews were made up of three men at a daily cost per man of $3.93 for wages and $1.69 for food (n = 5 interviews), for 180 work days.

[c] Cost of capital was calculated at a rate of 6% per year on initial investments described above in footnote a.

[d] Raft speed was 4 km/hr (est. tidal velocity according to the Pará State Port Authority), at a mean diesel oil consumption rate of 10.7 l/hr (n = 5) and motor oil consumption rate of 0.05 l/hr (n = 3), at a cost of $0.23/l and $1.87/l, respectively.

Logs are also transported in the eastern Amazon by truck (most common) and by large barges, or balsas, used for wood that is not buoyant. In the lower Amazon and lower Tocantins region, yearly log transportation costs for a distance of 100 km between a medium-sized sawmill and the logging site were around $30/m³ by truck, and $8/m³ by barge (Barros & Uhl, 1995). In other words, a mill that used river transportation could get its wood supply at the same cost from more-distant sites than one that trucked in logs; those that used wood brought in by raft from the floodplains could go even farther afield for their wood supply. These data illustrate the fact that for each additional dollar of revenue, a mill could traverse an extra 516 km by raft, 98 km by balsa, but only 7 km by truck.

Wood Processing in the Estuary

In 1991, there were 1026 sawmills in the Amazon estuary that used timber from the floodplain forests. These mills were classified as 972 small mills using circular saws, 48 mills using band saws (usually medium-sized), and 6 large plywood and laminated-wood factories (Fig. 2).

WOOD PROCESSING BY SMALL SAWMILLS   Most sawmills in the study area were small (95%) (Fig. 2) and were owned locally (100%, $n = 66$). These mills used logs mainly from floodplain forests, used circular saws to split the logs, and depended on family labor (53%, $n = 60$ interviews). An average of 5 people worked in each mill ($n = 60$).

The installation of these small mills in the Amazon estuary was a phenomenon of the 1980s which intensified after 1985 (Fig. 3) due to the existence of both domestic and foreign markets for low-cost sawnwood. Lumber was in great demand for building low-income housing projects and for public works in Belém (capital of the state of Pará) and in the Brazilian northeast. Export companies also bought virola wood from the small mills in order to increase the volume of sales, making this species the second highest in wood volume of timber exports from Pará. Market demand for wood plus low production costs led to the installation of many small mills in the region.

The reduced sawnwood prices of these small mills was due to the simplicity of installation plus the low cost of floodplain-forest logging and transportation. A boat motor ran the saw that was housed in a shed made of wood taken from the forest. The entire installation cost around $3000 (Fig. 3). Wood buyers often financed the installation of small sawmills by advancing capital and lending equipment or goods. This financial assistance gave the local people the opportunity to get involved in the timber industry even if they had no capital of their own (33% of the mills interviewed had received financing for installation, $n = 42$).

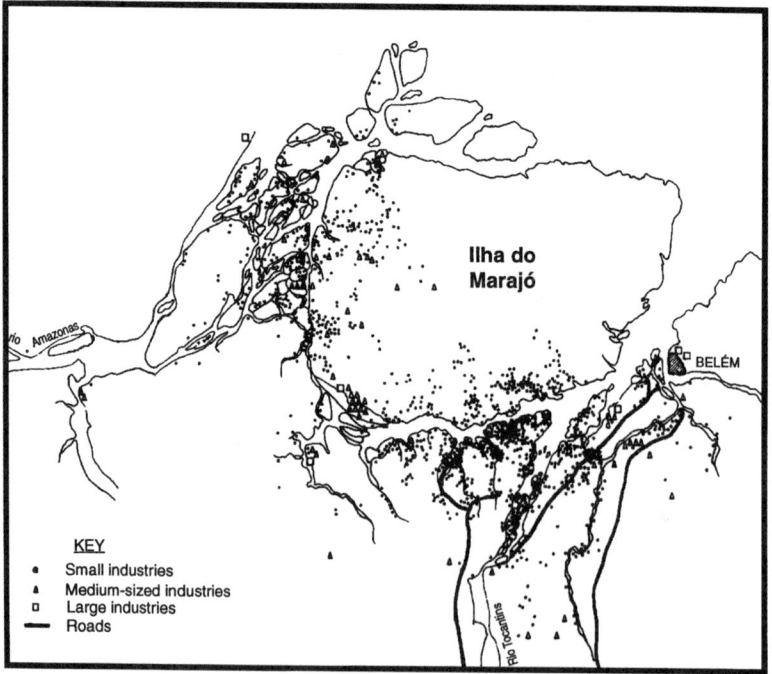

Figure 2. Distribution of sawmills and laminated-wood factories in the Amazon estuary.

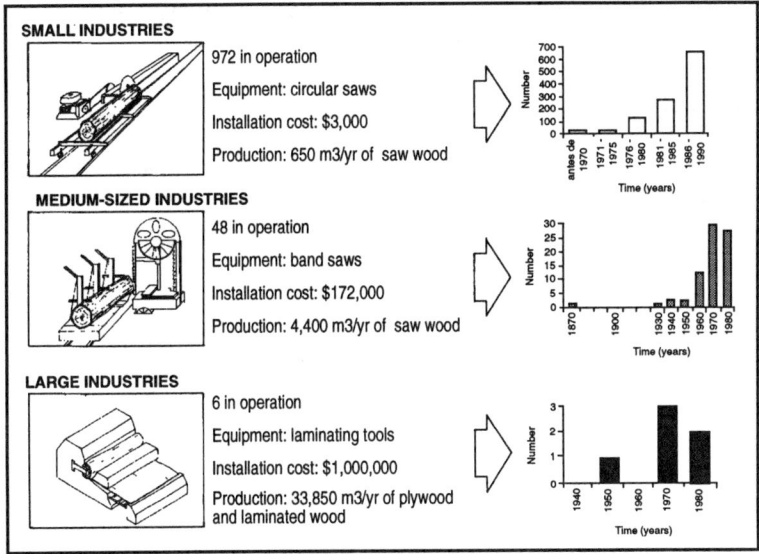

Figure 3. Characterization and installation period of small, medium-sized, and large timber industries that operate in the Amazon estuary.

A small sawmill produced an average of 650 m³/yr of sawnwood ($n = 61$) using 1850 m³ of logs. The price paid for this raw material varied from $1.95 to $4.40/m³ at the forest site, depending on wood type (soft or hard).

The purchase of the annual log supply represented 40% of total operational cost (Table III). Other expenses included transportation, fuel for the sawmill motor, maintenance and depreciation of the shed and the machinery, and, primarily, the payroll, when labor was hired from outside the family. The payroll for five employees based on wages of $3.93/man/day ($n = 28$) plus $1.69/man/day for food, totaled $5058 for 180 work days. Total production costs were around $22.80/m³, or $14,800/yr.

Wood was sold at an average price of $27/m³, with softwood bringing in $21/m³ ($n = 5$ interviews with middlemen for local businesses) and hardwood, $33/m³ ($n = 12$). In the early 1990s, wood buyers ($n = 83$) consisted of local customers (37%), middlemen who bought from the sawmills and resold to dealers in big cities (33%), big-city buyers who dealt directly with the mills (17%), and mills that exported softwood from the floodplains (11%). Wood sales at the mill (to local buyers, middlemen, and exporters) produced an annual net income of $2755. This represents a capital investment with an internal rate of return of 124% and net present value of $20,754 at an annual rate of 12% (Table III).

Annual income could go as high as $8500 in family-run businesses (Table III, excluding wages and food for employees, and log transportation). For a family with five workers, then, the sawmill would provide an income of about $1700 per worker, and was one of the best employment options around. Wages paid in other small sawmills

Table III. Production costs and annual income (US$) of small, medium-sized, and large timber industries in the Amazon estuary

	Small	Medium-sized	Large
Annual production (m³)			
Sawnwood	650	4400	0
Laminated wood	0	0	13,305
Plywood	0	0	10,450
Production costs[a]			
Maintenance and depreciation	$905	—	—
Fuel	$1139	—	—
Labor	$5058	—	—
Cost of capital	$89	—	—
Total	$7191	$260,360	—
Costs of raw material			
Purchase of logs	$5883	$113,400	—
Log transportation	$1721	$11,357	—
Total	$7604	$124,757	—
Total cost	$14,795	$385,117	$5,341,393
Production value	$17,550	$466,400	$5,932,356
Net income	$2755	$81,283	$590,963
Profit margin	17%	17%	10%
Net present value			
6% rate	$34,044	$777,168	—
12% rate	$20,754	$437,022	—
Internal return rate	124%	50%	—

Source: Adapted from Barros & Uhl, 1995.

[a] Economic analyses of small and medium-sized industries were based on interviews (see Barros & Uhl, 1995; Verissimo et al., 1993). For the large industries, we used the 1992 accounting records of a company from the estuary. Costs on capital invested in the industry were calculated at a rate of 6% per year on the value of the initial investment. Net present values and the internal return rate on the invested capital for small industries were based on a period of 20 years. Data from different sources explains information gaps in the table.

were around $707/yr, whereas other jobs in the region that paid minimum salary yielded $1176/yr.

In spite of the low operating costs and revenues generated, in the early 1990s, the working capital of small sawmills was provided by wood buyers in 42% of the cases ($n = 45$). This means that if the price of sawnwood were to go up, or if renewed demand for wood were to occur, as happened in the 1980s, the potential exists for installation of new small sawmills and increased output of those that are already in operation, just as happened in the past.

WOOD PROCESSING BY MEDIUM-SIZED SAWMILLS   Medium-sized sawmills in the Amazon estuary used band saws, produced an average wood volume of 4400 m^3/yr ($n = 17$), and consumed an average of 12,600 m^3 of logs, or a yield of around 35% (Fig. 3). These mills employed an average of 36 workers ($n = 14$) and 82% had only one set of band saws.

The owners of these businesses either were from the state of Pará (56%), were migrants from other states (24%), or were foreigners (20%). Initially, an investment of $170,000 was needed to install sawmills with band saws (Veríssimo et al., 1992), and in 32% of the cases surveyed some sort of financing was obtained from banks or sawn-wood buyers.

The 48 sawmills that used wood from the floodplains dealt with more than 50 tree species, mainly virola (*Virola surinamensis*) and andiroba (*Carapa guianensis*). This raw material was purchased from loggers at $9/$m^3$ ($n = 4$ suppliers) and was shipped to the mill in rafts, at a cost of $0.9/$m^3$ (Table II). Over a period of one year, then, a mill that used 12,600 m^3 of logs spent 29% of total expenses in the purchase of logs and 3% for transportation. Most of the production costs came from processing the logs at the saw-mill (66%).

The final sawmill product was lumber for domestic and international markets: one-third of the sawnwood was sold to foreign markets at $185/$m^3$ ($n = 5$ species); one-third (second-grade wood) was also exported, at $103/$m^3$ ($n = 7$); and one-third (third-grade wood) was sold to local and domestic markets at $30/$m^3$. Therefore, the average price of wood from the floodplain was $106/$m^3$, and the mill made an annual net income of around $80,000, or $18/$m^3$ of sawnwood. The internal rate of return on the invested capital was 50%, with a net present value of $437,022 at a discount rate of 12%/yr (Table III).

In the early 1990s, eight medium-sized sawmills that used floodplain wood in the Amazon estuary began to diversify their operations by exploiting upland timber. Four of these worked the upland forests only during summer and continued to cut trees in the floodplain forests during winter when water levels were high and very large logs could be floated out. The other four mills switched over completely to upland timber. This changeover required a $220,000 investment in a barge (including tugboat) to transport the logs, but that was gained back in less than two years of exploiting upland timber. So the sawmill that ceases to exploit the floodplain forests will have a greater cash flow with operational costs going up to $100,000, but will also see an increase in net annual income of $120,000, thus making three times the profit they earned with floodplain timber (Barros & Uhl, 1995).

WOOD PROCESSING BY PLYWOOD AND LAMINATED-WOOD FACTORIES   In 1991, there were six plywood and laminated-wood factories in the Amazon estuary, three belonging to foreigners and three to owners from other parts of Brazil. In these factories, about 15 tree species were used to make the plywood core, which would be covered by sheets of locally preferred species in the buyer countries, or to make sheets from Amazonian species for veneering.

Average output of these factories was 33,850 m³ of processed wood per year (*n* = 5), or almost 10 times the production of medium-sized sawmills. An average of 656 employees worked in these factories (*n* = 5) and annual wood consumption was 91,000 m³ of logs (*n* = 5). Because the lamination process requires larger logs, these businesses had to fetch raw material from much farther away and received shipments from the floodplains of the upper Solimões River, a distance greater than 2000 km.

We estimated the initial investment in a laminated-wood factory to be around $1 million, or six times the cost of a medium-sized sawmill. Production costs were assessed using data from one of these companies that produced 23,755 m³ of processed wood; costs were around $5.3 million, 92% of which went for production costs and 8% for commercialization.

The plywood and laminated board were sent south as well as to overseas markets, and, based on production costs (Table III), this company had a net annual income of around $590,000. However, the production value reported in the accounts of this firm was $15–20/m³ less than the mean price cited in the interviews for this study. If we base our calculations on a mean price of $268/m³ (*n* = 5), production values would go up to $6,366,340 (23,755 m³ @ $268) and the factory's net annual income would surpass $1 million ($1,003,548).

Like the medium-sized sawmills, large laminated-wood factories that used wood solely from the Amazon estuary also experienced supply problems because of seasonal flooding. During the summer months, low water levels made it harder to obtain trees with diameters large enough for lamination. These businesses therefore invested in transportation, bringing in the logs from farther away. They also invested in mechanized harvesting in the floodplains by financing the use of chain saws to fell trees and of winches and buffaloes to skid the logs. Mechanized removal in the floodplain forests could make the work productive on a commercial scale and allow logging to penetrate deeper into the forest, away from the watercourses.

The Importance of Timber Exploitation in the Amazon Estuary

CREATING JOBS   In 1991, timber harvesting in the Amazon estuary was directly responsible for 21,671 jobs, divided almost equally between loggers and sawmill employees (Table IV). A total of 23,939 logging jobs were created by floodplain forest exploitation, including the 2268 jobs related to log transportation (from Barros & Uhl, 1995). This represents 44% of all timber industry jobs in the state of Pará (IMAZON, unpubl. data).

Small sawmills alone were responsible for hiring 13,601 workers, or 57% of the total for all floodplain industries (Table IV). These mills created 25% of timber-industry related jobs in the state of Pará.

In our study area, the total number of people with jobs was 28.5% of the resident municipal population in 1989 (FIBGE, 1991). Using the same percentage for 1991, the timber industry accounted for 12% of the jobs in the Amazon estuary. Based on the same line of reasoning, and taking as an example the Breves municipality, where there

Table IV. Number of jobs created by timber industries that used wood from the floodplains of the Amazon estuary in 1991

Type of work	Employees	Loggers	Total number	Percent
Small sawmills	4860	6797	11,657	53.8
Medium-sized sawmills	1728	2286	4014	18.5
Laminated-wood factories	3936	2063	5999	27.7
Total	10,524	11,146	21,670	100
Percent of total	49	51	100	—

Source: Adapted from Barros & Uhl, 1995.

is a greater number of industries (175 small sawmills, 30 medium-sized sawmills and 1 laminated-wood factory), almost 30% of the total number of jobs in the area are directly related to the timber industry.

GENERATING INCOME   Total wood production of Amazon estuary industries that used floodplain timber was 1,050,000 m³ in 1991, of which sawnwood accounted for 85,000 m³ and laminated wood, 200,000 m³. Most of this output was from small sawmills (60%). Medium-sized mills produced 20% of this volume, and the remainder was attributed to laminated-wood factories (19%) (Table V).

Although the 6 laminated-wood factories produced 19% of the total volume of floodplain wood, they accounted for 58% of production value because of the relatively high price of the finished product. Small sawmills, on the other hand, produced more than one-half of the floodplain wood but represented only 18% of production value, due to the lower quality of the finished product and its low market value locally.

In 1991, there were 1874 timber industries registered in the state of Pará, with an annual sawnwood output of 4,300,000 m³ (Verissimo & Souza for IMAZON, unpubl. data). Floodplain wood accounted for one-quarter of production volume (24%) and 16% of the total production value in the state. The laminated-wood factories contributed 9% of statewide production value, while small sawmills contributed only 3%.

Sustainability

Forestry management data is lacking for floodplain forests and current research is focused primarily on virola. However, after more than three centuries of timber exploitation, and with some 1000 sawmills utilizing more than 50 tree species, the question arises of whether these industries are sustainable when using only wood from floodplain forests.

To assess the sustainability of this activity, we estimated the region's floodplain-forest area, collected data on wood consumption by industries, and simulated the volume of wood harvested per hectare in order to project the area of forest needed to sustain this activity over an extended period of time. We then compared these data to cutting cycles recommended for upland forests (Barreto et al., 1993).

Table V. Number of timber industries using wood from the floodplains, annual wood volume production and production value for each type of industry in the Amazon floodplain compared to total numbers for Pará state

Timber industries[a]	Number	%[c]	Annual production (m³)	%[c]	Value (US$)	%[c]
Small sawmills	972	52	631,800	18	17,058,600	3
Medium-sized sawmills	48	3	211,200	8	22,387,200	4
Laminated-wood factories	6	0.3	203,100	5	54,430,800	9
Total	1026	55	1,046,100	24	493,876,600	16
Rest of Pará[b]	848	45	3,253,900	76	494,626,854	79
All of Pará	1874	100	4,300,000	100	588,503,454	100

Source: Adapted from Barros & Uhl, 1995.
[a] See text for number of industries, production, and wood prices.
[b] To estimate sawnwood production for the rest of Pará (PA-150, Belém–Brasília, southern Pará, lower Amazon and the Transamazon Highway), we considered Pará state to have 1874 timber industries that produced 4,300,000 m³ of sawnwood in 1991. The sawmills in the estuary produced 1,046,100 m³, so industry production from other areas of Pará was assumed to be 3,253,900 m³. Value estimates for this volume of wood were based on the following: 3,013,017 m³ sold in the domestic market; 94,674 m³ of mahogany exports; and 146,209 m³ of sawn and laminated wood, also for the foreign market (export data furnished by the Association of Wood Export Industries in Pará State, or AIMEX). The prices used were $136/m³ for the domestic market; $433/m³ for mahogany, according to AIMEX; and $300/m³ for the remaining portion of sawnwood and laminated wood for export (Barros & Uhl, 1995).
[c] Percent of total in Pará.

To this day, we do not know what the total floodplain-forest area is in the Amazon estuary. Our estimates were based on data for forested areas in all estuarine municipalities (SUDAM/IBDF, 1988) plus the probable limits of floodplain forests traced on a map of sawmill sites. We arrived at an estimate of 3,700,000 ha from which we subtracted 500,000 ha of nonexploitable forest reserves. There are 972 small sawmills, 48 medium-sized sawmills, and 6 laminated-wood factories in this area, and these consume 1850 m³, 12,600 m³, and 91,000 m³ of logs per year, respectively. Together these industries use 2,949,000 m³ of log wood per year.

The sustainability of this activity was assessed by first assuming that the industries would harvest the maximum usable-wood volume: 56 m³/ha according to a survey of the commercially valuable timber trees in a native forest in the southern part of Marajó Island (Arima et al., 1997). This wood volume represents the most intensive model of exploitation possible and corresponds to the smallest area that the industries would require annually. According to this model, all estuarine businesses would have to log 52,661 ha to obtain their annual supply. This means that within the total estimated area of floodplain forest (3,700,000 ha minus 500,000 ha of forest reserve, or 3,200,000 ha), harvesting would be based on a cutting cycle of 60 years, the time allowed for forest recovery and stock replacement, after which the area would once more be ready for exploitation.

In the event of more-selective cutting (28 m³/ha, or one-half the maximum volume), the industries would require a larger area in order to obtain their annual supply. In this case, 30-year cutting cycles should be applied to the entire forested area of the estuary in order to maintain sustainable exploitation. A third case of even more-selective harvesting (14 m³/ha, or one-fourth the usable volume per hectare) would demand a 15-year minimum cutting cycle in order to sustain the industry in the estuary.

In upland forests, where damages are greater than on the floodplain due to heavy logging machinery, a minimum cycle of 30–35 years is recommended for intensive extraction (30–40 m³/ha), 20–30 years for the removal of 20–30 m³/ha, and 15–20 years for 15–20 m³/ha (Barreto et al., 1993). These recommended time intervals for timber extraction in upland forests are similar to those found in our simulations for the floodplain, and we therefore conclude that the estuarine region has enough timber to adequately supply the wood industry on a long-term basis.

In areas where there were many sawmills, however, logging practices did not appear to be sustainable. Breves municipality, for example, had 271,503 ha of forest (SUDAM/IBDF, 1988) and 175 small sawmills, 30 medium-sized mills, and 1 laminated-wood factory. If these businesses harvested a maximum volume of 56 m³/ha, sustainable exploitation within the municipality would require forest recovery in 19 years. Furthermore, if the Breves sawmills based their operations on species that furnish only 14 m³/ha, the entire forest would be exploited in 5 years. The timber industry must increase the number of species used if exploitation is to continue at the same site. Assuming that, under heavy exploitation, it would be impossible for native forests to recover their initial volume in 19 years (or in 5 years of more-selective logging), the pressure brought to bear on floodplain timber resources in areas such as Breves municipality is predictable.

These simulations are, of course, grossly oversimplified. Our estimates of forested area and wood-volume removal per hectare were based on assumptions, not actual measurements. We also did not take into consideration the accessibility of the forest, its diversity in terms of stock, growth and mortality rates, or the current state of the forest after decades of logging. However, the data give a good picture of timber abundance in the study area as related to current industry demand, in a situation where forestry management is lacking.

Conclusions: The Present State of the Logging Operations in the Amazon Estuary and Prospects for Change

Based on the current model, timber exploitation in the floodplain forests of the Amazon estuary appears to be a good investment with high return rates on low capital investment. The local population is involved in both harvesting and processing, in contrast to what usually happens in Amazonia where economic activities are concentrated in the hands of a few timber barons or large landholders. An increase in wood prices or in the demand for wood, such as took place in the 1980s, could bring about renewed construction of small sawmills, resulting in increased participation of the local population in the floodplain timber industry. There is no guarantee, however, that the forest can sustain this growth.

Our simulations showed that in areas with a higher concentration of sawmills, the entire forest or selected species were already feeling the impact of logging activities. Consequently, industries with more capital had begun to exploit upland and floodplain forests of the upper Solimões River, diverting their investments from the estuary. The lack of raw material forced these industries to move on to new logging sites, lured by the low cost of river transportation. The timber industry was shifting away from the estuary instead of becoming a development model for the region.

In order to reverse this trend and convince the floodplain industries that they will be able to operate in the estuary for a long time, forestry management must be brought to bear on the problem. In upland forests, management should be based primarily on careful harvesting methods that reduce environmental impact and on map-oriented construction of highways, railroads, and timber yards. The IMAZON pilot project in the municipality of Paragominas, Pará, is a good example of the advantages this form of forestry management brings to upland forests (Barreto et al., 1998), and also demonstrates its viability.

Although practical experience is lacking, there is greater potential for floodplain-forest management than for upland-forest management. Logging on the floodplains causes less damage because no heavy machinery is used and there are fewer vines entangled in the tree crowns. Floristic composition of the floodplain forest is simpler and trees grow more quickly. For these reasons, management techniques in the Amazon estuary should place special emphasis on forest enrichment and treatments that accelerate the growth of suitable trees. These kinds of initiatives also have at their disposal the local people's knowledge of plants including seed germination, growing conditions, and location of species.

Management techniques, however, are only part of sustainable utilization of the forest. In order for these methods to be accepted and put into practice by the local population, it is also essential to make changes in existing legislation, establish a monitoring system, and set up guidelines for the logging sector that include forestry extension and law enforcement; moreover, property titles should be normalized, to allow river dwellers to invest in their own lands, manage their forests, and properly observe environmental control regulations.

Forestry legislation was revised in late 1994 by IBAMA (Brazilian Institute for the Environment and Renewable Natural Resources) in an effort to overhaul national forestry policy. A new law was passed (Act 1282) and its regulatory counterpart came under discussion in early 1995. The most important change at the moment has been the elimination of the requirement to present a management plan in order to exploit forests on land holdings of less than 500 ha.

In the Amazon estuary, where forests are exploited by the local people, the exploitation-without-management clause for "small properties" is simply a way of bringing reality into the law. However, minimum felling cycles have not been established, which means that successive cuttings are allowed in the same area until there is nothing left. These new measures have legalized nonsustainable harvesting in the estuary, an area which supplies up to one-quarter of Pará's wood production. Anderson et al. (1994) presented several concrete proposals for a forestry policy in the Amazon estuary.

If the new rules laid down by IBAMA do not produce the desired effect, the state government may take matters into its own hands through the state legislature and the state environmental secretariat. The development of logging operations in the Amazon region must be based on research and extension work, but it must also include the participation of local politicians who are interested in, and well-informed on, the issues involved in land use and its environmental, social, and economic consequences.

Another important issue to be dealt with by the government in the Amazon estuary is the normalization of land ownership. The agency responsible for property titles in this region (ITERPA, or Pará State Land Institute) should reexamine all deeds and land titles and update legal papers that, even today, often date from the land grant period of colonial Brazil. An alternative would be the concession of extensive forested areas by the government to organized community groups that would exploit timber using sound forestry management practices. INCRA (National Institute for Colonization and Land Reform) has already gone through an experience of this type; 260,000 ha of forest were donated to settle communities near the Tucuruí dam in southern Pará. Social organization is an essential component of this process, to enable local inhabitants to work together with the government and control the marketplace. The logging issue in Amazonia is still an area where much work remains for political and social scientists.

Logging activities can create jobs and income for the estuarine population, thus guaranteeing development and well-being. But in order for this to take place, research institutions must play an important role through studies of the forest, the instruments and methods of government action, and the successes and failures of community organizations. Nongovernmental organizations that have access to the community are also important, to pass along the knowledge acquired through research, in a system of forestry extension that deals with management techniques, land-title normalization, and promoting social organization. And finally, government leaders should plan their actions to support the efforts of extension agents and researchers by passing laws to encourage forestry management for sustainable and continuous use of the forest.

Literature Cited

Anderson, A., I. Mousasticoshvily Jr. & D. Macedo. 1994. Impactos ecológicos e sócio-econômicos da exploração de virola no estuário amazônico. Boletim da WWF. World Wildlife Fund, Washington, DC.

Arima, E., N. Maciel, A. C. Barros, H. Pollak, M. Piedade & C. Uhl. 1997. Oportunidades para o desenvolvimento sustentado do estuário amazônico a partir de reservas extrativistas. Series Amazonia. IMAZON, Belém.

Barreto, P., C. Uhl & J. Yared. 1993. O potencial de produção sustentável de madeira em Paragominas–PA; na Amazônia oriental: Considerações ecológicas e econômicas. Pages 387–392 in Congresso Florestal Brasileiro, 7, 1993. Vol. 1. Sociedade Brasileira de Silvicultura and Sociedade Brasileira de Engenheiros Florestais, São Paulo.

————, E. Vidal, P. Amaral & C. Uhl. 1998. Costs and benefits of forest management in eastern Amazonia. Forest Ecology & Management (in press).

Barros, A. C. & C. Uhl. 1995. Logging along the Amazon River and estuary: Patterns, problems, and potential. Forest Ecology and Management 77: 87–105.

Bruce, R. W. 1976. Produção e distribuição da madeira amazônica. PRODEPEF Série Estudos, no. 4. Projeto de Desenvolvimento e Pesquisa Florestal (IBDF/FAO), Rio de Janeiro.

Fundação Instituto Brasileiro de Geografia e Estatística (FIBGE). 1991. Sinópse preliminar do censo demográfico do Pará. Vol. 5. FIBGE, Rio de Janeiro.

Gentil, J. 1988. A juta na agricultura de várzea na área de Santarém – Médio Amazonas. Boletim do Museu Paraense Emílio Goeldi, Antropologia 4(2): 118–199.

Santos, R. 1980. História econômica da Amazônia: 1800–1920. Editorial T. A. Queiróz, São Paulo.

Silva, M. 1987. Os trabalhadores da várzea no serviço da madeira. Tese de Mestrado, Núcleo de Altos Estudos Amazonicos (NAEA), Universidade Federal do Pará, Belém.

SUDAM/IBDF. 1988. Alteração da cobertura vegetal primitiva do estado do Pará. Programa de monitoramento da cobertura florestal do Brasil. Relatório técnico. SUDAM/IBDF, Belém.

Veríssimo, A., P. G. Barreto, M. Mattos, R. Tarifa & C. Uhl. 1992. Logging impacts and prospects for sustainable forest management in an old Amazonian frontier: The case of Paragominas. Forest Ecology and Management 55: 169–199.

Miriti (*Mauritia flexuosa*) Palms and Their Uses and Management among the Ribeirinhos of the Amazon Estuary

Mario Hiraoka

Introduction

An increasing number of studies has focused attention on traditional resource-management practices as an alternative to the continued deforestation in Amazonia. Indigenous systems, adjusted to the diverse environments and the cultural needs of the inhabitants, are complex but are believed to be ecologically sound and economically viable. Although Amerindian solutions to forest resource management are proven to be conserving of forest resources, they are often found to be incompatible with contemporary market-oriented economies. Indigenous groups are marginally integrated with the market economy, and the majority of their diverse forest products are place and culture specific, so that extralocal demands are generally limited (Anderson, 1990; Hladik et al., 1993; Redford & Padoch, 1992; Posey & Balée, 1989).

Research on the forest management practices among the floodplain inhabitants of Amazonia, collectively known as *ribeirinhos* in Brazil and *ribereños* in Peru, offers encouraging results. Consisting of a melange of ethnic groups—detribalized Amerindians, offspring of European-African-Amerindian unions, and others—the floodplain inhabitants practice a livelihood based largely on that of their aboriginal forebears (Chibnik, 1994; Hiraoka, 1985; McGrath, 1989; Padoch, 1988). Their familiarity with várzea or floodplain and adjoining terra firme or interfluvial highland resources, and their settling on the waterways led to their early integration with extralocal economy. The economic booms based on Amazonian resources—salted fish, turtle oil, animal pelts, rubber and other resins—relied largely on ribeirinho expertise and labor (Smith, 1985; Parker, 1985). External ties increased with the improvements in transportation and communication technologies and the expansion of regional, national, and international markets. The market-oriented economy practiced largely with traditional resource-management practices serves to make the ribeirinho model an alternate model of rural development and environmental management for the region.

Contemporary ribeirinho livelihood is based on small-scale farming, fishing, hunting, and extraction of forest products. Floodplain agriculture is risky in areas other than the estuary, where seasonal flood duration and level are variable. However, the diverse biotopes formed by the rise and fall of the water and the sediments deposited by the river offer possibilities for the sustained cultivation of a large number of crops. The varied aquatic environments and the rich and reliable ichthyofauna are for the riverine people a rich source of animal protein. The seasonal fluctuations in fish catch, caused by variations in hydrochemistry and water level changes, are supplemented by dwindling but still available land fauna. The harvest of a vast number of forest products, ranging from construction materials to food and from handicraft materials to medicinals, constitutes an important segment of ribeirinho subsistence and cash-earning activities. These items are obtained from diverse flooded-forest ecosystems, varying from highly anthropogenic forests to those with limited human interference.

A major feature of flooded forest is the abundance of a restricted number of products caused largely by the ecological conditions of the floodplain. The extended or frequent inundation of the low-lying terrain eliminates species that do not tolerate anaerobic conditions. The limited number of species increases labor productivity and reliability of harvests of the economically useful plants. The frequent addition of plant nutrients through floods also contributes to reduce risks in continuous output. The production reliability and species abundance of selected plants have led ribeirinhos to rely on few products as staples and cash-earning sources. Among the valuable species palms figure prominently.

In the floodplains, palms are especially prized since the economically important species occur in dense stands. For example, palms like açaí (*Euterpe oleracea*), miriti (*Mauritia flexuosa*), and yarina (*Phytelephas macrocarpa*) form dense concentrations and have played significant roles in the market and subsistence economies of floodplain inhabitants (Peters et al., 1989). Reflecting the importance of these plants, palms are increasingly managed or incorporated as part of the agroforests in the riverine communities. Investigations in the estuarine region note the elaborate care given to the expanding açaí forests. Several distinctive attributes make palms a major adjunct to ribeirinho economy. As palm seeds and trunks are able to withstand fire relatively well, they survive the swidden land-clearing process in larger numbers than non-palms. The crowns of tall species like pataua (*Oenocarpus bataua*) and miriti, when widely spaced in the fields, do not interfere with the growth of crops or agroforest species (Anderson, 1988; Johnson, 1983). Further, many valuable floodplain palms, including açaí, miriti, pona (*Socratea exorrhiza*), muru muru (*Astrocaryum murumuru*), jupati (*Raphia taedigera*), and ubucu (*Manicaria saccifera*) are found on agriculturally marginal, periodically to permanently flooded terrain. Such characteristics, combined with proper management, offer potential for increased income from the bottomland palm forests while maintaining the vegetation cover (Kahn & Granville, 1992).

Present knowledge indicates that palms play other important roles in the lives of the inhabitants and the ecology of the floodplains, including native cosmology, folklore, plant succession, and geomorphological changes (Anderson, 1988, 1990; Heinen & Ruddle, 1974; Kahn & Granville, 1992; Ruiz Murrieta & Levistre Ruiz, 1993; Wilbert, 1976). However, our understanding concerning the management of major floodplain palms is limited. Distribution maps of most palms are often inaccurate, except for those in a recent work by Henderson (1995), and areal estimates are restricted to a few areas. Ethnobotanical studies of main floodplain palms exist, but research on their management is limited to the vicinity of main urban centers like Belém, Manaus, and Iquitos. Except for *Euterpe oleracea*, practically no detailed economic data exist for other palms (Anderson, 1988). Likewise, ecological research on densely occurring palms, particularly those subject to human intervention, are wanting.

Among the estuarine inhabitants, açaí and miriti constitute the two most important palms. The fruit of the palms is consumed as a dietary staple, and there are numerous uses for diverse parts of the plant. The rapid urban growth of the past three decades has also created a large market for açaí and miriti products. The increased demand has been met by an expansion of fruit output, especially of açaí, through improved management of natural forests and plantings. Major contributions on diverse aspects of açaí, such as, management, spatial structure of production areas, ecology, and economic significance for ribeirinho households in the estuary of the Amazon near Belém, have been made by Anderson and his colleagues (Anderson, 1988; Anderson & Ioris, 1992; Anderson & Jardim, 1989; Anderson et al., 1985).

The objective of this paper is to relate the use and management of miriti (*Mauritia flexuosa*) palms by the tidal floodplain inhabitants in the Amazon estuary. This case study illustrates the economic potentials and limitations of miriti palms that form dense populations in agriculturally marginal habitats of the floodplain. Data are based on miriti forests and their users in the Abaetetuba Islands, near the confluence of the Tocantins and Para Rivers, about 80 km southwest of Belém, Brazil (Fig. 1).

The Estuarine Floodplain

The Amazon estuary is characterized by a lowland consisting of Holocene sediments bordered by older and slightly higher Tertiary deposits of the Barreiras Formation. Two distinctive levels exist within the low-lying terrain: *ica* and *várzea*. The former appears to be the remnants of old Holocene deposits which stand 2–3 m above the várzea. This slight rise prevents the icas from flooding. The well-drained but weathered ica surfaces, consequently, support a diverse vegetation akin to the terra firme forest. Icas are used mainly for manioc swiddens. The term *várzea da mare* is applied to tidal floodplain that fringes the Tertiary uplands and icas. Landforms in the várzea are dynamic and—reflecting past and present eustatic changes, shifts in stream channels, and tidal movements—are constantly reworked. The resulting microterrain varies from poorly to well-drained lands. Floodplain instability brings corresponding changes in vegetation formations. Plant communities are particularly susceptible to the ongoing tidal activities, such as in-filling of shallow depressions, bank erosion, and slumping of margins.

Unlike the seasonal floods caused by the rainy season's swelling of river level, the estuarine floods are tide driven. Twice a day, the incoming tide raises the river level and briefly submerges the várzea depressions. At the Ilhas de Abaetetuba the average tidal range is about 2.7 m. The high tides, lasting 2–3 hours, inundate the low-lying várzeas to varying degrees. The greatest amplitude is associated with the equinoctial spring tides. Flooding is especially frequent and longer lasting during the rainy months of February–April. The erosive and depositional power of spring tides, especially those of the rainy season, serves to maintain the equilibrium of resources in the várzea. Silting of depressions and lowering of natural levees, removal of organic debris, and seed dispersal are among the outcomes ascribed to the high tides.

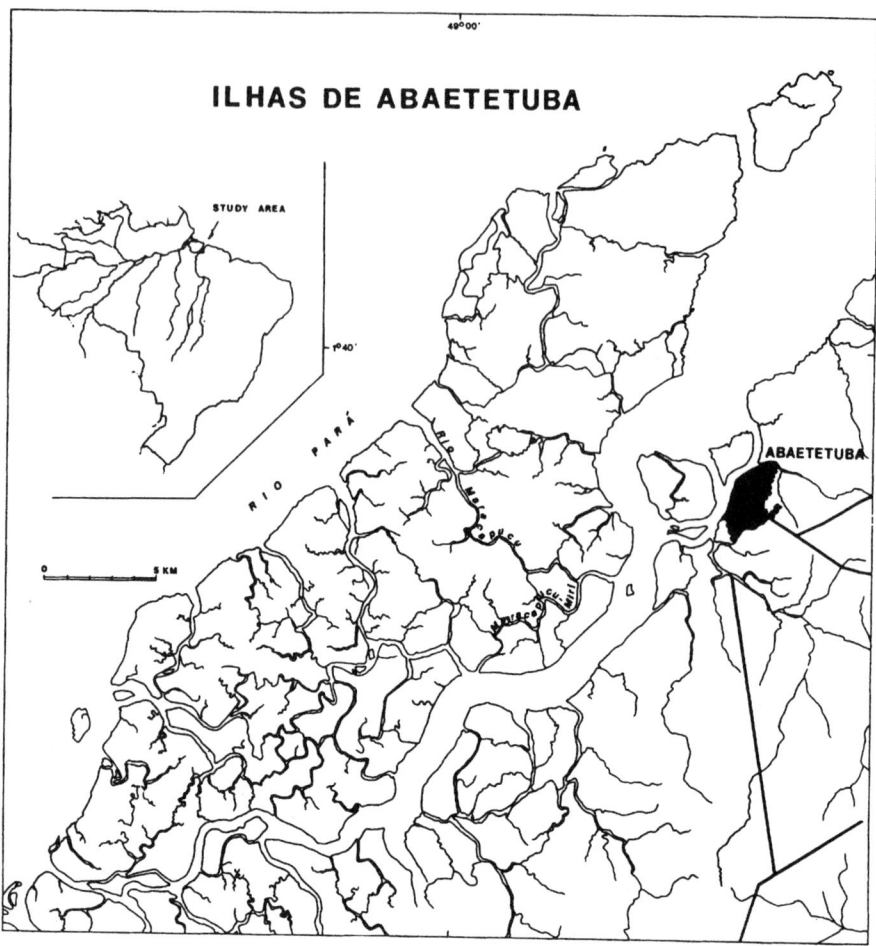

Figure 1. The Ilhas de Abaetetuba are located on the southern estuary of the Amazon, at the confluence of the Tocantins and Para Rivers. Abaetetuba, the main urban center of the region, is situated about 70 km west of Belém, the largest metropolitan area of Amazonia.

The combination of low relief and the tidal cycle in the várzea is responsible for the high water table and an abundance of poorly drained ground. Although the water table varies with tide levels, it is usually within a meter of the surface even on the better-drained portions. Aside from waterlogged conditions, heavy mottled clays followed by gley underlie a thin layer (10–20 cm) of organic surface materials. The low levels of soil oxygen, the frequent flooding, and the sterile clays a few centimeters below the surface affect plant growth. To cope with the ecological limitations of the várzea, plants developed several adaptations, including lenticels, pneumatophores, aerial roots, and other mechanical adaptations to clay soils. Since few plants survive in such a stressful environment, the várzea is noted for its low plant diversity. Examples of economically useful palms that successfully adapted to the estuarine floodplain include açaí (*Euterpe oleracea*) for food and palm hearts; jupati (*Raphia taedigera*) petioles for the manufacture of mats, fishing traps, and other traditional utensils; miriti (*Mauritia flexuosa*) for a great range of items; muru muru (*Astrocaryum murumuru*) fruit for fattening pigs; pataua (*Oenocarpus bataua*) as a source of vegetable oil; and ubucu (*Manicaria saccifera*) for thatching. The flood forests are also noted for economically important trees like the andiroba (*Carapa guianensis*) for timber and medicinal oil; cedro (*Cedrela* spp.), a valuable cabinetwood; mututi (*Pterocarpus officinalis*), used widely as firewood; the latex-yielding rubber (*Hevea brasiliensis*); and ucuuba (*Virola surinamensis*) for export-oriented timber (Anderson, 1988; Anderson et al., 1994). The concentration of valuable species and the ease of extraction from a dense network of waterways led to an early and intensive harvest of selected products for sale. Probably the best-known product extracted from the flood forest is the latex of *Hevea* trees, but the collection of oil seeds, the extraction of timber, and other products have been important for brief periods in the economic history of the estuary. The most recent of such booms is the harvest of the fruit and hearts of açaí palm. However, as is the case with timber species like andiroba and ucuuba, the heavy pressure on the exploitation of a few abundantly occurring species within a short time span has led to their virtual demise in parts of the region.

The unplanned extraction of a limited number of species, particularly trees, impoverished the floristic composition of the easily accessible flood forests, but the continued demographic growth, with consequent pressures on estuarine resources, expanding markets, and the long experience with extractivism were instrumental in the development of alternative land use– management practices (Hiraoka, in press). The recent "açaízation" of the estuary offers the best example. In response to a number of factors—e.g., demise of small-scale farming based on sugar cane and rice cultivation in the várzea, increasing demands for açaí fruit in the regional urban centers as a result of massive urban migrations and the consequent transfer of rural dietary patterns to the urban setting, supply of domestic and foreign markets with palm hearts following the exhaustion of naturally occurring *Euterpe edulis* in southeast Brazil, and the familiarity with the management of the palm—açaí stands began to expand after the mid-1970s in estuarine areas with easy access to Belém and other regional urban centers.

Although *Mauritia flexuosa* does not have the same market value as the açaí, it has traditionally been a palm as important as açaí for the ribeirinhos. Miriti and açaí are the two most commonly found palms in the tidal lowlands and share essentially the same environments. However, as miriti can withstand a terrain with poorer drainage than açaí can withstand, it is often found on the swampy portions where no competing land uses occur. As such, it is illustrative of a tree with major economic potentials that could be conserved in situ and be managed in association with other land uses.

The Miriti Palm

Miriti is one of the most abundant palms of the Amazonian floodplain. It is usually associated with seasonally perennially inundated lowlands at <800 m elevation, and an average precipitation of 2000 mm. Its distribution ranges from east of the Andes to Maranhão (Brazil), south to Mato Grosso and Goiás (Brazil) and the Bolivian lowlands, and north to the Llanos of Colombia and Venezuela. Within its range, miriti is·often found in quasi-monospecific stands, as in the floodplains of the Amazon, and the lower Huallaga and Ucayali basins of Peru (Peters et al., 1989). In the Peruvian Amazon alone, the palm covers an estimated 6–8 million ha. Dense concentrations are also found on the low islands and inner depressions of the estuarine floodplain. On the mud banks of islands, for example, solid stands of miriti grow on slightly higher ground inland from the water's edge, where the lower margins are colonized and consolidated by the aninga (*Montrichardia arborescens*) (Strudwick & Sobel, 1988). Floodplain depressions with waterlogged soils, away from tidal margins, are also sites of high miriti density.

Within areas of human-modified flora in the estuary, the *Mauritia flexuosa*, with its straight and tall trunk and the gracious crown projecting above the neighboring vegetation, is one of the most majestic plants. Miriti is a dioecious, monocaulous, and arborescent palm reaching a height of up to 35 m when mature. The 20–25 palmate leaves are supported by petioles reaching 3.5 m in length. There seem to be some conflicting views concerning the phenology of *Mauritia flexuosa*. In the Orinoco delta and elsewhere, pistillate or female tree flowering occurs annually at the end of the rainy season (Heinen & Ruddle, 1974; Gonzalez Boscan, 1987; Pedersen & Balslev, 1990; Ervik, 1993). In the estuarine floodplain, however, residents report that the miriti flowers biennially in the dry season and that the flower-to-fruit cycle requires more than 18 months. The particular phenology perhaps reflects the ecological conditions of the region. An average of 5.67 interfoliar inflorescences are produced during the flowering season. Insects of the *Derelomini* and *Alticinae* families appear to be the pollinators of *Mauritia flexuosa* (Ervik, 1993). Globose to ovoid mature fruits, up to 7 cm in length, fall between January and June. An infructescence may yield 1100 fruits, each tree producing between 1500 and 7000 units per season, or approximately 200 kg of fruit (Pedersen & Balslev, 1990; Gonzalez Boscan, 1987).

Miriti, similar to açaí, is an early successional species. Since germination appears to occur under exact light and moisture conditions, not all wetlands are colonized by the palm. It germinates on open ground such as the newly built mud bars above the daily tides, on gaps along stream banks, and on periodically drained edges of swamps. Seed dispersal is accomplished by animals, humans, and especially by the spring tides. Reproductive age is reached in 8–10 years. To thrive in the waterlogged soils of the várzea, miriti assimilates plant nutrients in an anaerobic environment through a specialized root structure. Part of the root is in the soil but the rest is aerial. Oxygen is absorbed by the pneumatophores on the aerial portions of the root (Granville, 1974; Kahn et al., 1993). Thus, the mass of underground root, located mostly within 20–30 cm of the surface, depends on the aerial roots for the absorption activities. The adult individuals, which survive up to 60 years, do not appear to have natural enemies. Parrots often make their nests in the stem when trees begin to die. In the estuary, the fruit is consumed by numerous frugivorous birds, land animals, fish, and humans, but the trunks and the coriaceous evergreen leaves are not attacked by herbivores. The oxygen-deficient substrate also slows down biomass decomposition. Consequently, most of the net primary production, estimated at 10–16 metric tons per year, accumulates in situ, forming a dense layer of organic matter. Over the years the detritus, broken up by bacteria and fungi, rapidly drains the nutrient-poor and acidic hydromorphic soils. Edaphic changes and the gradual canopy closure by *Mauritia* bring autogenic alterations to the *miritizal*, or miriti forest. As shade-tolerant species begin to emerge, *Mauritia* seedlings are unable to compete and the succession to a dicotyledonous flooded forest ensues (Gonzalez Boscan, 1987).

The Tree of Life

Mauritia flexuosa is probably the most useful palm of Amazonia. Recognizing the importance of the plant in the inhabitants' economy and society, Spruce (1908) called it the "tree of life." Reports from diverse parts of the basin indicate the great number of uses among the Amerindians and mestizos. For example, the Waraos of the Orinoco delta extracted a storable starch from the trunks of the *Mauritia* (Heinen & Ruddle, 1974; Wilbert, 1976). Among several indigenous groups, wine is made from the sap of felled trunks or stumps (Braun, 1968). Ribereños of Peru and Amerindians east of the Andes in Ecuador collect the larvae of the beetle *Rhynchophorus palmarum* from the pith of freshly felled trunks, for both consumption and sale (Hiraoka, 1985; Pedersen & Balslev, 1990). The mature trunks of the palm are often cut to make bridges over swampy terrain and streams, or walkways on the muddy floodplain. The mesocarp of ripe fruit is eaten as a snack over much of the basin by both rural and urban residents. In the Peruvian Amazon, the fruit is an important commodity and serves as a major source of seasonal cash income to the floodplain people. In Iquitos, the mesocarp is made into ice cream and icicles. A drink is made from a mixture of mashed pulp and water (Padoch, 1988). Indigenous inhabitants in Ecuador and in parts of Venezuela are reported to ferment the pulp to make a mild alcoholic beverage (Beckermann, 1979; Vickers, 1976). The strong and flexible fiber obtained from young leaf blades is widely utilized for the manufacture of ropes, hammocks, and bags (Schultes, 1977). Such objects have been observed among the Waraos of Venezuela and the contemporary inhabitants of Barreirinhas, in northeast Maranhão, Brazil (Table I). *Mauritia flexuosa* has also played important roles in the rituals and folklore of indigenous people in diverse parts of Amazonia (Heinen & Ruddle, 1974; Ruiz & Ruiz, 1993).

Miriti's Uses in the Estuary

The use of miriti among the ribeirinhos of the Ilhas de Abaetetuba illustrates the range of products the estuarine Amazonians have been able to obtain from *Mauritia flexuosa*. Concurrently, the management practices provide insights on the economic possibilities and limitations of miriti palm forests in the densely inhabited (>100 inhabitants/km^2) estuary, situated within easy access of Belém, the largest metropolitan region of Amazonia.

The miriti palm is used for a number of products and purposes. Although most uses are traditional, devised long ago to meet the subsistence needs, some have been devised to take advantage of new market opportunities and demands. In rural areas, the mesocarp serves as one of the main staples during the rainy season. Ribeirinhos depend on the mesocarp of the açaí as a basic component of the diet during most of the year, and in the rainy season, when its production declines, they switch to miriti pulp. The

Table I. Uses for different parts of miriti palm, *Mauritia flexuosa*

Portion of palm Use	Estuary	Elsewhere in Amazonia
Trunk		
Bridge, pontoon	X	X
Carbohydrate source		X
Wine		X
Insect repellant	X	
Leaves		
Thatch		X
Petiole		
Basket	X	
Fish trap	X	
Fan	X	
Sail, curtain	X	
Mat	X	X
Wall	X	
Float	X	
Toy	X	X
Cork	X	X
Toilet paper	X	
Leaf blade		
Hammock	X	X
Handicrafts	X	X
Rope	X	X
Fruit		
Food/snack	X	X
Animal feed	X	X
Oil		X
Flower		
Shrimp feed	X	
Root		
Medicine		X

pulp is mixed with water and consumed with the addition of *farinha* (manioc flour), or it is made into a mush by adding rice or farinha. Called *mingau*, the miriti mush becomes the main source of carbohydrates between January and May.

Despite its importance in the rural diet, the fruit has marginal cash value. Since Belém is supplied throughout the year with açaí, miriti has only limited market potential. In a regional city like Abaetetuba, where açaí supply is unreliable because of urban size and purchasing power, miriti finds a niche during the rainy season. The wide use of miriti pulp as discussed by Padoch (1988) for Iquitos does not occur in the estuarine region. Until about 1970, prior to the wide-scale availability of vegetable oil in the region, the oily mesocarp was a source of cooking oil. About 12% of the reddish yellow mesocarp is oil, and the oil that rose to the surface from the boiling process was collected for cooking purposes (Balick, 1979). Currently, the time-consuming procedure is no longer practiced; thus, the contribution of miriti fruit to the rural household income is minimal. A year-long income survey (June 1992–May 1993) of 12 households at the Ilhas de Abaetetuba indicated that *Mauritia flexuosa* contributed, on average, 1.05% of the total annual income, while açaí's share amounted to 64.91% (Fig. 2).

On the other hand, the fruit is an important feed source for domestic animals such as pigs and fowl. Ribeirinhos in the survey area keep an average of 4.1 pigs and 11.8 fowl per household. Chickens and their eggs are raised mostly for family consumption, but pigs and ducks are raised for sale. Sale of domestic animals produced an average of 10.66% of the annual income; of this portion, the sale of pigs amounted to 74%. Several factors lead small-scale farmers of the várzea to keep hogs: Pigs can be readily converted to cash, their price remains relatively stable during the year, they require minimum care, and, most important of all, they are fattened mostly with flooded-forest fruit, especially that produced by palms. Although muru muru, inaja (*Attalea maripa*), ubucu (*Manicaria saccifera*), and açaí are significant in the diet of semi-feral pigs, the main source is the fruit of the miriti. The miriti's copious fruit production (up to 350 kg/tree) is seasonal, and pigs are synchronized to be fattened during the fruit-fall period. Piglets are bought by estimating the level of fruit production. Two- to three-month-olds, weighing 3–4 kg, are acquired in the beginning of the rainy season (December or January), when the fruit begins to fall. The semi-feral pigs forage on the forest floor, and by the end of the fruiting season in June or July, when they are marketed, the animals have gained 30–35 kg. The small-scale rearing of hogs is an important adjunct to subsistence, especially toward the end of the rainy season when alternative sources of income are unavailable (Fig. 3).

The petiole of miriti has been used traditionally for a number of purposes. The long petiole (3–3.5 m), consists of a hard but flexible outer layer and a spongy inner material. Having a high buoyancy, the light and soft tissues, locally called *buxo*, are employed as flotation material. For instance, the petiole is commonly used to extract nonbuoyant logs from the flooded forest. During high tide, rafts of timber tied with miriti petioles are floated and moved from the forest floor. The easily worked buxo is also used for other purposes, such as boat caulking, sail for small watercraft, toy making, cork, walls, window blinds, and toilet paper substitutes in rural areas (Table I). The thin epidermis (1–2 mm), however, is perceived as the most useful part of the petiole. Several products are made by removing the epidermis in 5–7-mm strips. These are woven into manioc-processing tools, such as the circular sieve, the *tipiti* (elongated press), and as-sorted containers. In addition, the pliable and resistant epidermis is woven to produce diverse products, including the standardized açaí container, the sieve for extraction of açaí pulp, and protective covers for boats and canoes. The reliance on miriti epidermis increases as other basketry materials from the flooded forest—the *aruma* (*Ischnosiphon* spp.), *ambe* (*Philodendron spruceanum*), and *jacitara* (*Desmoncus polyacanthos*)—become de-pleted. Many of the basketry items are sold occasionally, but no product rivals the

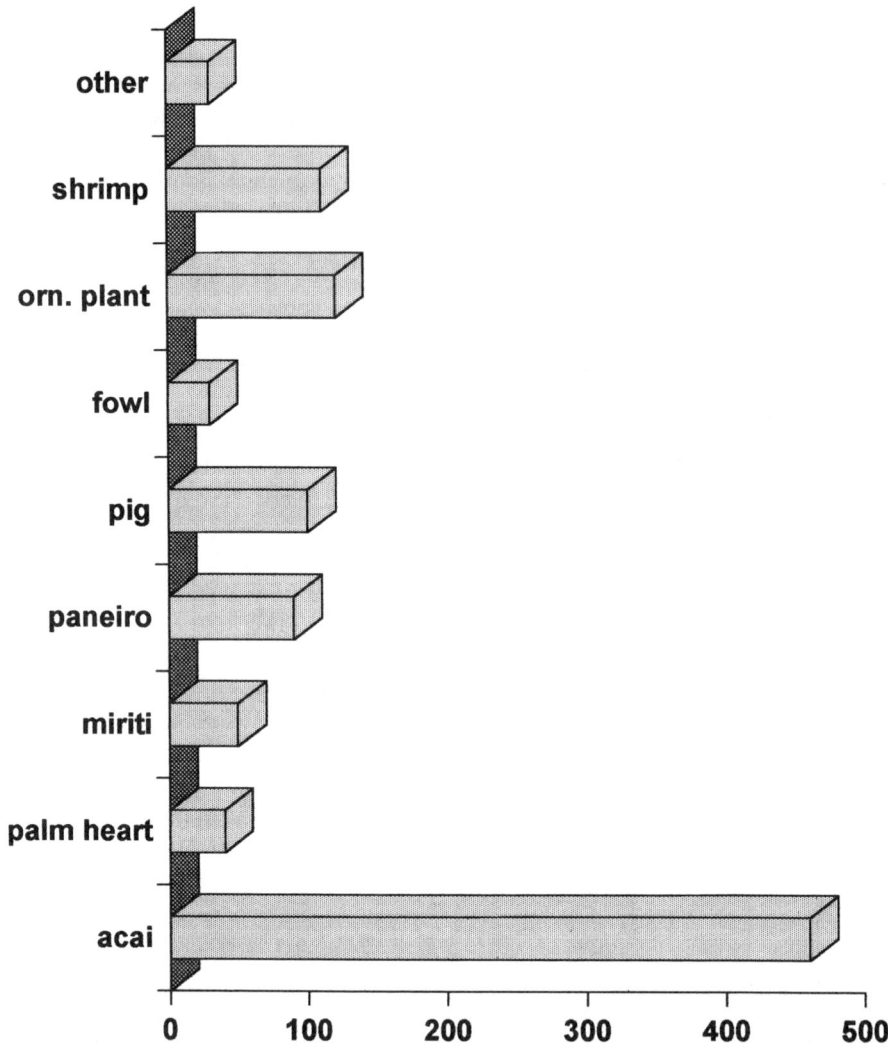

Figure 2. Annual income composition of sample households. Miriti products and by-products, such as miriti fruit, paneiros, and pigs, constitute a small proportion of the total income.

coarsely woven *paneiros* as a source of cash income. During the past 15 years, a regional market developed for the locally made paneiros. Used as biodegradable, disposable containers in the supermarkets and wholesale fruit and vegetable distribution centers, a major demand was created for the competitively priced paneiros. The ribeirinhos of the Ilhas de Abaetetuba recognized the niche and began to produce for the Belém metropolitan market. In our survey, a family produced an average of 3125 units/yr, contributing 5.2% of the total income (Fig. 3). Baskets are made by women during

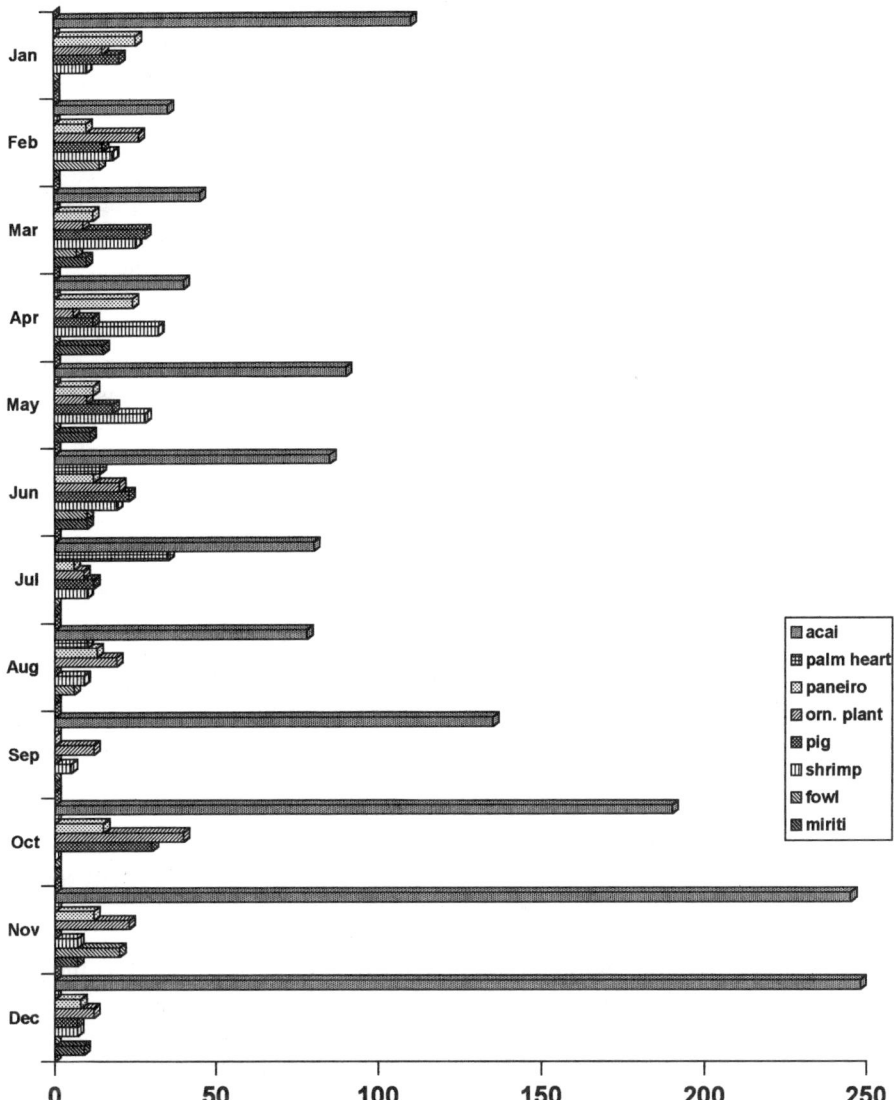

Figure 3. Açaí is the main source of income during the dry season and provides the bulk of the annual income. Inhabitants rely on diverse products for income during the rainy season. Miriti palm–derived products are important during this season.

their spare time. Petioles are harvested from acaulescent to mature trees. The demand has created a cash value for the petioles, so that landless residents are required to purchase the material. Cutting and transporting of the petioles are done either by men or women, but the removal of the epidermis and the weaving are performed by women. Paneiro-making has expanded the activities of women, but concurrently it offered them an independent source of income. Although the cash receipts are limited, the paneiros are sold year-round and provide the women with a continuous flow of disposable income. Partial economic independence, on the other hand, contributes to increase the self-esteem of the female population, while opening the possibilities for other employment and entrepreneurial opportunities.

Other parts of miriti serve the ribeirinhos in significant ways. The mature trunk, lasting as long as five years in water, is the favorite material for floating docks. The highly buoyant trunk, secured at both extremities by vertical poles of açaí, moves with the oscillating tidal levels of the estuary. The long, durable, uniform diameter and the readily available trunk is important not only for giving access to watercraft at low tide, when large parts of the shoreline become exposed, but also for bathing and fetching cooking and drinking water from the river. Dead trunks often serve as nesting sites for parrots. The baby parrots, in turn, are avidly caught around mid-February, when they are said to be ideal for raising as pets or for sale. Opossum (*Didelphis marsupialis*) is hunted on the high canopy of miriti, where dens are found. The organic soil resulting from pith of dead trunks, and the partially decomposed organic matter deposited around the base of miriti trees, is collected as potting soil for the raised-platform gardens of the ribeirinhos.

Tender leaves of miriti are sources for strong fibers. All-purpose tying material is extracted by splitting the young leaf blades. A superior and softer fiber, believed to be even more durable than *Astrocaryum chambira*, is prepared from the epidermis of young leaf spikes (Schultes, 1977). The fiber, locally termed *carua* or *linho*, is prepared by separating the epidermis from the leaf blades, boiling the fiber, and sun-drying it. Hammocks, belts, cordage, and fishing nets used to be manufactured from the carua, but more recently, placemats, hats, handbags, and other handicraft items are beginning to be made for the urban clientele.

In the estuarine floodplain, as elsewhere in Amazonia, *Mauritia flexuosa* continues to be important in the household economy. Its diverse parts are utilized throughout the year, but it is during the rainy season that ribeirinhos become virtually dependent on the plant for their subsistence. During the lean months of February and June, miriti becomes a valuable source of sustenance, because few crops are produced on the waterlogged tidal floodplains and few income-generating opportunities exist. Miriti has been insignificant in cash generation because of its limited prestige as a regional staple, especially in comparison to açaí, but the increased demand for petiole and leaf epidermis-based products appears to offer rewarding prospects. Since the baskets and handicrafts are not influenced by rains, they utilize the idle rainy-season labor and thus balance income distribution during the year. Valuation of female labor, on the other hand, not only increases household income but also is leading to empowerment of women.

Management of Miriti Forests

The existing miriti stands of the Ilhas de Abaetetuba are highly anthropogenic. Extractivism and agriculture prior to the mid-1970s led to the removal of most old-growth miriti from the better-drained areas. Today, mature stands are confined to *baixas*, poorly drained areas unfit for swiddens. Following the rapid decline of the sugar cane industry in the region after the late 1960s, and the subsequent contraction of farmlands, miriti began to recolonize the floodplain. Young trees, usually less than 20 years old, are common in the secondary growth, açaí agroforests, new *roçados* (swiddens), and orchards near dwellings. Therefore, the present distribution of miriti is a strongly human-modified one. On the other hand, the increasing valuation of the palm and the pressure on the land caused by rapid population growth, are forcing ribeirinhos to devise new management practices. These also contribute in modifying the *Mauritia flexuosa* distribution patterns.

Miriti management practices vary within the region and between ribeirinho holdings. Differences in ecological characteristics, economic history, access to market, and demographic pressures, among other factors, are responsible for the regional variations. At the farm level, management differences are influenced mainly by land use types.

In preparing the rocado, most of the standing vegetation is removed. Care is taken to save the economically valuable species, such as rubber trees (*Hevea brasiliensis*), tapereba (*Spondias mombin*), açaí, and andiroba, from the ax and the fire. Virtually the entire miriti population reestablished in the *capoeiras* (secondary forests) is eliminated at deforestation. Only female trees in production, i.e., those more than 8–10 years old, and those not likely to interfere with swidden crops are left standing. Qualities like sizes of fruit and infructescence, color, taste, and thickness of mesocarp are taken into account in choosing the trees to be spared. No more than 15–20 adult female trees per hectare remain in the new rocados. When the rocados are converted later to açaí agroforests, as is happening with most swiddens, miriti is often seen as a liability. Consequently, they are generally cleared.

On the other hand, miriti in the acaulescent stage are beginning to be protected in some swiddens. Ribeirinhos without miriti groves in swampy terrain often save acaulescent individuals for extraction of petioles in the rocados. They begin to be cut when the plants reach 3–4 years of age. Although petioles from juveniles (usually <3 m) hardly reach the dimensions of adult palms (3.5–4 m), the ease of harvests more than compensates for the shorter length. The annual returns (about 1% of gross receipts) from petiole sales are minuscule, but they serve as an additional income source to the household. Once the miriti reaches the arborescent stage and begins to shoot upward, it is cut to limit shade from the crown.

Ribeirinhos have few miriti trees in their *terreiro*, or house garden/orchard. Some acaulescent trees, germinating from seeds discarded on the terreiro by humans and animals following the consumption of the mesocarp, are intermixed with other planted or protected species. The young miriti is protected primarily for the extraction of petioles and tender leaf blades. The palm is removed when it begins to compete with açaí and other fruit trees for light. Mature miriti, on the other hand, is not present because of the ribeirinhos' fear of lightning and tree falls. Even before the açaízation of the estuary, the terreiro was generally dominated by açaí. It is even more so today, with increasing commercialization of the fruit and palm heart. In this sort of an economic environment, except for a few shade-tolerant plants like cacao (*Theobroma cacao*) and aruma (*Ischnosiphon* spp.), most competing species are slashed at the time of annual weeding.

Dense stands of miriti are increasingly restricted to the baixas. The concentrations, known as *pontas de miriti*, are managed differently. Unlike the floodplains upriver from Iquitos, Peru, where the commercial harvests have led to the virtual disappearance of mature female trees from the easily accessible parts of the floodplains, ribeirinhos of the estuary have devised ways to enhance and conserve the miriti trees. Production is increased by changing the sex ratio of trees. The practice of removing the male trees is common among the inhabitants. The outcome of such a procedure, where the sex ratio is unbalanced, is unknown. Local people relate that they have not seen changes in the production and productivity of remaining trees. But the biennial flowering of *Mauritia flexuosa* in the region—a phenomenon ribeirinhos began to notice within the past decade or so—may be related to the novel procedure. Another possibility for the change in the phenology may result from the damming of the Tocantins River at Tucuruí, for hydroelectric purposes, in 1984. The control of discharge, resulting in lower flood ranges during the rainy season, may be correlated with the changed phenological behavior. Another management procedure common for açaí, the thinning of all trees regardless of sex ratio, is still rarely practiced for miriti. Such a practice, a form of intensification, is likely to be adopted with the valuation of the fruit or petioles.

The mounting pressure on várzea lands, caused by the expansion of açaí in the estuarine areas neighboring the two major markets of Belém and Abaetetuba and the increasing population, are leading to the reclamation of baixas. Ribeirinhos without sufficient land for açaí replace the miriti in the baixas by broadcasting açaí seeds or by planting açaí seedlings. Thus, if the present açaízation trend continues, the tendency is for the replacement of miriti forests by the economically more desirable açaí.

Sustainable practices have been developed for the extraction of açaí fruit. Ripe fruit fallen on the base of trees, or those flushed out by the high tides, are collected for home consumption or animal feed. Fruit destined for sale is harvested by climbing the trees. Açaí fruit is harvested by climbing the slender trunks and removing the infructescences, but the height of the mature miriti and its trunk diameter of up to 60 cm do not make for easy climbing. Moreover, a single miriti infructescence weighs more than 50 kg. Ribeirinhos have solved the problem by gaining access to the crown from neighboring trees. Easily climbed trees that are commonly associated with the miriti in the baixas—trees like anani (*Symphonia globulifera*), jenipapo (*Genipa americana*), and mututi (*Pterocarpus officinalis*)—are protected to provide access to the miriti crown. The harvest of leaf blades and petioles, although not common from mature miriti, is also accomplished from neighboring trees.

Sustainable procedures are likewise practiced in harvesting the petioles and leaf blades. The petioles are harvested, usually from juveniles, in such a way as to avoid permanent damage and to allow the plants to regenerate new petioles. By leaving a minimum of 3–4 mature petioles, 6–8 leaf stems can be extracted twice a year from each juvenile. Likewise, leaf spikes are removed semiannually, without apparent damage to the plant.

Current practices of extracting diverse parts of miriti for economic purposes appear to be sustainable, yet the long-range tendency is for continued elimination of the palm from the várzea, unless an increase in value is created for the palm. As demonstrated, *Mauritia flexuosa* continues to be extremely useful to life in the tidal floodplain, but its ubiquity leads to its successive removal with each land use cycle. For the small-scale farmers of the estuary, struggling to improve their economic well-being while attempting to feed a rapidly expanding population, the conservation of a plant with limited cash value is increasingly difficult. The inhabitants also recognize the significance of the miriti stands as important feeding, nesting, and breeding sites for various species of animals, including rodents and marsupials such as agouti, paca, and opossum. These animals, though increasingly scarce, are hunted and serve as occasional sources of income and animal protein. The fruiting season of miriti attracts a number of species of aquatic fauna to the tidal forest (Goulding, 1980, 1985). For example, ribeirinhos see a correlation between bountiful freshwater shrimp harvests and miriti fruit. Shrimp size and population appears to be related to fruiting levels of *Mauritia flexuosa*.

Conclusions

The study of contemporary Amazonian floodplain inhabitants and their resource management practices offers important insights into sustainable methods of resource use. In addition to the fruit that serves as one of the primary staples among the estuarine inhabitants of Ilhas de Abaetetuba, the miriti palm is used for diverse purposes by these rural households. The management techniques devised by the inhabitants to utilize the palm appear to be sustainable. Reflecting the importance of the palm, suggestions have been made for in situ conservation of the monospecific miriti forests (Kahn et al., 1993; Kahn, 1991, 1993; Peters et al., 1989). The miriti forests in the swampy terrain would be set aside for the extraction of its varied products, while better-drained land would be placed into crops. Despite the miriti's multiple uses and its relevance in the rural economy, miriti-dominated forests are on the decline. Miriti in the Brazilian Amazon, unlike that in the Peruvian Amazon, does not enjoy much acceptance in the estuarine markets. The competition from açaí progressively pushes miriti away from its habitat. As expansion of açaí agroforests occurs, miriti stands are cleared and replaced with açaí. The decline of miriti-dominated forests is akin to that occurring to the babaçú (*Attalea speciosa*). The babaçú, common in the eastern and southern peripheries of Amazonia, has many uses and contributes to income, but cash returns are minor. When babaçú begins to compete with crops and cattle pastures in areas of more intensive land use, it usually loses ground to other land uses (May et al., 1985; Anderson et al., 1991).

The present case study points out the difficulty of harmonizing the dichotomous concepts of development and conservation even in societies supposedly knowledgeable about resource-conserving practices. On-farm management of native flora is a challenging task, even without considering the biological implications of small-scale conservation attempts. The experience of the Ilhas de Abaetetuba suggests that a combination of factors may in time deflect the pressures from the miriti forests. For example, as the young ribeirinhos' behavior becomes increasingly urban oriented, an increasingly small proportion of the youth chooses to remain in the rural estuarine areas. Also, in the rural areas, the proportion of ribeirinhos incorporating nonfarm activities—e.g., retailing of diverse merchandise, off-farm work, transportation—is on the rise. As nonfarm income-generating activities become common among the ribeirinhos, there is a consequent decline in the expansion of farmlands. Finally, a proven strategy is to increase the value of miriti-derived products. As the example of açaí in the region illustrates, the valuation of the product as a result of increasing demands for its fruit and palm hearts is seen not only in the in situ management but also in the expansion of cultivated areas (Anderson & Ioris, 1992). However, the prospects for conservation of *Mauritia* stands in densely occupied areas are limited. Drastic changes in markets are unlikely in the near future, both locally and regionally. Intensification of land uses (e.g., wet rice, fish culture, and açaízation) resulting from improved access and population growth is likely to cause further incursions into the baixas.

Acknowledgments

My sincere thanks to the numerous individuals who assisted me in understanding the *Mauritia flexuosa* palm. I am particularly indebted to Flor Chavez for sharing her references and explaining the natural history of the palm; to Li Yun Hsu, Shiro Hiraoka, and Ana Maria Dias for important insights concerning the diverse uses; and to my ribeirinho neighbors of Maracapucu and Maracapucu-Miri who shared their knowledge of the use and management of miriti. Funding for this study was provided by the National Geographic Society, U.S. National Science Foundation, Ministry of Education (Japan), CNPq, the Heinz Foundation, the Ford Foundation, and the Rockefeller Foundation.

Literature Cited

Anderson, A. B. 1988. Use and management of native forests dominated by *açaí* palm (*Euterpe oleracea* Mart.) in the Amazon estuary. Pages 144–154 *in* M. J. Balick, ed., The palm—Tree of life: Biology, utilization and conservation. Advances in Economic Botany 6. New York Botanical Garden, Bronx.

———, ed. 1990. Alternatives to deforestation: Steps toward sustainable use of the Amazon rainforest. Columbia University Press, New York.

——— & E. M. Ioris. 1992. Valuing the rain forest: Economic strategies by small-scale forest extractivists in the Amazon estuary. Human Ecology 20: 337–369.

——— & M. A. G. Jardim. 1989. Costs and benefits of floodplain forest management by rural inhabitants in the Amazon estuary: A case study of açaí palm production. Pages 114–129 *in* J. O. Browder, ed., Fragile lands of Latin America: The search for sustained uses. Westview Press, Boulder, CO.

———, A. Gely, J. Strudwick, G. Sobel & M. G. Pinto. 1985. Um sistema agroflorestal na várzea do estuario amazonico (Ilha das Oncas, Municipio de Barcarena, Belém, Para, Brasil). Acta Amazonica, Suppl. 15 (1–2): 195–224.

———, P. H. May & M. J. Balick. 1991. The subsidy from nature. Columbia University Press, New York.

———, I. Mousasticoshvily Jr. & D. S. Macedo. 1994. Impactos ecologicos e socio-economicos da exploracao seletiva de virola no estuario amazonico. World Wildlife Fund, Washington, DC.

Balick, M. J. 1979. Amazonian oil palms of promise: A survey. Economic Botany 33: 11–28.

Beckermann, S. 1977. The use of palms by Bari Indians of the Maracaibo basin. Principes 21(4): 143–154.

Braun, A. 1968. Cultivated palms of Venezuela. Principes 12, nos. 2–4.

Chibnik, M. 1994. Risky rivers: The economics and politics of floodplain farming in Amazonia. University of Arizona Press, Tucson.

Ervik, F. 1993. Notes on the phenology and pollination of the dioecious palms *Mauritia flexuosa* (Calmoidae) and *Aphandra natalia* (Phytelephantoideae) in Ecuador. Pages 7–12 *in* W. Barthlott et al., Results of the annual meeting of the German Society for Tropical Ecology, held at Bonn, 13–16 February 1992. Zoologisches-Forschnungsinstitut und Museum Alexander Koenig, Bonn.

Gonzalez Boscan, V. 1987. Los morichales de los Llanos Orientales: Un enfoque ecologico. Ediciones Corpoven, Caracas.

Goulding, M. 1980. The fishes and the forest. University of California Press, Berkeley.

———. 1985. Forest fishes of the Amazon. Pages 267–276 *in* G. T. Prance & T. E. Lovejoy, eds., Amazonia. Pergamon Press, New York.

Granville, J.-J. de. 1974. Apercu sur la structure des pneumatophores de deux especes des sols hydromorphes en Guyane *Mauritia flexuosa* L. et *Euterpe oleracea* Mart. (Palmae). Generalisation au systeme respiratoire racinaire d'autres palmiers. Cahiers ORSTOM, Ser. Biol. 23: 3–22.

Heinen, H. D. & K. Ruddle. 1974. Ecology, ritual, and economic organization in the distribution of palm starch among the Warao of the Orinoco delta. Journal of Anthropologic Research 30: 116–138.

Henderson, A. 1995. The palms of the Amazon. Oxford University Press, New York.

Hiraoka, M. 1985. Mestizo subsistence in riparian Amazonia. National Geographic Research 1(2): 236–246.

———. 1995. Land use changes in the Amazon estuary. NATO ASI Series I, Global Environmental Change 5(4): 323–336.

Hladik, C. M., H. Pagesy, O. F. Linares, A. Hladik & M. Hadley, eds. 1993. Food and nutrition in the tropical forest. UNESCO-MAB, Paris.

Johnson, D. 1983. Multi-purpose palms in agroforestry: A classification and assessment. International Tree Crops Journal 2: 217–244.

Kahn, F. 1991. Palms as key swamp forest resources in Amazonia. Forest Ecology and Management 38: 133–142.

———. 1993. Amazonian palms: Food resources for the management of forest ecosystems. Pages 153–162 *in* C. M. Hladik et al., eds., Food and nutrition in the tropical forest. UNESCO-MAB, Paris.

——— & J.-J. de Granville. 1992. Palms in forest ecosystems of Amazonia. Springer-Verlag, Berlin.

———, K. Mejia, F. Moussa & D. Gomez. 1993. *Mauritia flexuosa* (Palmae), la mas acuatica de las palmeras amazonicas. Pages 287–322 *in* F. Kahn et al., comps., Las plantas vasculares en las aguas continentales del Peru. L'Institut Français d'Études Andines, Lima.

May, P. H., A. B. Anderson, M. J. Balick & J. M. F. Frazao. 1985 Subsistence benefits from the babassu palm (*Orbignya martiana*). Economic Botany 39: 113–129.

McGrath, D. G. 1989. The Paraense traders: Small-scale, long-distance trade in the Brazilian Amazon. Ph.D. dissertation, University of Wisconsin, Madison.

Padoch, C. 1988. Aguaje (*Mauritia flexuosa* L.f.) in the economy of Iquitos, Peru. Pages 214–224 *in* M. J. Balick, ed., The palm—Tree of life: Biology, utilization and conservation. Advances in Economic Botany 6. New York Botanical Garden, Bronx.

——— & W. de Jong. 1990. Santa Rosa: The impact of the forest products trade on an Amazonian village. Pages 151–158 *in* G. T. Prance & M. Balick, eds., New directions in the study of plants and people. Advances in Economic Botany 8. New York Botanical Garden, Bronx.

Parker, E. P. 1985. Caboclization: The transformation of the Amerindian in Amazonia 1615–1800. Pages 1–49 *in* E. P. Parker, ed., The Amazon caboclo: Historical and contemporary perspectives. Studies in Third World Societies 32. Department of Anthropology, College of William and Mary, Williamsburg, VA.

Pedersen, H., H. Borgtoft & H. Balslev. 1990. Ecuadorean palms for agroforestry. AAU Reports 23. Botanical Institute, Aarhus University, Risskov.

Peters, C., M. Balick, F. Kahn & A. Anderson. 1989. Oligarchic forests of economic plants in Amazonia: Utilization and conservation of an important tropical resource. Conservation Biology 3: 341–349.

Posey, D. A. & W. Balée, eds. 1989. Resource management in Amazonia: Indigenous and folk strategies. Advances in Economic Botany 7. York: New York Botanical Garden, Bronx.

Redford, K. & C. Padoch. 1992 Conservation of neotropical forests: Working from traditional resource use. Columbia University Press, New York.

Ruiz Murrieta, J. & J. Levistre Ruiz. 1993. Aguajales: Forest fruit extraction in the Peruvian Amazon. Pages 797–804 *in* C. M. Hladik et al., eds., Food and nutrition in the tropical forest. UNESCO-MAB, Paris.

Schultes, R. E. 1974. Palms and religion in the northwest Amazon. Principes 18: 3–21.

———. 1977. Promising structural fiber palms of the Colombian Amazon. Principes 21: 72–82.

Smith, N. J. H. 1985. The impact of cultural and ecological change on Amazonian fisheries. Biological Conservation 32: 355–373.

Spruce, R. 1908. Notes of a botanist on the Amazon and Andes. Macmillan, London.

Strudwick, J. & G. L. Sobel. 1988. Uses of *Euterpe oleracea* Mart. in the Amazon estuary, Brazil. Pages 225–253 *in* M. J. Balick, ed., The palm—Tree of life: Biology, utilization and conservation. Advances in Economic Botany 6. New York Botanical Garden, Bronx.

Vickers, W. T. Cultural adaptation to Amazonian habitats: the Siona-Secoya of eastern Ecuador. Ph.D. dissertation, University of Florida.

Wilbert, J. *Manicaria saccifera* and its cultural significance among the Warao Indians of Venezuela. Botanical Museum Leaflets, Harvard University 24: 275–335.

An Overview of the Palms of the Várzea in the Amazon Region

Francis Kahn and Andrew Henderson

Introduction

A traveler going by river from the city of Belém, at the mouth of the Amazon, to the city of Iquitos, almost 2500 km to the west (as the crow flies), would see almost all of the palms of the várzea. This route, or at least part of it, was taken by all the early travelers and naturalists in the Amazon (e.g., Carl von Martius, Richard Spruce, Henry Walter Bates, Alfred Russell Wallace), and many of them commented on the abundance and usefulness of the palms they saw. These palms are still as abundant and useful now, in the late twentieth century, as they were in the mid-nineteenth century—more so, in some cases. In this paper we give an overview of the economically important genera and species of várzea palm.

The taxonomy and general ecology of várzea palms is now relatively well understood. Henderson (1995) has reviewed taxonomy of all Amazonian palms, and Kahn and de Granville (1992) have done the same for their ecology. Approximately 20 species in 14 genera of palm commonly occur in the várzea. Nine genera, the most economically important ones, are reviewed below. Where palms are concerned, várzea forests are usually characterized by few species with high density and by high fecundity. Phillips (1993) showed that despite its general low diversity in tree species, in terms of fruit production the várzea was more productive than terra firme forests. This abundance and fecundity is the main reason for the usefulness of várzea palms. They also tend to be widespread and, in general, occur in various habitats—not only in floodplains but also on nonflooded soils. For example, both *Socratea exorrhiza* and *Euterpe precatoria* are common várzea palms, but both also occur on steep Andean slopes to 2000 m elevation. Their adaptation to flooded soils seems to be physiological rather than morphological.

Euterpe

The most prominent palm—and the one any Amazon traveler first notices, occurring in great abundance along the banks of the numerous channels of the Amazon estuary—is the açaí (*Euterpe oleracea*). *Euterpe* is a genus of seven species (Henderson & Galeano, 1996) that are widespread in tropical America. *Euterpe oleracea* is abundant from the Amazon estuary northward along the coast of Brazil (Amapá), the Guianas, and Venezuela, and just reaches Trinidad. It then disappears and reappears on the Pacific coast of Colombia and Ecuador (and just enters Panama). *Euterpe oleracea* is a clustered-stemmed species harvested principally for palmito and edible fruits. Both of these are important items of commerce in Belém. There are also a host of minor uses (Strudwick & Sobel, 1988). Because of the apparent sustainability of its harvesting, it has attracted much interest from researchers interested in nontimber forest products (e.g., Anderson, 1988). Hiraoka (in this volume) has shown how açaí stands have expanded considerably in recent years, having been planted in response to an increased demand for fruits in Belém. He calls this the "açaízation" of the estuary (also known as the "l'euterpisation"; Kahn & Moussa, 1995). A curious feature of useful palms, exemplified by *E. oleracea*, is the parochial nature of their use. This species is very important in the city of Belém but is far less used in other parts of its range. Here we will concentrate more on the lesser known *E. precatoria*.

Near Santarém, at the mouth of the Rio Tapajós, near the limit of tidal influence, the Amazon traveler will notice that *Euterpe oleracea* becomes less common and then is finally replaced by another species, *E. precatoria*. This species differs in several respects, most obvious of which is that it has a solitary stem. When it is cut for its palmito, the plant is killed. Nevertheless, *E. precatoria* is used in the same way as *E. oleracea*—for palmito and fruits.

Peña (1996) studied palmito extraction from wild populations of *Euterpe precatoria* in várzea areas of eastern Bolivia, where more than 2 million hectares have been allocated for palmito concessions. Palmito is the third largest forest product exported from Bolivia (timber and Brazil nut are first and second, respectively) and was worth US$2 million in 1993. Although management plans are required under Bolivian law, Peña found no cases where they were in operation, and she described palmito extraction in Bolivia as "a mining activity." Although a harvesting cycle of 8 years has been proposed, Peña found that, based on her demographic studies, a harvesting cycle of 400 years would still not be sustainable. There seems no reason to doubt that palmito extraction from *E. precatoria* in Bolivia will go the same way it went for *E. edulis* in coastal Brazil: massive destruction of wild stands followed by collapse of the industry.

De Castro (1993) commented on the collection of fruits of *Euterpe precatoria* for sale in the markets of Manaus. Because fruits deteriorate quickly, the effective harvesting zone is within a 350 km radius of Manaus. All fruits are transported by river. The fruiting season, from January to August, varies from place to place over the 700 km harvest region along the main river. Upstream, fruiting begins earlier, in January, and the season "moves" downstream until September. This staggered fruiting seems to be associated with high water on the river. Once in Manaus, the fruits enter an intricate marketing system and pass through a hierarchy of distributors. Although de Castro believed the harvesting of fruits did not jeopardize natural populations of *E. precatoria*, there are as yet no demographic studies with which to show that this is the case.

Euterpe precatoria is extremely widespread, perhaps the most so of neotropical palms; even local overharvesting will not place it in any danger of extinction. It can occur in various habitats, from the várzea forests of the Amazon region to the steep 2000 m slopes of the Andes.

Raphia

As one travels from Belém upstream along the river, another palm soon becomes evident: the jupatí (*Raphia taedigera*). This species was described by Martius (1823–1837), who had firsthand knowledge of the palm. Soon after he arrived at the mouth of the Amazon, in 1819, Martius got lost in the forest:

> The joys of agreeable contemplation of nature were replaced by terror; for I came to a swampy area where I found myself surrounded by impenetrable groves of prickly palms. . . . It was starting to get dark. . . . I clambered up the trunk of a *jubatí* palm (*Raphia taedigera*), several footstalks of which formed a kind of stairway. I was surely safe from wild animals in the thick branches of this tree. [Spix & Martius in Gheerbrant, 1992]

Raphia taedigera occurs in the Amazon estuary, and also in Central America and the Pacific coast of Colombia. It is one of the few neotropical genera of palms that is not endemic. In fact, most species of *Raphia* are African, and only one neotropical. This has led to considerable speculation that even the neotropical species was introduced from Africa. It is the kind of question that cannot be easily resolved—at least, not without resorting to molecular techniques. Ripe fruits sink in water, so presumably the species would have to have been introduced by humans. Ecological studies have been equivocal (discussed in Henderson, 1995). Tuley (1994) suggested that pre-Columbian Moorish expeditions across the Atlantic introduced this palm to the New World. *Raphia* has few uses in the Amazon estuary, and these were described by Wallace (1853). Currently it seems most important in the manufacture of shrimp traps.

Mauritia

On the afternoon of 29 August 1848, the English naturalist Henry Walter Bates, at the start of his trip up the Amazon, arrived at a place near the mouth of the Tocantins. He wrote:

> We went ashore on an island covered with palm-trees, to make a fire and boil our kettle for tea. I wandered a short way inland, and was astounded at the prospect. The land lay below the upper level of the daily tides, so that there was no underwood, and the ground was bare. The trees were almost all of one species of Palm, the gigantic fan-leaved *Mauritia flexuosa*; on the borders only was there a small number of a second kind, the equally remarkable Ubussú palm (*Manicaria saccifera*). The Ubussú has erect, uncut leaves, twenty-five feet long, and six feet wide, all arranged round the top of a four-feet high stem, so as to form a figure like that of a colossal shuttlecock. The fan-leaved palms, which clothed nearly the entire islet, had huge cylindrical smooth stems, three feet in diameter, and about a hundred feet high. The crowns were formed of enormous clusters of fan-shaped leaves, the stalks alone of which measured seven to ten feet in length. Nothing in the vegetable world could be more imposing than this grove of palms. [Bates, 1863]

Mauritia flexuosa, the miriti of Brazil and aguaje of Peru, is indeed an imposing palm. Bates was impressed by the size of the palms, but he could not have imagined the sheer numbers of individual plants that exist (or, rather, existed) in the Amazon várzea. Estimates of area made from Landsat satellite images suggest that stands *of M. flexuosa* cover 6–8 million hectares in the Peruvian Amazon (Ruiz & Ruiz, 1993), and more than 2 million hectares of this are high-density stands. Kahn and Mejia (1990) found 645 individual *M. flexuosa* trees in one hectare of permanently flooded forest in Peru. *Mauritia flexuosa* is dioecious and has very tall stems. Fruits are a popular item in Iquitos markets (Padoch, 1988). The mesocarp is either eaten directly or made into various food items. Harvesting of fruits is carried out by cutting down female trees. This has led to large-scale damage to populations, which now consist of purely male trees (Vásquez & Gentry, 1989).

As with *Euterpe*, uses of *Mauritia flexuosa* tend to be local. In eastern Brazil the fruits are an important source of vegetable oil (Altman, 1964), whereas in Iquitos, as we mentioned, the mesocarp of the fruits is eaten.

Manicaria

The other palm that impressed Bates, the ubussú (*Manicaria saccifera*), does not generally occur in such large stands, and is rather patchily distributed in the Amazon (Henderson, 1995). It also extends into Central America, and thus has a distribution surprisingly similar to both *Euterpe oleracea* and *Raphia taedigera*. It was formerly important to the Warao Indians of the Orinoco delta as a source of sago (Wilbert, 1976), as was *Mauritia flexuosa* (Heinen & Ruddle, 1974). Currently, leaves of *Manicaria saccifera* are important as a source of thatching, especially when a "rustic" look is desired.

The last sentence of Bates's (1863) book, describing his departure by sea from Brazil in 1859, after 11 years of fieldwork, was this: "Among the masses [of floating grass] I espied many fruits of that peculiarly Amazonian tree the Ubussú palm; this was the last I saw of the Great River."

Socratea

Alfred Russell Wallace was not so lucky, nor so nostalgic, in his departure in 1852 from the Amazon (Wallace, 1889). The ship on which he was traveling caught fire and sank in the Atlantic. Wallace lost all his specimens, except for, among other things, some sketches of palms. With these he wrote, from memory, his classic account of the ethnobotany of Amazon palms (Wallace, 1853). He was particularly taken with the stilt-rooted palm *Socratea*.

Socratea is a genus of five species (Henderson, 1990), primarily of Andean distribution. One, *S. exorrhiza*, is very widely distributed throughout moist-forest areas of Central America and the Amazon region, and is very common in the lower Amazon. It occurs on nonflooded soils in Central America, as it does on Andean areas. In the Amazon region, however, the species almost always occurs in várzea or other temporarily inundated areas. This led some authors to suppose that stilt roots are an adaptation to inundated soils (Bouillenne, 1924), although the palm most often occurs on non-flooded soils. Its adaption to flooded soils is physiological.

In the Brazilian Amazon, *Socratea exorrhiza* is known as "paxiuba," from the Tupi words *paxi* (a thin palm) and *yúa* (trunk or stem). In the Peruvian Amazon it is known as "casha pona." In both countries the basal parts of the split stem are used in house construction, especially floors. So common is this use in Peru that "el emponado" has come to mean floor. This kind of construction is the most important use of *Socratea*, as it is of *Iriartea*.

Iriartea

As Bates traveled upstream from Manaus to Tefé ("In passing slowly along the interminable wooded banks week after week"), he saw his first paxiuba barrigouda, (*Iriartea deltoidea*). Like *Socratea*, a related genus, *Iriartea* has stilt roots, and often a swollen stem. It has several uses, the most important of which is planks for walls and floors of houses. It is extensively used on a local scale in the western Amazon region. A floor made from *Iriartea* planks is reported to last for 20–25 years. Pinard (1993) has recently studied demography of this species in Acre, Brazil. She asserted that rubber tappers' use of *I. deltoidea* was ecologically sustainable. Only the tallest palms are felled for planks.

Astrocaryum

Astrocaryum is a moderately large genus most easily recognized, and remembered, by its large, black spines. Richard Spruce (1871) wrote that he once got a spine from an *Astrocaryum* in his finger joint, and even 16 years later his finger would occasionally be paralyzed. Two species are common in the várzea: *A. murumuru* and *A. jauari*. *Astrocaryum murumuru* is a thorny taxonomic problem. Kahn and Millán (1992) split this species complex into 13 separate species, a scheme not followed by Henderson (1995).

Astrocaryum jauari is common in both whitewater floodplains and blackwater river margins. Young plants can be inundated for up to 300 days per year (Schlüter et al., 1993) and have both physiological and anatomical adaptations to flooding, but they are better adapted to blackwater. Indeed, *A. jauari* is probably most abundant on the Rio Negro, where there is interest in its potential as a source of palmito.

Phytelephas

Richard Spruce traveled extensively in the Amazon region in the mid-nineteenth century. He was especially interested in palms (Henderson, 1996)—in particular, the phytelephantoid palms.

> I regret that an accident deprived me of the means of drawing up a botanical description of this species [*Phytelephas tenuicaulis*]. On my voyage up the Huallaga in May 1855, I gathered one morning some fully formed fruits of *Yarina*, and, as they were infested by stinging ants, I laid them near the fire, where our breakfast was being cooked, to disperse the ants, and then plunged into the forest in quest of other objects. During my abscence the Indians, not knowing I wanted to preserve the fruits, struck their cutlasses into them, and finding the seeds still tender enough to be eaten, munched them all up, and thus destroyed my specimen. [Spruce, 1871]

Even though the endosperm of *Phytelephas* fruits is edible while immature (as Spruce found), it ripens into one of the hardest substances in the plant world, the well-known vegetable ivory, or tagua. This product was once an important export from Ecuador, mostly for making buttons, until replaced by plastics (Barfod et al., 1990). There is now again great interest in tagua as a nontimber forest product. It has innate appeal to Westerners, for several reasons. The product is unusual, comes from a palm tree, is compared to elephants, and appears to be sustainable. This leads to such headlines in local papers as "Buttons could save rain forests, elephants." Most of the current export is from Pacific coastal areas in Colombia and Ecuador, but large stands also occur in the western Amazon region.

Oenocarpus

Oenocarpus is a genus of nine species, all of which occur in the Amazon region, and only two of which extend beyond (Henderson, 1995). The genus now includes *Jessenia* (i.e., *Oenocarpus bataua*). This latter is a widespread and important várzea species and occurs in large stands. It is of interest as a source of high-quality oil, extracted from the mesocarp of the fruits (Balick, 1986). Some demographic data has been supplied by Sist (1987). Fruits are often harvested destructively by cutting down trees. Vásquez and Gentry (1989) have commented on how this practice is leading to destruction of populations of *O. bataua* near human settlements.

Conclusions

Though there are few demographic studies on várzea palms, we do know that these palms are widespread. They are usually physiologically adapted to life in the floodplain, and elsewhere occur on nonflooded soils. They are usually abundant and fecund, and so are economically important, both historically and currently. Várzea palms are extensively harvested, but usually not in a sustainable manner. Because they are widespread outside the várzea, they are not immediately threatened.

Literature Cited

Altman, M. 1964. A industrialização do fruto de Buriti. INPA, Manaus.

Anderson, A. 1988. Use and management of native forests dominated by açaí palm (*Euterpe oleracea* Mart.) in the Amazon estuary. Pages 144–145 *in* M. Balick, The palm — Tree of life: Biology, utilization, and conservation. Advances in Economic Botany 6. New York Botanical Garden, Bronx.

Balick, M. 1986. Systematics and economic botany of the *Oenocarpus–Jessenia* complex. Advances in Economic Botany 3. New York Botanical Garden, Bronx.

Barfod, A., B. Bergmann & H. Pedersen. 1990. The vegetable ivory industry: Surviving and doing well in Ecuador. Economic Botany 44: 293–300.

Bates, H. 1863. The naturalist on the river Amazons. J. Murray, London.

Bouillenne, R. 1924. Les racines-échasses de *Iriartea exorrhiza* Mart. (Palmiers) et de *Pandanus* div. sp. (Pandanacées). Mémoires de l'Académie Royale des Sciences, Belgique, ser. 2, 8: 1–45.

De Castro, A. 1993. Extractive exploitation of the açai (*Euterpe precatoria*) near Manaus, Amazonia. Pages 779–782 *in* C. Hladik et al., eds., Tropical forests, people and food: Biocultural interactions and applications to development. UNESCO, Paris; Parthenon Publishing, Carnforth, UK.

Gheerbrant, A., trans. 1992. Excerpts from J. Spix & C. Martius [1831], Reisen in Brasilien in den Jahren 1817 bis 1820. *In* The Amazon: Past, present, and future. Harry N. Abrams, New York.

Heinen, H. & K. Ruddle. 1974. Ecology, ritual, and the economic organization in the distribution of palm starch among the Warao of the Orinoco delta. Journal of Anthropological Research 30: 116–138.

Henderson, A. 1990. Arecaceae. Part I. Introduction and the Iriarteinae. Flora Neotropica 53. New York Botanical Garden, Bronx.

———. 1995. The palms of the Amazon. Oxford University Press, New York.

————. 1996. Richard Spruce and the palms of the Amazon and Andes. Pages 187–196 *in* M. Seaward & S. Fitzgerald, eds., Richard Spruce: Botanist and explorer. Royal Botanic Gardens, Kew.

———— & G. Galeano. 1996. A revision of *Euterpe*, *Prestoea*, and *Neonicholsonia* (Palmae). Flora Neotropica 72. New York Botanical Garden, Bronx.

Kahn, F. & K. Mejia. 1990. Palm communities in wetland forest ecosystems of Peruvian Amazon. Forest Ecology and Management 33/34: 169–179.

———— & J.-J. de Granville. 1992. Palms in forest ecosystems of Amazonia. Springer-Verlag, Berlin.

———— & B. Millán. 1992. *Astrocaryum* (Palmae) in Amazonia: A preliminary treatment. Bulletin, Institut Française d'tudes Andines 21: 459–531.

———— & F. Moussa. 1995. Les migrations de palmiers provoquees par l'homme en Amazonie et sa peripherie—Un premier constat. Biogeographica 71: 161–177.

Martius, C. 1823–1837. Historia Naturalis Palmarum. Vol. 2. Genera et species. T. O. Weigel, Leipzig.

Oficina Nacional de Evaluación de Recursos Naturales (ONERN). 1976. Inventario, evaluación e integración de los recursos naturales de la zona Iquitos, Nauta, Requena y Colonia Angamos. Oficina Nacional de Evaluación de Recursos Naturales, Lima.

Padoch, C. 1988. *Aguaje* (*Mauritia flexuosa* L. f.) in the economy of Iquitos, Peru. Pages 214–224 *in* M. Balick, The palm—Tree of life: Biology, utilization, and conservation. Advances in Economic Botany 6. New York Botanical Garden, Bronx.

Peña, M. 1996. Ecology and socioeconomics of palm heart extraction from wild populations of *Euterpe precatoria* Mart. in eastern Bolivia. M.Sc. thesis, University of Florida, Gainesville.

Phillips, O. 1993. The potential for harvesting fruits in tropical rainforests: New data from Amazonian Peru. Biodiveristy and Conservation 2: 18–38.

Pinard, M. 1993. Impact of stem harvesting on populations of *Iriartea deltoidea* (Palmae) in an extractive reserve in Acre, Brazil. Biotropica 25: 2–14.

Ruiz, J. & J. Ruiz. 1993. *Aguajales*: Forest fruit extraction in the Peruvian Amazon. Pages 153–162 *in* C. Hladik et al., eds., Tropical forests, people and food: Biocultural interactions and applications to development. UNESCO, Paris; Parthenon Publishing, Carnforth, UK.

Schlüter, U., B. Furch & C. Joly. 1993. Physiological and anatomical adaptations by young *Astrocaryum jauari* Mart. (Arecaceae) in periodically inundated biotopes of central Amazonia. Biotropica 25: 384–396.

Sist, P. 1987. Regéneration, dynamique des populations et dissémination d'un palmier de Guyane française *Jessenia bataua* (Mart.) Burret subsp. *oligocarpa* (Griseb. & Wendl.) Balick. Bulletin du Muséum d'Histoire Naturelle, sect. B, Adansonia 3: 317–336.

Spruce, R. 1871. Palmae Amazonicae. Journal of the Linnean Society, Botany 11: 65–183.

Strudwick, J. & G. Sobel. 1988. Use of *Euterpe oleracea* in the Amazon estuary, Brazil. Pages 225–253 *in* M. Balick, The palm—Tree of life: Biology, utilization, and conservation. Advances in Economic Botany 6. New York Botanical Garden, Bronx.

Tuley, P. 1994. Columbus and *Raphia taedigera*. Principes 38: 99.

Vásquez, R. & A. Gentry. 1989. Use and misuse of forest-harvested fruits in the Iquitos area. Conservation Biology 3: 350–361.

Wallace, A. 1853. Palm trees of the Amazon and their uses. Van Hoorst, London.

————. 1889. A narrative of travels on the Amazon and Rio Negro. Ward, Lock & Co., London.

Wilbert, J. 1976. *Manicaria saccifera* and its cultural significance among the Warao Indians of Venezuela. Botanical Museum Leaflets 24: 275–335.

Section 3: Conservation

Introduction

John G. Robinson

The conservation of the flooded forests (*várzea*) of Amazonia represents both an imperative and a challenge. An imperative because flooded forests have been largely destroyed in other parts of the world. These forests are characterized by species diversity, a distinct fauna and flora, and a high productivity important to resident people. And it is a challenge because their productivity attracts outside capital and development schemes that lead to system simplification so that biological, resource, and sociocultural diversity is lost. The agricultural potential of the soils has attracted the attention of the international development community, and projects and schemes abound to extensively convert the várzea for large-scale agricultural and agroforestry production (Goodland, 1980). One estimate is that 75% of the flooded forests in the lower Amazon have already been lost (Alexander, 1994). The five papers in this section examine different approaches to conserving these unique forests.

Conserving this ecosystem requires, at a minimum, maintaining the forest cover over a significant portion of the area. How to do this? Creating localized protected areas, such as parks and reserves, would do little to conserve an ecosystem whose functioning is dispersed across the landscape. For instance, many fish species migrate seasonally many hundreds of kilometers (e.g., Goulding, 1980), whereas others migrate among widely dispersed habitats over the course of their lifetimes (e.g., Queiroz, 1995a). Protecting areas, or indeed regulating any land use, must take into account the fact that many people live in or exploit this ecosystem, and governmental or regulatory presence is minimal. The success of any conservation effort in flooded forests therefore will depend on the active support of the local people. Accordingly, the more successful efforts have turned to community-based conservation—vesting local peoples in conservation efforts by allowing them access to the natural resources.

The traditional caboclo and ribeirinho communities depend for their subsistence on a mixture of floodplain agriculture and resource extraction from the surrounding forest and water (Padoch, 1988; Padoch & de Jong, 1992). These activities are generally compatible with maintaining forest cover. Floodplain agriculture by its very nature has an ephemeral impact on the forest. Low-lying sites are the most prone to flooding and thus do not long remain under agriculture. Agricultural sites on higher ground are less frequently flooded, but even here, fields are transformed to agroforestry stands and eventually to forest after some years. The extraction of products from the forest—timber, various nontimber products, and wildlife—generally maintains forest cover. However, when forest people live at significant densities, or when external commercial markets drive extraction rates, then these activities are frequently not sustainable. Natural resources can be locally extirpated, with concomitant declines in wild species diversity and household income. If the goal is the conservation of várzea, we must understand the impact of harvest on different resources and thus define the limits for extractivism.

The extraction of nontimber forest products has traditionally determined the cash economy of people living in várzea. The great rubber boom at the end of the nineteenth century and the beginning of the twentieth was perhaps the most dramatic example of this phenomenon, but the extraction of a succession of forest products—sarsaparilla, spotted cat skins, peccary hides, cacao, Brazil nuts, mahogany—has been an omnipresent characteristic of the region. Each has gone through a "boom and bust" cycle, defined in part by changes in external markets and in part by the exhaustion of the resource. As eloquently described by Browder (1992: 176): "The historical record of extractive economies in Amazonia consistently demonstrates that such economies tend to self-destruct. If the prices of extracted products rise (usually due to scarcity), further depletion from the wild may follow as rational extractors (or their bosses) seek quick profits, or as corporate groups elsewhere developed cultivated or synthetic substitutes that usually displace the product extracted from the wild."

Browder concludes: "I suspect that there has never been even one renewable tropical forest resource, once introduced into commerce, that has provided the basis for sustained economic activity without external subsidization."

Yet because extractivism offers perhaps the only mechanism to integrate use of the forest with its conservation, it has been promoted (e.g., Peters et al., 1989b; Allegretti, 1990). Can the extraction of forest products be done on a sustainable basis? If it can be, then and only then will extractivism be compatible with conservation. For resources to be used sustainably, the following conditions must apply (Robinson, 1993):

- Resource extraction must be constrained by the biological limits on productivity. If extraction rates exceed production rates, the resource will be exhausted.
- Losses in biodiversity associated with extraction must be acceptable. Species loss and environmental degradation are associated with resource extraction, and these losses must be acceptable to all interested parties (which include local peoples, national governments, and international groups).
- People with rights to extract resources from the forest must be limited. Certain people must be given rights to resources, others must be denied them. Otherwise, you have open access and the failure of users to conserve the resource.
- Income produced from the extractive activities must meet the needs and aspirations of the people. Otherwise, people will be forced to overexploit the resource base.

The projects outlined in the papers in this section seek to define these conditions for specific resources within the different flooded forests. Human endeavors in the Mamirauá Sustainable Development Reserve, as described in the chapters by Ayres et al. and Lima, flood for a larger portion of the year, on average, and residents rely disproportionately on aquatic resources, such as fish and caiman. In somewhat drier areas, such as the Pacaya-Samiria Reserve (see chapters by Durand and McCaffrey and by Bodmer), while fishing still dominates, the local economy depends on a higher proportion of nontimber products and wildlife harvesting. This is even more the case in *terra firme* areas such as the Tamshiyacu Tahuayo Communal Reserve (mentioned in

Bodmer's chapter). For the rest of this introduction, I will first briefly note what we know of the biological characteristics of some of these resources and their potential as a basis for sustainable extraction. Then I will outline some of the social and economic conditions that affect the sustainability of resource extraction in the várzea.

Wildlife is harvested for food, hides, and fur throughout the region. Extrapolating from available case studies, Redford and Robinson (1991) provide one estimate that some 19 million mammals, birds, and reptiles are consumed in rural communities throughout the Brazilian Amazon every year. Much of this is consumed directly by rural peoples. Some is sold in local markets as meat or enters the commercial hide trade. Padoch et al. (1985) estimate that in the Tamshiyacu area southeast of Iquitos, wildlife provides on average about 10% of the cash income of rural families and over 60% of the income from forest products. Bodmer's estimate, in this volume, of 11% of cash income from the same area accords with these numbers. Wildlife is less important in the more frequently flooded forests (e.g., Hilger, 1995). Historically, during times when spotted-cat fur and peccary hides were demanded by international markets, the average contribution of wildlife to cash income was certainly much higher. Yet wildlife harvesting is frequently not sustainable in neotropical forests (e.g., Robinson & Redford, 1994), with resulting declines in wildlife populations and local extirpations.

Fish and other aquatic species are significant resources to people throughout the várzea, providing the main source of animal protein consumed and income earned (Shrimpton & Giugliano, 1979). As an example, in Mamirauá, initial estimates are that the 672 family units within the main focal area of the project annually consume some 1800 metric tons of fish and sell an additional 500 metric tons to local markets (Queiroz, 1995b). Yet, despite the importance of aquatic resources to local economies, we still know little about the impact of fisheries on fish populations, although some stocks are showing signs of overexploitation (Goulding, 1980; Goulding et al., 1988; see also Queiroz, this volume). The additional impact on fish populations of large-scale mining and dam-building in the Amazon basin is even more incompletely understood (Bayley & Petrere, 1989).

While timber production contributes the largest percentage to the value derived from renewable natural resources in the Amazon, and the flooded forests contribute some 75% of this commercial wood production, the importance of timber production to the economies of local extractivists is relatively minor. Most timber extraction is highly capitalized, with companies tending to employ experienced outsiders (Schmink, 1988). There is some timber extraction practiced by local peoples that provides direct income. For instance, wood production by local extractors from the main focal area of Mamirauá annually exceeds 15,000 m^3 (Albernaz & Ayres, 1995, this volume). While the impact of local timber extraction on forest structure still needs to be understood, the system might be more sustainable than commercial operations, which have done little to further forest conservation (Fearnside, 1989).

Nonwood plant resources, especially latex, gums, fiber, and oil, have traditionally been important to local extractivists (Fearnside, 1989, 1992). The most famous product has been rubber, which at the present time is still the most valuable product extracted by local peoples from the forest. Allegretti (1990) estimates that in the Brazilian Amazon, 340,000 people still largely depend for their livelihood on the tapping of native rubber trees. Yet as the sole basis of a sustainable extractive economy, rubber is limited: Rubber tapping does not meet the economic needs and expectations of many, and the forest can be destroyed by extractivist communities to provide additional income (Browder, 1992).

Fruit from the forest is another nonwood product that increasingly has been promoted (e.g., Padoch et al., 1985). Many flooded forests are dominated by one or two tree species, and if the fruits are of economic importance, this concentration of resources makes extraction more feasible (Peters et al., 1989a). Açaí palms (*Euterpe oleracea*) and aguaje palms (*Mauritia flexuosa*) are noteworthy examples. In theory, fruit harvesting could have a relatively light impact on the forest, but, as Bodmer notes in his paper here, fruit harvesting is frequently destructive, involving the felling of trees, and present practices are generally not sustainable (Vasquez & Gentry, 1989).

The biological productivity of these different várzea resources limits the size of the human population that can engage in extractive activities. With increases in human population density extractive resource economies lose their sustainability. For instance, based on less productive terra firme sites where human communities depend on wildlife for meat, Robinson and Redford (1994) estimated that catchment areas of about 8 km^2 per person are necessary to support a human community while having a minimal impact on the biological community. Communities with access to less than 1 km^2 per person have depauperate faunas and low-density wildlife populations (e.g., Peres, 1993). These numbers can be compared to the population densities found in extractive reserves in more productive flooded forests where people depend more on aquatic resources (Ayres et al., this volume). Mamirauá, with an overall human population density of about 10 km^2 per family, still retains populations of wildlife at or near carrying capacity, yet when human populations have access to less than 1 km^2, wildlife densities are again very low (e.g., Martins, 1992). Many variables, of course, will affect the relationship between human population density and the integrity of the biological community, but these numbers are indicative of the range of variation.

What social and economic conditions need to be encouraged in order to increase the sustainability of resource extraction? One need is to promote social and governmental mechanisms that act to maintain the productivity of natural resources. While some local management traditionally characterizes caboclo and ribeirinho resource use (e.g., Padoch et al., 1985; Anderson, 1990), additional mechanisms need to be established as specific resources increasingly enter the market economy. Bodmer (this volume) identifies management programs being established for palms and game hunting in the areas around the Pacaya-Samira Reserve. The challenge is to help establish management that, beyond simply restricting the harvest of resources, provides incentives for people to increase resource productivity.

Another need is to diversify the number of forest resources being exploited. The core lesson from the history of extractivism in the Amazon is the vulnerability of local economies that depend on a single resource. The papers in this section identify additional resources that have the potential to enter the commercial market. The potential of agricultural intensification by rural peoples should also not be ignored (e.g., Subler & Uhl, 1990). As part of a mixed strategy of land use, this would in effect diversify the items contributing to the cash income. Defining who owns and can harvest resources remains a need in many areas. To the extent that extractivists have no secure land title, there will be little incentive to manage resources sustainably (Browder, 1990). Extractive reserves, a legal land category in Brazil, and indigenous reserves in Colombia and Peru offer a mechanism to achieve this end. Another example is provided in the papers by Ayres et al. and Lima, who describe the recently instituted categories of lake use in Mamirauá. Access to certain lakes termed "subsistence lakes" is restricted to local communities. Regulating open-access resource systems, such as frequently characterize fisheries, is important to avoid "scramble competition" and the resulting degradation of the resource base.

Extractivism as a conservation strategy remains a hypothesis. The historical record is not encouraging, yet extractivism still offers one of the few ways, in theory, to support local development while maintaining the integrity of the biological community. As a strategy, it has the highest potential at relatively low population densities in generally productive natural systems—a situation characteristic of the várzea of Amazonia.

Literature Cited

Albernaz, A. & J. M. Ayres. 1995. Extração seletiva de madeira na EEM. Pages 143–162 *in* Relatório Semestral #6. Projeto Mamirauá, Belém.

Alexander, B. 1994. People of the Amazon fight to save the flooded forest. Science 265: 606–607.

Allegretti, M. H. 1990. Extractive reserves: An alternative for reconciling development and environmental conservation in Amazonia. Pages 252–264 *in* A. B. Anderson, ed., Alternatives to deforestation: Steps toward sustainable use of the Amazon rain forest, Columbia University Press, New York.

Anderson, A. B. 1990. Extraction and forest management by rural inhabitants in the Amazon estuary. Pages 65–85 *in* A. B. Anderson, ed., Alternatives to deforestation: Steps toward sustainable use of the Amazon rain forest. Columbia University Press, New York.

Bayley, P. B. & M. Petrere. 1989. Amazon fisheries: Assessment methods, current status and management options. Pages 385–398 *in* D. P. Dodge, ed., Proceedings of the International Large River Symposium. Canadian Special Publications in Fisheries and Aquatic Science 106.

Browder, J. O. 1992. The limits of extractivism. BioScience 42: 174–181.

Fearnside, P. M. 1989. Extractive reserves in Brazilian Amazonia. BioScience 39: 387–393.

————. 1992. Reservas extrativistas: Uma estratégia de uso sustentado. Ciência Hoje: 14(3): 15–17.

Goodland, R. 1980. Environmental ranking of Amazonian development projects in Brazil. Environmental Conservation 7: 9–26.

Goulding, M. 1980. The fishes and the forest. University of California Press, Berkeley.

————, **M. Leal de Carvalho & E. G. Ferreira.** 1988. Rio Negro: Rich life in poor water. SPB Academic Publishing, The Netherlands.

Hilger, B. R. 1995. The use of fish and game resources among caboclos of the Estação Ecológica Mamirauá. Pages 259–284 *in* Relatório Semestral #6. Projeto Mamirauá, Belém.

Martins, E. 1992. A caça de subsistência de extrativistas na Amazônia: Sustentabilidade, biodiversidade e extinção de espécies. M.S. thesis, University of Brasília, Brazil.

Padoch, C. 1988. People of the floodplain and forest. Pages 127–140 *in* J. S. Denslow & C. Padoch, eds., People of the tropical rainforest. University of California Press, Berkeley.

———— **& W. de Jong.** 1992. Diversity and change in ribeirenho agriculture. Pages 158–174 *in* K. H. Redford & C. Padoch, eds., Conservation of neotropical forests: Working from traditional resource use. Columbia University Press, New York.

————, **J. Chota Inuma, W. de Jong & J. Unruh.** 1985. Amazonian agroforestry: A market-oriented system in Peru. Agroforestry Systems 3: 47–58.

Peres, C. A. 1993. Biodiversity conservation by native Amazonians: A pilot study in the Kaxinawá Indigenous Reserve of Rio Jordão, Acre, Brazil. Report to World Wildlife Fund, Washington, DC.

Peters, C. M., M. J. Balick, F. Kahn & A. B. Anderson. 1989a. Oligarchic forests of economic plants in Amazonia: Utilization and conservation of an important tropical resource. Conservation Biology 3: 341–349.

————, **A. H. Gentry & R. O. Mendelsohn.** 1989b. Valuation of an Amazonian rainforest. Nature 339: 656–657.

Queiroz, H. L. 1995a. Sub-projeto Pirarucu—Considerações gerais após mais um ano e meio de ativadades. Pages 505–522 *in* Relatório Semestral #6. Projeto Mamirauá, Belém.

————. 1995b. Alguns indices relevantes acerca de pesca, das pescarias e dos pescadores na EEM segundo dados disponiveis até o momento. Pages 560–612 *in* Relatório Semestral #6. Projeto Mamirauá, Belém.

Redford, K. H. & J. G. Robinson. 1991. Subsistence and commercial uses of wildlife in Latin America. Pages 6–23 *in* J. G. Robinson & K. H. Redford, eds., Neotropical wildlife use and conservation. University of Chicago Press, Chicago.

Robinson, J. G. 1993. The limits to caring: Sustainable living and the loss of biodiversity. Conservation Biology 7: 20–28.

———— **& K. H. Redford.** 1994. Community-based approaches to wildlife conservation in neotropical forests. Pages 300–319 *in* D. Western et al., eds., Natural connections: Perspectives in community-based conservation. Island Press, Washington, DC.

Schmink, M. 1988. Big business in the Amazon. Pages 163–171 *in* J. S. Denslow & C. Padoch, eds., People of the tropical rainforest. University of California Press, Berkeley.

Shrimpton, R. & R. Giugliano. 1979. Consumo de alimentos e aliguns nutrientes em Manaus, Amazonas, 1973–1974. Acta Amazonica 9: 117–141.

Subler, S. & C. Uhl. 1990. Japanese agroforestry in Amazonia: A case study in Tomé-Açu, Brazil. Pages 152–166 *in* A. B. Anderson, ed., Alternatives to deforestation: Steps toward sustainable use of the Amazon rain forest. Columbia University Press, New York.

Vasquez, R. & A. H. Gentry. 1989. Use and misuse of forest-harvested fruits in the Iquitos area. Conservation Biology 3: 350–361.

Mamirauá: The Conservation of Biodiversity in an Amazonian Flooded Forest

*J. Márcio Ayres, Ana Rita Alves, Helder Lima de Queiroz, Miriam Marmontel,
Edila Moura, Deborah de Magalhães Lima, Aline Azevedo, Marise Reis, Pedro
Santos, Ronis da Silveira, and Donald Masterson*

Introduction

There are few areas on this planet that remain seven to fifteen meters under water for six months of the year. The few flooded forests that still exist in tropical regions are located on the margins of several rivers in Southeast Asia and west Africa, and mostly along the rivers of Amazonia. It is estimated that the Amazon basin, mainly in Peru and Brazil, includes over 180,000 km² of flooded forests. While flooded savannas are most common in the lower part of the basin and in the Amazon River's estuary, forests predominate along the upper stretches of the main river. The flooded forests are locally known as either "várzea" or "igapó," according to their origin. Igapó areas are formed by blackwater rivers poor in nutrients (usually originating in the Amazon basin). The várzea areas are associated with the whitewater rivers that flow from the Andes and that, in contrast to blackwaters, carry large amounts of sediment.

The várzeas have considerable biological significance. The plant and animal species of the várzeas have to adapt themselves to the variations caused by the annual flood of several meters; consequently, there is high degree of endemism among species in várzea areas (Ayres, 1986, 1993). Due to the annual renewal of nutrients by the Andean rivers that bring sediments from the mountains, the várzea areas are also highly productive. Thus, these lands have become a focus of economic exploitation by the human population of the region. It is believed that at least 80–90% of the population in the Amazon basin lives near these flooded areas, mostly on the margins of the large rivers. Reports from the famous expedition that discovered the great river, led by the Spaniard Francisco Orellana some 450 years ago, described large Amerindian populations living and exploiting the várzea along almost the entire Amazon River (Carvajal in Medina, 1988). Today these areas are used for intensive fishing and logging and some seasonal agriculture. Until the 1980s there was no conservation unit located entirely in the Brazilian Amazon várzea.

The Mamirauá Reserve

The Estação Ecológica do Lago Mamirauá (Mamirauá Ecological Station, or MES) was created in 1990 by the government of the state of Amazonas, on the basis of work done by a team of sceintists headed by J. M. Ayres. The total area of the station is around 11,240 km², comprising the region between the Japurá, Solimões, and Auati-Paraná Rivers. In July 1996, the Amazonas government transformed the MES into the first Brazilian sustainable development reserve (SDR). This new category of conservation unit allows for the coexistence of a human population with the protection of local biodiversity.

Mamirauá is the only conservation unit in Brazil located entirely in the Amazonian flooded forests, or várzea. The Ecological Station of Anavilhanas and the Jaú National Park include areas of flooded forest; these, however, are flooded by blackwater rather than whitewater rivers. In the Peruvian Amazon, much, although not all, of the Pacaya-Samiria Reserve's 20,000+ km² is seasonally flooded.

The Mamirauá Sustainable Development Reserve (MSDR) várzea dates mainly from the Holocene period. Most of the area is very recent in origin, less than 10,000 years old (Klammer, 1984). The region between the Solimões and Japurá Rivers is characterized by hundreds of oxbow lakes, many of which originate from abandoned channels, streams classified as paranás and canos, small islands, restingas (levees) along the channels and lakes, and large swamps that are connected during floods. A recent inventory identified 499 lakes in the focal area of the reserve, which is 260,000 ha in size (Schuster, pers. comm.). The Mamirauá Reserve experiences an average 12-m difference between low water level (September/October) and high water level (May/June).

The várzea of the lower Japurá River region is composed of a great number of different environments, several of which are forests. The higher sections of the restingas are subject to annual floods lasting 2–4 months, during which they are covered by 1–2.5 m of water. These areas represent roughly 12% of all Mamirauá forests. They are structurally similar to upland (terra firme) forests. Their species composition, however, is very different. In the várzea, the higher restingas have the highest species diversity among arboreal environments. The lower restingas are a transition between forested and scrub areas of the várzea and represent about 50% of the land area. The low restingas remain under water for 4–6 months each year; in some spots the floodwaters may reach depths of 5 m. The restingas are located, in general, along channels and lakes (Ayres, 1993).

The habitat known as "chavascal" is a backswamp on lower-lying ground; it probably represents the largest portion of the Mamirauá Reserve's várzea. It consists of low, scrubby, swampy vegetation, almost impossible to cross. The chavascal remains under water for 6–8 months of the year, with a water level of 6–7 m. In the middle of this scrub vegetation there are some emergent trees and some islands of restingas. The chavascais are usually located behind the restingas.

Due to the annual floods of 11–12 m that cover all the várzea lands of the middle Solimões River and the lower Japurá River, the terrestrial fauna characteristic of the neighboring terra firme forests does not occur in the Mamirauá Reserve. For instance, among mammals we find only aquatic species (e.g., river dolphins, manatees, otters), good swimmers (e.g., jaguar), arboreal mammals (e.g., primates), and flying mammals (e.g., bats). Terrestrial mammals such as armadillos (*Dasypus*), agoutis (*Dasyprocta*), pacas (*Agoutis*), tapirs (*Tapirus*), and peccaries (*Tayassu*)—all common species in Amazonian upland habitats—are not found in the area. The várzea abounds instead in aquatic fauna (especially fish) that enter the forest with each flood.

Mamirauá is the habitat of many rare and endangered species of the Amazonian fauna, including manatees (*Trichechus inunguis*), black caimans (*Melanosuchus niger*), white uakari monkeys (*Cacajao calvus calvus*), blackish squirrel monkeys (*Saimiri vanzolinii*), and Amazonian turtles (*Podocnemis* spp.), all of which are officially listed as endangered species in Brazil.

Among the eight known species of primates that are found in the Mamirauá Reserve, only the white uakari monkey and the blackish squirrel monkey are endemic to the reserve. Though the latter type of monkey is the most abundant in numbers (approximately one individual per hectare of forest), it is found only in the extreme east of the reserve between the channel of the Jarauá River and the mouth of the Japurá River. On the other hand, white uakaris are found almost everywhere in the reserve, but they occur in lower densities (approximately one individual per 14 hectares of forest).

The black caiman is the largest predator in the Amazon; it can reach a length of more than 5 m. This type of caiman almost disappeared between the 1940s and the 1970s due to illegal hunting; its skin was highly prized in the international trade. The Mamirauá Reserve now harbors the largest known population of this type of caiman in the entire Amazon basin (Silveira & Thorbjarnarson, pers. comm.). This population, however, is still under hunting pressure, because its meat is marketed as pirarucu, a valuable fish, to buyers in the State of Pará (Brazil) and Colombia.

Among the five known species of turtles (quelônios) in the reserve, the giant Amazonian river turtle (tataruga, *Podocnemis expansa*) has been the most exploited over the last several centuries. Today it is much rarer than the tracajá (*Podocnemis unifilis*) and the iaçá (*Podocnemis sextuberculata*). In the middle Solimões River region, these two smaller species of the *Podocnemis* are also threatened due to their commercial value.

So far, approximately 290 fish species and 310 bird species have been identified in the focal area of the reserve. Many of the birds are aquatic and migratory. There are also several other species of reptiles, a great diversity of amphibians, and an abundance of different types of fish with high commercial value, including several, such as the tambaqui (*Colossoma macropomum*) and the pirarucu (*Arapaima gigas*), that are threatened by intensive fishing. In general, the lakes serve as a refuge for young tambaquis during the dry season, and the shallower lakes give shelter to the pirarucus so they can build their nests during the flood. The forested areas of the reserve supply food during the flood to the many different types of fish that are caught for food and sold in the neighboring towns (e.g., the matrinchã [*Brycon* sp.], several species of pacu [*Mylossoma* spp., *Myleus* sp., *Metynnis* sp.], pirapitinga [*Pyaractus bidens*], several types of sardine [*Triphorteus* spp.], and several types of aracu [*Leporinus* spp. and *Schizodon fasciatus*]).

The tree diversity of the Mamirauá várzea is superior to that of other várzea regions of the middle and lower Amazon River (Ayres, 1993). More than 250 species of trees with diameters larger than 10 cm have been identified so far. The populations of many of these species have been greatly reduced by selective logging. The samaúma (*Ceiba pentandra*) is an example. Despite these losses to logging, there are still substantial populations of açacu (*Hura crepitans*), muiratinga (*Maquira coriacea*), and ucuuba (*Virola sur-*

inamensis). These are all timbers that have been extensively extracted from the forest to supply the plywood industry of Manaus. There are also some valuable hardwoods, such as louro-inamuí (*Ocotea cymbarum*) and mulateiro (*Calycophyllum spruceanum*). Logging companies from towns neighboring the reserve have been extracting timbers from Mamirauá for several decades. The timber is used in house construction, boatbuilding, and furniture making. Recently, envira-vassourinha (*Xylopia frutescens*) has been in great demand by brickmakers in the nearby city of Tefé. Logging activities not only destroy trees but may also pose a major threat to the regional fauna. Many species, especially the arboreal animals and fish, depend on fruits or seeds of these trees for their nourishment.

The Mamirauá Project

The founding of the Mamirauá Ecological Station in 1990 brought many challenges. First, national conservation legislation had to be changed. The residents of the reserve needed to be assured of their right to stay and their right to use and market natural resources of the region in sustainable ways. The communities in the reserve have a historical right to own the lands they occupy. Second, there was a need to carry out an integrated program of interdisciplinary research that combined scientific and theoretical approaches with the needs of furthering nature conservation and as well as enhancing the well-being of the families that live in the várzea. Third and most important, there was a need to strengthen the participation of local communities in the project. This last consideration has always been regarded as absolutely fundamental to the viability of Mamirauá as a conservation unit. Along with the strengthening of communities' participation, the creation of a nongovernmental organization to manage project activities was deemed necessary. This special organization complements government activities, secures funding, helps in the development and maintenance of a specialized technical team, and guarantees continued long-term functioning of the structures developed during the reserve's implementation period.

The Wildlife Conservation Society and the CNPq (National Council of Research of Brazil) financed the preliminary studies that served as the basis for the proposal to create the reserve. In 1991, a project was devised and sent to national and international financial institutions, proposing the setting up of an experimental area of 260,000 hectares between the Aranapu, Japurá, and Solimões Rivers. This focal area is being used to implement pilot reserve management activities as well as ecological research; these activities will later be expanded to other areas. To date, approximately 80 researchers and extension agents have contributed their efforts to the Mamirauá Project. About 30 research institutions and funding agencies from Brazil and abroad have contributed financially. The most important objective of the project has been to devise and implement a management plan for the reserve. The project is divided into five complementary programs that cover administration and development, research, and extension of the project.

Parallel to the activities developed by the Mamirauá Project, in 1992 the Sociedade Civil Mamirauá (SCM) was created. According to its charter, the SCM's objective is to contribute to the conservation of the renewable natural resources, especially in the areas of flooded rain forests. The creation of this society is one of the alternatives that the project has devised to guarantee the implementation and maintenance of the reserve in the long run.

The implementation model of the Mamirauá Reserve is based on a broad definition and vision of nature conservation. It is informed by the most recent national and international discussions on the issue. It aims to avoid environmental sectarianism and acknowledges the importance of integrating conservation efforts with processes of social development (Lima-Ayres, 1994; see also Lima, this volume).

The Use of Natural Resources and Human Life in the Mamirauá Flooded Forests

As in other areas of rural Amazonia, there were neither census data, nor maps of the settlements in the Mamirauá region when the project began in 1991 (Lima-Ayres, 1992). Much field research went into collecting basic data on the socioeconomic reality of the region and mapping and identifying the communities that use the reserve. The results of this research, which made adequate planning for the project's extension activities possible, will constitute the baseline for future evaluation of the impact of the reserve's implementation. The combination of data obtained through in-depth synchronic sampling and through monitoring of changing economic activities and health conditions of the human population throughout the year allows an adequate understanding of the social reality, including the identification of the effects of seasonal variation. In the management plan, data on communities and their distribution were integrated with the results of biological research, to create a geographical information system for zoning of the reserve. In October 1991 a demographic census and a survey of economic activities was carried out. This research identified and quantified human population dynamics in the reserve and adjacent areas, as well as the communities' main sources of income and household survival strategies.

The settlements located within the reserve are all located in the várzea, though they vary somewhat in the elevation of the lands used for housing and resource-management activities. The people who live in these settlements have adapted to the flood season, when each year their land disappears under water for 4–6 months. Due to this environmental fluctuation, the várzea presents certain challenges to human occupation and generally reduces the lifetime of settlements and restricts the available methods of exploitation of natural resources. In general, very high floods are an important factor in encouraging emigration and, in general, limiting human settlement in the várzea. Geomorphological modifications caused by the floods—especially the formation of beaches and the collapse of banks—lead to the moving and extinction of most settlements. According to data obtained on the history of settlements within this area, the mean time that any community has remained in one place is 41 years.

Despite the importance of environmental change in determining the patterns of the human occupation of the várzea of the Mamirauá Reserve, the characteristics of the economic production and the social organization—land tenure, kinship, economic activities, politics, religion—are the factors that historically have defined the patterns of settlement (Lima-Ayres & Alencar, 1994). The main economic activities of Mamirauá Reserve users are agriculture, fishing, and logging, each adapted to seasonal changes in the várzea as well as to alterations in the composition of domestic groups that are the basic units of production and consumption. Manioc (cassava) is the main agricultural product and, along with fish, the main source of food for the local human population. Manioc is planted in the varzea as floodwaters recede and is harvested six months later, before the next inundation. Families regularly lose a significant part of their manioc crop to the unusually high or early floods. Other important agricultural products include many varieties of cultivated bananas and plantains.

For the human population of the reserve area, fish is by far the most important source of animal protein. Per capita fish consumption is very high, estimated at 500 g/day. This adds up to an annual consumption rate of approximately 240–300 tons of fish by the reserve's focal area population. This is equivalent to more than 12% of all the fish consumed in the city of Tefé (and the reserve's population is only about 5% of the size of Tefé's). During the dry season, when the water level is low and the pirarucu and migrant catfish (Siluriformes) season begins, fishing becomes even more important economically. Fish that are caught are dried and salted and sold to the regatões (local boats that trade goods in the Amazon) by the remote communities. Villages located closer to the urban areas may sell their dried and salted fish, or sometimes even fresh fish, directly to the market. The chance to sell fresh fish allows the marketing of some species, such as tambaquí and tucunaré (*Cichla* sp.), that are kept fresh inside small ice boxes, and some species, such as acari-bodó (*Pterygoplichthys* sp.), that are sold alive in the local markets.

Fish become more vulnerable to fishermen when they abandon the flooded forests during the ebb of the floodwaters. Many species migrate to the river during the dry season, while others stay in the lakes, where they are easily captured. When the flood begins, the fish return to the flooded forests and search for shelter and food. During this period, night fishing of tucunaré in shallow waters, with a facho (torch) and a trident spear, becomes more intense. When the forests are flooded, fish become more difficult to catch because they scatter. The human population's knowledge of the environment helps them considerably in catching some species of fish during the flood: Local fishermen know which fruits are important in some frugivorous fish species' diet, and they use these fruits as bait.

The pirarucu is a species of great economic importance within the reserve, and its catch is highly seasonal: More than 85% of the annual catch is taken between September and December. The size and weight of captured pirarucu are in the ranges 80–240 cm and 5–135 kg, respectively, the average being 134.5 cm and 13.4 kg. In 1993, 1994, and 1995, the catch of pirarucu in the reserve was estimated at 63, 115, and 70 tons, respectively. The frequent use of gill nets (introduced into the Amazon basin during the early 1970s) is leading to a decline in the sizes of caught fish and may eventually cause a production collapse (Queiroz & Sardinha, 1996; also see Queiroz, this volume, for an extensive discussion of this fishery).

Ornamental fish were once intensively harvested in the reserve area. In 1991, fish tanks used for keeping ornamental fish were first noticed in the reserve. But intensive fishing appears to have diminished the numbers of the sought-after acará-disco (*Symphysodon* sp.) in previous years. Today, the number of individuals of this species in the reserve is very much reduced. At present, commercial fishing of ornamental species is not done in the reserve (see Crampton, this volume).

Timber is cut at the end of the summer (low-water) season and is taken out and sold in the winter (high-water) time. In years with exceptionally high floods (e.g., 1993 and 1994) logging is especially intensive. In 1993 and 1994, for example, the extraction of timber reached approximately 20,000 m³. Extraction is highly selective; nine species account for 85% of all individual trees extracted. The soft white timbers that represent 80% of the total volume extracted are destined for plywood mills in Manaus, Itacoatiara, and Belém. The heavy woods and the firewood that make up a smaller portion of the volume extracted (though in terms of numbers of trees, they make up a high portion of the total) are locally consumed, mainly in Tefé. Approximately 330 villagers annually participate in logging in the focal area of the reserve. Those who are able to sell their product directly to buyers from Manaus and Itacoatiara earn about 50% more than those who sell to middlemen from Tefé (see Albernaz & Ayres, this volume).

The main animal species hunted are caimans. Caiman hunting takes place during the dry season, normally after the fishing of pirarucu—that is, between December and March. The intensively hunted species are the jacaré-açu (black caiman) and the jacaré-tinga (spectacled caiman, *Caiman crocodilus*). There is a third species, the jacaré-paguá (Cuvier's dwarf caiman, *Paleosuchus palpebrosus*), that occurs in lower densities and has no commercial value. Tons of caiman meat from the Mamirauá Reserve and periphery are sold each year. Caiman is sold as pirarucu meat in Pará (Brazil) and in Colombia, which makes this activity much more profitable, considering that, at the source of production, the price of pirarucu is much higher than the price of caiman.

Most of the marketed caiman meat is black caiman, the most seriously threatened species. Approximately 6500 kg of the caiman meat offered for sale in markets near the reserve between January and March of 1995 was from the Mamirauá Reserve. More than 50% of that meat was black caiman, 25% was spectacled caiman, and the rest was indistinguishable. The great majority of spectacled caiman that were hunted were adult males (in a proportion of 11 males per 1 female), while the proportion for black caiman was smaller (2 males per 1 female). This indicates that the hunting of the black caiman can more grievously affect the maintenance of this population. In a recent inventory of caiman offspring carried out at various points of the reserve, approximately 955 offspring of spectacled caiman and 14 offspring of black caiman were found. It is estimated that nearly 250 people in the reserve are caiman hunters.

The hunting of caiman is illegal in Brazil. It is, however, an important source of income for the riverine population of the middle Solimões River during certain times of the year. In the Mamirauá Reserve, caiman hunting is carried out very intensively in the side channels of the Aranapu and Panauá Rivers. Recently, some villagers have reported that Colombian merchants have offered the same prices for skins of large and medium-sized caimans. Consequently, caiman hunters have an incentive to intensively hunt and possibly exterminate populations of preadult, medium-sized caimans. On the other hand, in the meat business, the bigger the animal, the higher its value.

Subsistence hunting is less important in the várzea than in the neighboring terra firme forests, because fish contributes so much protein to the várzea diet. The communities' dependence on some wild animal resources, however, cannot be ignored. The manatee, for instance, is a large species that represents a good source of protein for local residents. Despite being protected by law since 1967, the manatee is still being hunted within the reserve. The meat is usually consumed locally and distributed among the members of community; all the parts are used, and often sausages are prepared. Small-scale marketing of manatee meat (fresh, salted, or preserved in its own fat in a form called "mixira") does occur in the bigger towns of the region, where it is highly valued. To be a manatee hunter requires a lot of patience and skill. Manatee hunters are few and ageing: Apparently the young are not interested in learning the capture techniques. This reluctance may help preserve and ensure the continuation of the Amazonian species (Marmontel, 1995).

Other species sought by subsistence hunters are red howler monkey (*Alouatta seniculus*), curaçao (*Crax globulosa* and *Mitu tuberosa*), and wild duck (*Cairina moschata*). Apart from these species, many others are caught opportunistically in the course of other activities such as fishing, farming, and logging. In three of the sampled communities in the focal area of the reserve, with 14 domestic groups in each village, 15 howlers, 8 curaçaos, and 2 wild ducks were hunted during one year's inventory (Santos, 1995). Of 27 families interviewed in the Mamirauá Reserve, only 3 (11%) had hunted in the week previous to the interview. However, 59% of households in the sampled communities possessed firearms. This contrasts with terra firme forest communities near the neighboring Amaná Lake, where 7 of 17 families (37%) had hunted in the previous week and 95% of the families had firearms (Ayres, 1990).

Game is occasionally sold in neighboring towns. An inventory done in the markets of Tefé and Alvarães, the two most important towns of the region, over a period of approximately 10 months found 42 curassows, 6 ducks, 5 howlers, and 1 manatee offered for sale. The young of many species of wild animals are captured to become pets, and these are also occasionally sold. Parrots (*Amazona aestiva*) and parakeets (*Brotogeris versicolorus* and *B. sanctithomae*) could become especially important trade items in the future (Santos, 1995).

The variety of economic activities described above is the basic source of both subsistence and market goods for the domestic groups in the reserve. The average annual income of a family is about US$907, representing an annual per capita income around $130; a figure much lower than the average national income. Due to the difficulty of reaching markets and the fact that the greater part of the diet is provided directly by the household, these monetary values should not be considered accurate measures of the well-being of reserve families. However, per capita income is useful as a baseline indicator for monitoring future changes in markets and commercialization of natural resources taken from the reserve area.

Data collected on total expenses incurred in one year by sampled households show that basic supplies needed to maintain the family account for most outflow: Almost 70% of income goes for purchase of food, and another 25% goes for buying tools and maintaining equipment. Only a very small remainder is used to acquire any items of enduring value (Lima-Ayres, 1992).

According to the most recent census, the population of the focal area of the reserve is 1668 people (295 families) distributed along the banks of the Japurá and Solimões Rivers in 23 small communities. The surrounding areas of the reserve have an additional population of reserve users of 3600 people (576 families) distributed in 37 communities. Thus the total population directly dependent on the resources of the focal area of the Mamiraua reserve is nearly 5300 people (Mamiraua, 1996). All the activities reported above generate at least $4.5 million per year, of which $2.05 million is fish production for subsistence purposes. Two species of fish, the pirarucu (*Arapaima gigas*) and the tambaqui (*Colossoma macropomum*), account for nearly 25% of the total income. This information, based on an average of three years' income, indicates that each hectare of the floodplain generates by "sustainable use of the resources" at least $17 per year (Mamiraua, 1996).

If all the resources taken from the reserve were to be used solely for the benefit of reserve residents, these resources would generate an income of at least $15,250 per family (with 6–7 persons per family). This value would be much greater than the average annual income of most families in the Amazon region. However, invasions by fishing boats from larger towns, timber exploitation by logging companies, and other forms of resource exploitation by nonresidents remove much of the income from the area, reducing the average annual income to much lower levels. Despite the fact that the residents of the reserve are unaware of these economic details, they are perfectly aware of the loss caused by outsiders' activities.

Since fish are the reserve's most important economic resource for the residents of the area (representing 78% of their annual income), the creation of the reserve in 1990 was widely accepted by local people but rejected by those nonresidents who also depend on the resources of the area (Mamiraua, 1996; see also Lima, this volume). The major problem faced by the residents of the reserve before its creation was the invasion of the lakes that had been set aside for protection and sustainable use. The idea of protecting lakes as resources in Amazonia doubtless dates from man's arrival in the flooded forests thousands of years ago. The idea of categorizing lakes for different uses, however, sprang from the activities of Tefé's Catholic Church working in cooperation with the communities of the reserve in the late 1980s (see also Lima, this volume). Today the practice of using some lakes for sustainable production and others for conservation has been adopted by almost all várzea communities along the Amazon River. The communities' right to guard these lakes gained legitimacy and became the major factor for communities' support of the creation of the reserve.

Project Results and Challenges

The forms of community participation in the Mamiraua Project were defined by the community members on the basis of their own previous experiences with Catholic Church–linked development efforts (see also Lima, this volume). The model they adopted specified that (1) neighboring communities constitute organized clusters that meet every two months, with each community sending two representatives to the meetings; (2) nine clusters of communities cover the entire focal zone; (3) each cluster elects a coordinator, who is responsible for organizing meetings; and (4) representatives from all clusters meet in annual general assemblies.

To date, five general assemblies have been held. During these assemblies, two new categories of commercial lakes were created: those for use by the communities and those for use by fishermen from the nearby municipalities of Tefé and Alvarães. These two new categories complement the original categories, of protected and sustainable-use lakes (Ayres et al., 1995; Reis, 1993).

Today the local population is primarily responsible for monitoring compliance with regulations. Whenever a protected or sustainable-use lake is invaded by an outsider, a group of members from that village or cluster get together and advise the interloper to leave the area. If he refuses to leave the area, the leader of the group radios project headquarters in Tefé, where functionaries of IBAMA (the Brazilian environmental agency) are briefed and two or three authorized guards are sent to the area of conflict. This practice has helped in banning from the area the large fishing boats that come from Manaus, Manacapuru, and Itacoatiara (three of the largest cities in Amazonas). In addition, there has been a substantial reduction—by at least two-thirds—in fish taken for sale in Tefé. In terms of the portion it represents of the total supply of fish in the Tefé market, the catch taken from the reserve dropped from 20% in 1992 to 6.9% in 1995 (Barthem, 1996). The practice of designating protected and sustainable-use lakes has also increased the biomass of some important commercial species of fish (Costa et al., 1996).

Although logging represents only a small fraction of the annual income of reserve residents, it is an activity that requires attention because of its possible widespread ecological consequences to the ecosystem. Many species of fish, invertebrates, and arboreal vertebrates depend on fruits, seeds, and leaves as important dietary items. Because of the potential severe impact of timber extraction, the Mamirauá Project is trying to promote conservation measures similar to those used to manage the fish stocks of the reserve. The logging situation, however, is more difficult than fishing. First, the results of tree conservation are not easily detected over a short period of time as is the case with fishing. Second, this activity is very profitable to the few individuals who control the logging industry.

Selective logging was discussed with the communities' leaders during the second general assembly in July 1993. It was decided then that no logging was to take place on the restingas along the banks of protected lakes, but that it was to be permitted on the sustainable-use lakes' restingas, for community use only. Commercial logging was permitted on the restingas of lakes that had already been designated for commercial use. In the third general assembly, in July 1994, permissible minimum diameters for logging different types of trees in the reserve were discussed. While the federal law sets a minimum of 45 cm diameter for logging of any type of tree, in the Mamirauá Reserve communities decided to vary the minimum diameter for logging according to the species.

In 1994, no logging took place in the areas of the protected lakes, with the exception of some communities' territories where the use of lakes was in dispute. The trees that were extracted in 1994 had larger diameters than those extracted in 1993. This rise in size occurred despite the communities not having reached a consensus on minimum diameters for different species.

Caiman, manatee, and turtle management issues were also introduced to reserve users during the fourth general assembly, where residents were invited to comment on and suggest recommendations in addition to those offered by the project staff. One of the communities recently decided to cease hunting caiman. Their reason was that since caiman hunting is, according to federal law, an illegal activity, they fear that outsiders may use this argument against them in order to gain access to fish in their area.

In addition to community participation, environmental education, health and nutrition, socioeconomic research, and monitoring (Albernaz, 1994; Maranhão & Lins, 1993; Maranhão & Silva, 1994; Moura & Lima-Ayres, 1994), a number of studies are being conducted by the project, including such topics as the exploitation of natural resources, ecology, and behavior of indicator species or economically important species such as dolphins, manatees, white uakaris, red howlers, and fish including pirarucu, tambaqui, and ornamental fish. (Some of the results are reported in this volume in articles by Queiroz and Crampton.) Other studies that have been done include vegetation surveys and research on annual patterns of plant primary production so that more is known about the seasonal distribution of herbivorous foods and the long-term sustainability of some logging practices. These key multidisciplinary studies will lead us to a better understanding of the flooded-forest habitat and will certainly help us make decisions concerning effective management of this important ecosystem.

After four years of scientific research and participatory community extension work, the Mamirauá Project produced a management plan based on the cumulative results. Further conservation measures are being discussed and negotiated with the residents and users of the reserve. The project proposes reserving two large areas as strict protection zones in the interior of the focal area. Additional critical habitat areas for the management of manatees, turtles, tambaqui, and pirarucu, as well as bird nesting sites, have also been identified within the sustainable-use areas. Outside the core protected zones, local villagers will continue to manage their lakes and the surrounding vegetation on the basis of principles employed over the past three years, i.e., by employing distinct management regimes for three distinct classes: preservation lakes, maintenance lakes, and commercialization lakes.

The results that the project and the local communities realized during these first four years led the government of Amazonas to create in July 1996 a new category of conservation unit for the Brazilian Amazon, a Sustainable Development Reserve. This new category envisions active community participation in biodiversity conservation and sustainable resource management (see also Lima, this volume). Although high human population densities are incompatible with biodiversity conservation in this ecosystem, the sustainable development reserve model seems to apply better to the Amazon flooded forests than do other conservation models. It may not be an appropriate model for protected areas in other habitats of the region, but so far it has proved to be effective under local conditions. It is important to create site- and situation-appropriate models if we wish to increase the amount of less-disturbed habitats in the region. Increasing local community participation in the conservation process will help to enlarge protected areas and as a result will help to maintain the biodiversity and its ecological and evolutionary processes.

Since the completion of the management plan, the project extension agents and researchers have been involved in discussing it with all the communities in the focal area of the reserve. These discussions, begun in the second half of 1996, raised many issues that will be quite important in directing phase two of the Mamirauá Project. It is clear, for example, that alternative economic activities will have to be implemented in order to compensate for economic effects of some of the management rules of the reserve. Among the economic activities planned for the second phase is the establishment of small ice factories for fish processing in some communities, the start of ecotourism at the eastern edge of the reserve, and the development of new agricultural activities. In addition to these economic alternatives, other priorities include the enhancement of health and sanitation and increased environmental education activities.

Today the Mamirauá Reserve has better-developed infrastructure than many other conservation units in the Brazilian Amazon. It is still insufficient, however, to deal with the need for continued research and surveillance. The project owns one stationary house and six floating houses located at different strategic surveillance points of the reserve. Each of these units is equipped with a solar energy system that provides energy for six light bulbs, one water pump, and one shortwave radio used to communicate within the reserve, with project bases in Tefé and Belém, and, on a specific frequency, with the IBAMA office in Tefé. Moreover, the reserve has about 15 aluminum motor boats and 5 on-board diesel-engine wooden boats (the motors vary from 22 to 124 horsepower). These boats are used for research, extension services, supplying the floating houses, and transporting community participants. There is also a boat for the exclusive use of residents and users of the reserve, so they can organize community and cluster meetings as well as carry out surveillance of the lakes. The base in Tefé has a small library, a computer room, a studio for the production of two weekly radio programs and videos for environmental education, lodging for researchers, an administrative section, and a warehouse to store equipment and supplies.

The founding of the Sociedade Civil Mamirauá (SCM) created a mixed system of management that joined the active participation of government agencies (CNPq, INPA, Fundo Nacional do Meio Ambiente, IPAAM/State of Amazonas and Museu Goeldi) with the more flexible operation of nongovernment agencies. The SCM allows foreign financial aid (currently from the Overseas Development Administration–UK, European Union, Wildlife Conservation Society, and World Wide Fund for Nature) to give the reserve's activities a guarantee of continuation without risk of short-term interruption. The foreign aid does not, however, guarantee the maintenance of the reserve and the project in the long run: Thus the need for an arrangement with the federal and state governments is important.

One of the SCM's objectives is to find new ways of obtaining funds, including the creation of an endowment to help continue research and extension activities in the Mamirauá area. The marketing of books, workbooks, and postcards is already underway. An agreement between the clothing manufacturer Aqualung and the SCM was reached to generate funds in the medium and long run. Other fundraising activities are in their initial phases of implementation; these include ecotourism and the establishment of an endowment for the Mamirauá activities.

One of the most important steps to guaranteeing the sustainability of the reserve in the long run was taken by CNPq/MCT. At the end of 1994, the CNPq bought about 13 hectares of land along the banks of Lake Tefé on the outskirts of the city. On that location the CNPq will build a research center (the Mamirauá Institute for Sustainable Development) geared toward the study and management of the flooded forests of the Mamirauá Reserve.

The Pilot Program to Conserve the Brazilian Rain Forest (a PPG-7/Ministry of the Environment program) is planning to establish a rain forest corridor in central Amazonia that links the conservation units of Jutaí-Solimões, Juami-Japurá, Mamirauá, Jaú, Anavilhanas, Uatumã, and several other reserves in the state of Amazonas. In addition, the state of Amazonas will create the Amanã reserve linking the Jaú National Park to Mamirauá, thus forming the largest continuous block (ca. 8 million ha) of tropical rain forest protected by law. The Mamirauá experience will be an important tool for the establishment of this unique conservation experiment in Brazil.

Acknowledgments

We thank all the supporters of the Mamirauá Project and of the Mamirauá Civil Society, especially the National Council of Scientific and Technological Development (CNPq/MCT), the Overseas Development Administration–UK, the Wildlife Conservation Society, the World Wide Fund for Nature, the European Union, the Institute for Environmental Protection of Amazonas State (IPAAM), the Aqualung Institute, the Brazilian Institute of Amazonian Research (INPA), the Goeldi Museum, and the Federal University of Pará. This paper is a revised version of a document presented at the Ramsar Convention held in Brisbane, Australia, in 1996.

Literature Cited

Albernaz, R. K. M. 1994. Relatório da pesquisa sócio-epidemiológica realizada com demanda espontânea de 20 comunidades da EEM. Pages 234–236 *in* Relatório Semestral no. 4, Setembro 1993–Outubro 1994. Projeto Mamirauá, internal document.

Ayres, J. M. 1986. Uakaris and the Amazonian flooded forest. Ph.D. dissertation, University of Cambridge.

———. 1990. Relatório anual de atividades do Wildlife Conservation Society. Wildlife Conservation Society, internal document.

———. 1993. As matas de várzea do Mamirauá. CNPq e Sociedade Civil Mamirauá, Rio de Janeiro.

———, **E. A. F. Moura & D. Lima-Ayres.** 1995. Estação Ecológica Mamirauá: O desafio de preservar a várzea amazônica. Pages 35–52 *in* Trópico em desenvolvimento: Alternativas contra a pobreza e a destruição ambiental no trópico úmido. Série POEMA. Universidade Federal do Pará, Belém.

Barthem, R. B. 1996. A pesca na região do médio Solimões e região de Tefé. Unpublished report for Mamirauá's Management Plan.

Costa, L., R. B. Barthem & M. V. C. Correa. 1996. Estudo de crescimento, reprodução e migração de tambaquis. Unpublished report for Mamirauá's Management Plan.

Klammer, G. 1984. The relief of extra-Andean Amazon basin. Pages 47–83 *in* H. Sioli, ed., The Amazon: Limnology and landscape ecology of mighty tropical river and its basin W. Junk, Dordrecht.

Lima-Ayres, D. 1992. The social category *caboclo* — History, social organization, identity and outsider's social classification of the rural population of a Amazonian region (the middle Solimões). Ph.D. dissertation, University of Cambridge.

———. 1993. Diagnóstico estatístico preliminar e sumário etnográfico da população da Estação Ecológica Mamirauá e áreas rurais adjacentes. Pages 491–500 *in* Relatório Semestral no. 3, Outubro 1992–Março 1993. Projeto Mamirauá, internal document.

———. 1994. A implantação de uma unidade de conservação em área de várzea: A experiência de Mamirauá.

Pages 403–409 *in* M. A. D'Incao & I. M. Silveira, organizers, Amazônia e a crise da modernização. Museu Paraense Emílio Goeldi, Belém.

———— **& E. Alencar.** 1994. Histórico da ocupação humana e mobilidade geográfica de assentamentos na área da Estação Ecológica Mamirauá. Pages 353–384 *in* Anais do IX Encontro Nacional de Estudos Populacionais. Vol. 2. Associação Brasileira de Estudos Populacionais, Caxambu, Minas Gerais.

Maranhão, S. & D. A. Silva. 1994. Boletim informativo sobre o trabalho de saúde. Ano I Número I. Pages 361–366 *in* Relatório Semestral no. 4, Setembro 1993–Outubro 1994. Projeto Mamirauá, internal document.

———— **& A. Lins.** 1993. Relatório do inquérito sócio-epidemiológico. Pages 378–397 *in* Relatório Semestral no. 3, Outubro 1992–Março 1993. Projeto Mamirauá, internal document.

Mamirauá. 1996. Plano de manejo de Mamirauá. Sociedade Civil Mamirauá (SCM) and Conselho Nacional para o Desenvolvimento Científico e Tecnológico (CNPq/MCT), eds.

Marmontel, M. 1995. O peixe-boi na Estação Ecológica Mamirauá. *In* II Congresso Internacional sobre Manejo de Fauna Silvestre en la Amazonia (resumenes). Iquitos, Peru.

Medina, J. T. 1988. The discovery of the Amazon according to the account of Friar Gaspar de Carvajal, and other documents. Dover Publications, New York.

Moura, E. & D. Lima-Ayres. 1994. Fecundidade e mortalidade infantil da população de usuários da Estação Ecológica Mamirauá. Relatório Semestral #4, Setembro 1993–Outubro 1994. Projeto Mamirauá, internal document.

Queiroz, H. L. & A. D. Sardinha. 1996. Plano de manejo dos pirarucus (*Arapaima gigas*) na Estação Ecológica Mamirauá. Unpublished manuscript for Mamirauá's Management Plan.

Reis, M. B. 1993. Relatório de atividades semestral, sub-programa de participação comunitária. Pages 319–333 *in* Relatório Semestral no. 3, Outubro 1992–Março 1993. Projeto Mamirauá, internal document.

Santos, P. 1995. Relatório de atividades semestral, sub-programa de sistemas terrestres: Caça de subsistência. Relatório Semestral 6, Abril 1995–Setembro 1995. Projeto Mamirauá, internal document.

Game Animals, Palms, and People of the Flooded Forests: Management Considerations for the Pacaya-Samiria National Reserve, Peru

Richard E. Bodmer, Pablo E. Puertas, Juan E. Garcia, Doris R. Dias, and Cesar Reyes

Introduction

Far from being homogeneous, the Amazon basin encompasses a variety of forest eco-systems. A relatively small but significant part of these ecosystems are the flooded forests of the upper Amazon basin. This unique ecosystem has an abundance of endemic wildlife and is fundamental to the ecology of the Amazon basin (Ayres, 1993). In addition, flooded forests constitute an extremely important resource for local human populations (Padoch, 1988). However, these forests are among the most vulnerable in Amazonia, because of their accessibility for resource extraction by fluvial transportation. Given the relatively small size of this ecosystem, it is undoubtedly the most threatened by unmanaged development.

Flooded forests of Amazonia may be conserved if biodiversity conservation and the needs of local people are joined through the efforts of natural-resource management. This, in turn, may avert degradation by unmanaged development, because forest resources can often be managed in accordance with biodiversity conservation. Natural-resource management can only link biodiversity conservation and the needs of local people if resources are not overexploited and if the economic, social, and political aspirations of local people are included in management programs (Levin, 1993; Ehrlich & Daily, 1993; Holling, 1993). Therefore, natural-resource management requires adequate information not only about species populations and ecosystems, but also about the people who most frequently use the resources (Browder, 1992).

Because of the ecological and economic situation in the flooded forests of the Pacaya-Samiria National Reserve, management programs must integrate game animals, palm trees, and rural communities. In lowland forests of the Peruvian Amazon, local people hunt large mammals for food and to sell (Redford & Robinson, 1991; Bodmer et al., 1994a; Alvard, 1995). These game species rely on palm fruits of the floodplain for food. However, people who live in these forests also use palm fruits (Padoch, 1987; Kahn, 1988). Thus, game animals, palms, and people are intricately linked. In this paper we examine how people use game animals of the flooded forests and how palms are connected ecologically to game animals and people, and we discuss how managing these resources must consider this interdependence.

Study Site

Studies on game animals, people, and palms were conducted within the Pacaya-Samiria National Reserve in the vicinity of Maipuco, Nueva Esperanza and San Antonio (Fig. 1). These three villages are situated on the south side of the Marañon River. People residing in these villages use the reserve to extract natural resources.

Maipuco is the district capital of Urarina, and many of its inhabitants have strong ties to the city of Iquitos. Thus, unlike many rural villages, Maipuco can be considered sort of an "urban extension." Nueva Esperanza is a more typical *ribereño* village inhabited by detribalized Amazonians. San Antonio, on the other hand, has declared itself a Cocama/Cocamilla Indian village. The three villages of Maipuco, Nueva Esperanza, and San Antonio make up a continuum of social organizations, proceeding from an urbanized setting to detribalized people to an indigenous community. These social organizations are typical of other villages found throughout the Pacaya-Samiria National Reserve. Maipuco has approximately 300 inhabitants; Nueva Esperanza, 460; and San Antonio, 300.

The forests surrounding the three villages are dominated by *várzea* (flooded forest) habitats with substantial *Mauritia flexuosa* swamps and *restingas* (floodplain levees). Várzea forests around the study site have experienced long- and short-term modifications as a result of sedimentation and erosion, and are rich in plant species and biomass. Changes in water level throughout the year are the most important causes of seasonal variation, and they affect cycles of local fauna and flora as well as activities of local people.

Data Collection

Information on hunting pressure was obtained by involving hunters in the process of data collection. This participatory method has been used successfully in other parts of the Peruvian Amazon and is a vital part of community-based wildlife management programs. The method specifically asks hunters to become involved in the research by having them collect skulls of animals they hunt. Hunters were informed that these skulls would become part of the collection of the Zoology Museum at the Universidad Nacional de la Amazonia Peruana.

Hunting pressure was determined by recording the number of animals hunted in each area from skulls collected by hunters. An error of 25% was added to account for animals that were hunted but whose skulls were not collected. The error was determined through informal interviews with local residents.

Hunting Pressure

During the year of August 1994 through July 1995, approximately 580 game mammals were hunted by residents of Maipuco, Nueva Esperanza, and San Antonio. The most

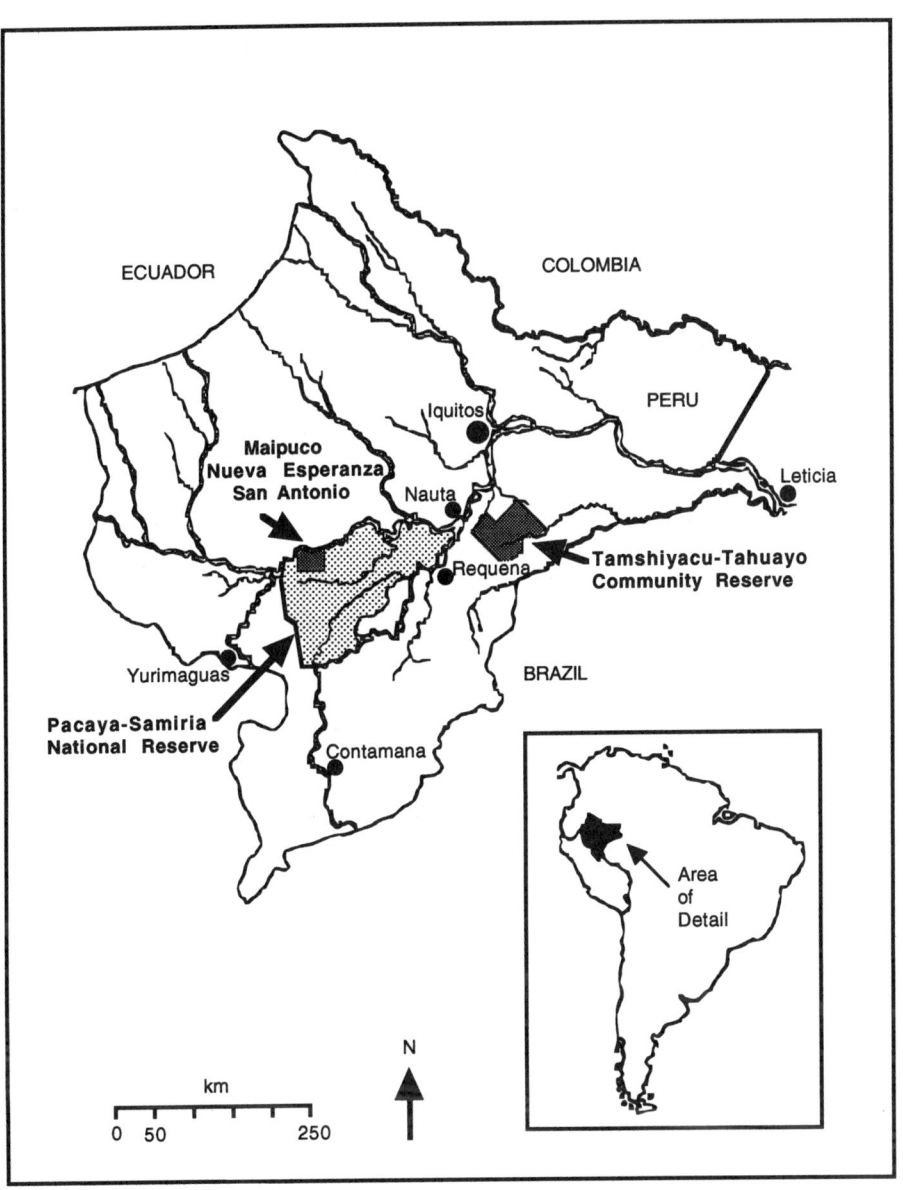

Figure 1. Map of northeastern Peru showing the study sites of Maipuco, Nueva Esperanza, and San Antonio within the Pacaya-Samiria National Reserve, and the Tamshiyacu-Tahuayo Community Reserve.

frequently hunted animal was the paca (*Agouti paca*), which made up 22% of the harvest, followed by white-lipped peccary (*Tayassu pecari*, 21%), agouti (*Dasyprocta fuliginosa*, 13%), brown capuchin monkeys (*Cebus apella*, 9%), howler monkeys (*Alouata seniculus*, 5%), kinkajou (*Potus flavus*, 5%), nine-banded armadillo (*Dasypus novencinctus*, 4%), and lowland tapir (*Tapirus terrestris*, 3%) (Table I).

In terms of the number of individuals harvested, large rodents made up 38%; ungulates, 28%; primates, 20%; edentates and marsupials, 7%; and carnivores, 6% (Table II). However, in terms of biomass extracted, which is a measure of the quantity of meat harvested, ungulates dominated, with 8255 kg of ungulate biomass extracted during one year, making up 73% of the total mammalian biomass extracted. Rodents made up 16% of the biomass extracted; primates, 5%; carnivores, 2%; and edentates and marsupials, 1.4%.

Hunters harvested the largest quantity of meat from lowland tapir, with 4000 kg of lowland tapir biomass extracted during one year, making up 48% of the ungulate biomass extracted and 35% of the total mammalian biomass extracted. The next most important game animal, in terms of meat harvested, was another ungulate, the white-lipped peccary, with 3630 kg of white-lipped peccary biomass extracted, making up 44% of the ungulate biomass extracted and 32% of the mammalian biomass extracted. The paca was the next most important animal hunted in terms of meat harvested, with 1170 kg of paca biomass extracted during the year, making up 66% of the large rodent biomass extracted and 10% of the mammalian biomass extracted.

Intervillage Variation in Game Harvests

Social differences between the villages appear to affect game harvests. As mentioned previously, the villages can be broadly categorized into an indigenous declared village, a ribereño village, and a more urban-connected village. The harvests in the indigenous declared village (San Antonio) and the ribereño village (Nueva Esperanza) were quite similar: 268 and 209 game mammals, respectively, during the year (ANOVA, N.S.).

Per capita, the indigenous declared village had the greatest hunting pressure of the three villages, with 0.68 mammals harvested annually per resident, or about 17 kg of game meat per person per year. The ribereño village hunted less per capita, with 0.34 mammals harvested annually per resident, or about 9 kg of game meat per person per year. The urban-connected village of Maipuco had a lower hunting pressure than the other two villages (ANOVA, F-ratio = 5.2, df = 1, $P = 0.03$), with 103 game mammals hunted by the residents of Maipuco during the year. This made up an annual per capita harvest of 0.26 mammals per resident in Maipuco, or about 7 kg of game meat per person per year.

The differences of game harvests between villages are also reflected in the number of hunters in each village: San Antonio had 27 hunters, Nueva Esperanza had 20, and Maipuco had 10.

Table I. Number of mammals hunted and biomass extracted from the vicinity of Maipuco, Nueva Esperanza, and San Antonio, August 1994 to August 1995

Latin name	Common name	Number hunted	Biomass extracted	Mean body weight (kg)
Artiodactyls				
Tayassu pecari	White-lipped peccary	121	3630	30
Tayassu tajacu	Collared peccary	7	175	25
Mazama americana	Red brocket deer	15	450	30
Perissodactyls				
Tapirus terrestris	Lowland tapir	20	4000	200
Primates				
Saguinus fuscicollis	Saddleback tamarin	3	1.2	0.4
Samiri spp.	Common squirrel monkey	4	3.2	0.8
Cebus albifrons	White-fronted capuchin	13	39	3
Cebus apella	Brown capuchin	53	185.5	3.5
Aotus nancymae	Night monkey	1	0.8	0.8
Alouatta seniculus	Red howler monkey	27	229.5	8.5
Lagothrix lagothricha	Common woolly monkey	11	121	11
Ateles paniscus	Black spider monkey	1	8	8
Pithecia monachus	Monk saki monkey	5	10	2
Callicebus cupreus	Titi monkey	1	1.2	1.2
Rodents				
Hydrochoerus hydrochaeris	Capybara	13	390	30
Agouti paca	Paca	130	1170	9
Dasyprocta fuliginosa	Black agouti	75	300	4
Sciurus spp.	Squirrels	3	2.1	0.7
Marsupials & edentates				
Didelphidae	Opossums	11	11	1
Dasypus novemcinctus	Nine-banded armadillo	23	115	5
Tamandua tetradactyla	Collared anteater	3	18	6
Bradypus variegatus	Three-toed sloth	4	16	4
Carnivores				
Potos flavus	Kinkajou	28	70	2.5
Felis yagouaroundi	Jaguarundi	1	6	6
Felis pardalis	Ocelot	1	10	10
Felis concolor	Puma	1	60	60
Panthera onca	Jaguar	4	320	80
Total		579	11,342.5	

Note: Values have been corrected by 25%.

Table II. Number of mammalian groups hunted and biomass extracted from the vicinity of Maipuco, Nueva Esperanza, and San Antonio, June 1994 to May 1995

	Number hunted	Biomass extracted
Ungulates	163	8255
Primates	119	599.4
Rodents	221	1862.1
Edentates & marsupials	41	160
Carnivores	36	466

Note: Values have been corrected by 25%.

Comparisons of Game Harvests between Flooded and Upland Forests

Flooded forests differ considerably from upland, or *altura* (terra firme), forests in Amazonia. Upland forests usually sustain greater levels of game harvests because these are the favored habitats of large-bodied mammals. Thus, hunting is common and productive in nonflooded habitats. In addition, upland forests are not productive fisheries, because they usually have small streams with swift currents that are less productive than the water bodies of flooded forests (Bayley & Petrere, 1989).

In contrast, large game animals are often scarce in flooded habitats, because they are unable to withstand long periods of inundation. Since hunters' preferences are for large-bodied game (Bodmer, 1995b), it is less profitable for people to hunt in seasonally flooded habitats than in upland habitats. However, flooded forests support most of the Amazonian fisheries and yield the greatest fish harvests of the upper Amazon (Bayley & Petrere, 1989).

Game hunting in flooded forests and in upland forests was examined further by comparing harvests in flooded forests of the Pacaya-Samiria National Reserve with studies of harvests in the nearby upland forests of the Tamshiyacu-Tahuayo Community Reserve. This comparison was done because differences between flooded and upland forests are so important for Amazonia. Methods used in both studies are the same, making a comparison of results possible.

Hunting pressure in upland forests was considerably greater than that in flooded forests. During a one-year period, a community residing in the vicinity of the Tamshiyacu-Tahuayo Community Reserve with 34 hunting households and approximately 50 hunters harvested 1267 game mammals in a 500 km² area of upland forests. This harvest is much greater than the 580 game mammals taken by 57 hunters in a similar-sized flooded forest in the vicinity of Maipuco, Nueva Esperanza, and San Antonio. This difference in game harvests probably reflects lower abundances of large game species in flooded forests. Interestingly, hunters did not show seasonal differences in either the flooded forests or the upland forests (ANOVA, N.S. in both cases).

The composition of game harvests differed between flooded and upland forests

(Pearson chi-square = 175, df = 13, $P < 0.001$) (Table III). In flooded forests large rodents were most frequently hunted, whereas in upland forests ungulates were most frequently hunted, followed by primates and then large rodents. This apparently reflects the lower number of ungulates in flooded forests.

The ungulate harvest differed between flooded and upland forests because of differences in the adaptations of ungulates to seasonal flooding. In upland forests the two species of peccary were harvested in similar proportions, whereas in flooded forests white-lipped peccary were hunted in much greater numbers than collared peccary (*Tayassu tajacu*). Likewise, tapir made up only 8% of the ungulate harvest in upland forests, whereas in flooded forests tapir made up 12% of the ungulate harvest.

White-lipped peccary and lowland tapir are hunted in greater numbers in flooded forests than are other ungulate species, because white-lipped peccary and lowland tapir are the ungulates least affected by inundation. Tapirs are semiaquatic and can stay submerged for long periods of time, commonly using water as an escape route from predators and hunters. The semiaquatic tapir can feed in waterlogged forests and move readily between floodplain levees. Tapirs are also less vulnerable to changes in fruit availability, because they consume large proportions of woody browse (Bodmer et al., 1994b).

White-lipped peccary can move in and out of inundated forests depending on the water level. Their large groups, comprising 50–300 individuals, enable them to traverse

Table III. Number of mammals hunted, in a one-year period, from the upland forests in the Tahuayo study site of the Tamshiyacu-Tahuayo Community Reserve

	Number hunted
Ungulates	
White-lipped peccary	166
Collared peccary	165
Red brocket deer	60
Grey brocket deer	28
Lowland tapir	38
Primates	335
Rodents	
Paca	174
Capybara	108
Black agouti	97
Others	36
Marsupials & edentates	71
Carnivores	98

Source: Data from Bodmer et al., 1994.

long distances. Thus, white-lipped peccary can move into the vast flooded forests during low water and exploit the rich fruit production (Bodmer, 1990b). During high water, white-lipped peccary can travel the long distances necessary to reach upland forests.

Arboreal mammals, such as primates, can also avoid inundations and are abundant in areas that are flooded for up to eight months a year. In contrast, collared peccary and brocket deer (*Mazama* spp.) are not well suited to flooded forests. Collared peccary and brocket deer are unable to move across long distances and therefore cannot easily move in and out of floodplain habitats. Indeed, the collared peccary and brocket deer that inhabit flooded forests must retreat to floodplain levees to avoid inundation during high water. Mortality is usually great on these levees because of food shortages.

Rodents also become trapped on floodplain levees. However, rodents can survive better on these levees than ungulates, because they usually consume smaller quantities of food. In addition, rodent populations, because of rodents' fast reproductive rates, can recuperate rapidly after high mortality from prolonged inundation.

Importance of Game for Rural People in Flooded and Upland Forests

The importance of game meat for rural inhabitants is also considerably different between upland and flooded forests. In upland forests, mammalian game meat is an important source of both subsistence food and cash income. In the Tamshiyacu-Tahuayo Community Reserve, hunters of the upland forests extracted 22,136 kg of mammalian biomass during a one-year period (Bodmer et al., 1994a). Ungulates made up 78% of this biomass. This made for an annual yield of about 74 kg of mammalian game meat per person. This is considerably greater than the per capita harvest of San Antonio, Nueva Esperanza, and Maipuco.

Differences in the availability of game between upland and flooded forests appear to influence how game meat is used. Much of the game meat harvested in upland forests is sold in city markets for cash income. In contrast, hunters of San Antonio, Nueva Esperanza, and Maipuco rarely sell game meat to city markets: Virtually all of the meat is consumed within hunting camps and villages.

Vulnerability of Game to Overharvesting

Wildlife management programs in Amazonia must consider the differences in vulnerability of species to overhunting. Studies in the Tamshiyacu-Tahuayo Community Reserve have shown that mammals with lower rates of r_{max} (intrinsic rate of natural increase) show greater biomass declines caused by hunting than species with higher rates of r_{max} (Bodmer, 1995a). Thus, species with lower rates of r_{max} are more vulnerable to overhunting than species with higher rates of r_{max}. In Amazonia, game species with high rates of r_{max} include brocket deer, peccaries, and large rodents, and species with low rates of r_{max} include lowland tapir, primates, and carnivores.

These results were used to examine the proportion of game harvested that are within the less vulnerable category (deer, peccaries, and large rodents) and those that are within the more vulnerable category (lowland tapir, primates, and carnivores) in the

Pacaya-Samiria National Reserve. In terms of the numbers of individuals hunted, 67% are within the category of less vulnerable to overhunting and 32% are within the category of more vulnerable to overhunting. However, 55% of the mammalian biomass extracted is within the category of less vulnerable to overhunting, and 45% is within the category of more vulnerable to overhunting. Thus, nearly half of the meat harvested by hunters in the vicinity of Maipuco, Nueva Esperanza, and San Antonio comes from mammals that are categorized as being vulnerable to overhunting.

As mentioned previously, game harvests in the vicinity of Maipuco, Nueva Esperanza, and San Antonio are lower than upland forests. This relatively low hunting pressure might be sustainable for some of the animals in the more vulnerable category. Even animals with a low r_{max} can be harvested sustainably if the hunting pressure is low enough for the given area (Robinson & Redford, 1991, 1994).

Animal populations can become more vulnerable to overhunting if they dwell in habitats not suitable for their species, because unsuitable habitat will result in lower reproduction rates and increased natural mortality. Flooded habitat is not suitable for some of the species in the less vulnerable category, such as brocket deer and collared peccary. Thus, these species might be more vulnerable to overhunting in flooded forests than in upland forests. However, since brocket deer and collared peccary were rarely hunted in the vicinity of Maipuco, Nueva Esperanza, and San Antonio, the current levels of hunting of these species may be sustainable.

Palms and Ungulates

Hunting is not the only threat to game animals of the Pacaya-Samiria National Reserve. Destruction of preferred feeding habitats, namely palm swamps, is also threatening animal populations. As noted previously, ungulates are the most important game animals in terms of the quantity of meat extracted by local people.

Ungulates rely primarily on fruit as their major food source. For example, red brocket deer consume 81% of fruit in their diet; grey brocket, 87%; collared peccary, 59%; white-lipped peccary, 66%; and lowland tapir, 34% (Bodmer, 1989, 1990a).

The most important fruits for these ungulates are from palm species (Table IV). For example, the fruit consumed most frequently by lowland tapir is the pulp of *Mauritia flexuosa* palms, which occurred in 76% of their diet samples. The next most common fruit consumed by tapir is the *Oenocarpus bataua* palm, which occurred in 24% of the diet samples (Bodmer, 1990a). The most common food items consumed by brocket deer are palm fruits of *Euterpe precatoria* and *Iriartea* sp. These palm fruits occurred in 80% of grey brocket deer diet samples and 59% of red brocket deer diet samples. Both peccary species also consume large quantities of palm fruits, which occurred in 91% of white-lipped peccary diet samples and in 43% of collared peccary diet samples (Bodmer, 1989).

Palms and People

People of the Peruvian Amazon also use palm fruits. Products from wild harvested palms are used for food products such as palm heats (*Euterpe precatoria*), ice creams, fruit snacks, and drinks (*Mauritia flexuosa* and *Oenocarpus bataua*) (Table V). Indeed, palm trees

Table IV. Rank of palm fruits in diets of Amazonian ungulates

Palm	Rank in diet[a]				
	Collared peccary	White-lipped peccary	Red brocket deer	Grey brocket deer	Lowland tapir
Astrocaryum sp.	3	4	28	n.e.	10
Euterpe precatoria	16	10	1	1	n.e.
Iriartea sp.	6	1	2	2	n.e.
Oenocarpus bataua	2	2	30	n.e.	2
Mauritia flexuosa	10	5	18	16	1

Source: Data from Bodmer, 1989, 1990a.
[a] Rank of 1 denotes food which is eaten most frequently; increasing numbers denote decreasing importance in the diet; ranks 1–5 denote foods that are very important; ranks >15 denote foods that are not important.

are the most important wild fruit resource in the Peruvian Amazon and contribute 61% of the market value for wild fruit production (calculated from Peters et al., 1989). These palms are important forms of economic income for both rural extractors and many urban families who work as middlemen and market vendors (Padoch, 1987).

The potential of wild harvested fruits as a conservation strategy is great, because natural forests will have true economic value as intact habitats, and this value would exceed that of clear-cutting and timber extraction (Peters et al., 1989). But this argument depends on the degree of destruction of trees during harvesting. Many rural extractors cut down palm trees to harvest fruits, making this potentially sustainable activity de-

Table V. Palm trees commonly used by ribereños in the Peruvian Amazon

Common name	Latin name	Part used	Product
Aguaje	*Mauritia flexuosa*	Fruit	Edible fruit, drink, ice cream
		Cortex	Starch
Aguajillo	*Mauritiella peruviana*	Fruit	Edible fruit, drink, ice cream
Chambira	*Astrocaryum* sp.	Fruit	Oil, edible fruit
		Bark	Fiber
Huasai	*Euterpe precatoria*	Apex	Palm heart
		Wood	Building material
		Fruit	Edible fruit
Irapay	*Lepidocaryum tessmannii*	Leaf	Roofing material
Pijuayo	*Bactris gasipaes*	Fruit	Edible fruit, oil
Pona	*Iriartea* sp.	Wood	Building material
Sinamillo	*Oenocarpus mapora*	Fruit	Edible fruit, oil
Ungurahui	*Oenocarpus bataua*	Fruit	Edible fruit, drink, ice cream, oil
Yarina	*Phytelephas macrocarpa*	Leaf	Roofing material

structive (Vasquez & Gentry, 1989; Phillips, 1993). The palm trees most commonly felled by rural people of the Peruvian Amazon are *Mauritia flexuosa, Oenocarpus bataua*, and *Euterpe precatoria*.

The use of wild palms by the three communities studied in the Pacaya-Samiria National Reserve varies considerably. For example, commercial use of *Mauritia flexuosa* palm fruit appears to depend on the proximity of productive palm swamps to the villages. San Antonio uses *Mauritia flexuosa* palm fruits most frequently, with the greatest harvests occurring from November to February. San Antonio has a *Mauritia flexuosa* palm swamp behind the village that is used for palm fruit harvests. During the harvest season, approximately 150 sacks of *Mauritia flexuosa* are sold per month from San Antonio. This results in the felling of around 50 trees per month or 200 trees per year.

Residents of Maipuco also sell *Mauritia flexuosa* palm fruits, but in fewer numbers than San Antonio. Residents from Maipuco use the same palm swamps as San Antonio and sell around 50 sacks a month during the harvest season, which results in the felling of approximately 17 trees per month or 68 trees per year. Residents of Nueva Esperanza do not use *Mauritia flexuosa* fruits for commercial sale but will occasionally use them for household consumption. Residents of Nueva Esperanza use *Mauritia flexuosa* palm swamps that are quite distant from the village, requiring several hours on foot path or canoe. Therefore, the use of *Mauritia flexuosa* fruits appears to depend largely on the proximity of productive palm swamps to the villages.

Palms, Ungulates, and People

The population growth of ungulates probably depends on the availability of palm fruits, since these fruits are their most important food resources. Unfortunately, both ungulates and people rely on the same palm fruits. However, while people often degrade palm swamps, ungulates help maintain palm productivity by dispersing seeds (Bodmer, 1991; Fragoso, 1994).

The carrying capacity of ungulates will depend in part on the quantity of palm fruits available. Assuming these ungulates have density-dependent relationships, it is possible to envision how altering the carrying capacity will effect the level of a sustainable hunt (Caughley, 1977). Density-dependence models would predict that the level of maximum sustainable hunting could be increased if palm resources were increased. Likewise, if palm resources are being depleted, then the carrying capacity of ungulates would be lowered, which in turn would lower the number of individuals that could be harvested at a sustainable level.

Approximately 268 trees of *Mauritia flexuosa* were removed from the vicinity of San Antonio and Maipuco during one year. This harvest rate will probably have an effect on the game animal populations. However, since the trees were felled close to the villages, the impact of their removal was probably greatest on smaller game species such as paca, agouti, kinkajou, and small-bodied primates. Since most of the larger game did not venture close to the villages, the felling was probably not affecting populations of lowland tapirs, peccaries, and deer.

If demand for *Mauritia flexuosa* fruits continues in city markets, and if no steps are taken to curb felling, then local residents will probably travel further into the forests to harvest fruits and subsequently impact populations of the larger game species. Indeed, long-term benefits local people can obtain from wildlife are linked directly to palm fruit harvests. If people maintain productive palm swamps in natural areas and allow this fruit production to be consumed by wildlife, then game harvests can become more sustainable. If felling of palms continue for short-term benefits, then both palm and game resources will become depleted and long-term benefits from these resources will not be realized.

Managing Palms in the Pacaya-Samiria National Reserve

Managing the buffer zone of the Pacaya-Samiria National Reserve requires programs for both the extraction of forest fruits and the management of game hunting. In accordance, management programs for palms are currently being set up in the buffer zone of the Pacaya-Samiria National Reserve. The fruit species that needs the most urgent attention is *Mauritia flexuosa*, because harvesting of this palm over the past decade has been extensive. Alternative management strategies for the harvesting of *M. flexuosa* palms include (1) no action, (2) improving harvest techniques, (3) setting up agroforestry production of palm species, and (4) combining agroforestry production and improved harvesting techniques.

No Action   If no action is taken, local extinction or near local extinction of *Mauritia flexuosa* will probably occur in many parts of the buffer zone of the Pacaya-Samiria National Reserve. As these palms become scarcer close to the villages, harvesting pressure will increase on palms further in the interior of the reserve.

Improving Harvest Techniques   Palms are harvested by climbing in many parts of the Amazon by both indigenous groups and ribereño/caboclo communities (Anderson, 1990; Vasquez & Gentry, 1989). Felling of palm trees in the Peruvian Amazon appears to be due to socioeconomics. Open-access resources, such as wild palms, have repeatedly been overused in the Peruvian Amazon, because the rights of users usually have not been established or are unclear. Therefore, local people will only accept more energy-consuming climbing techniques if palm trees have closed access with defined user rights. This will give local people confidence that the palm trees they climb will not be cut by other extractors.

Introducing improved harvesting techniques will help to maintain palm trees. However, improvements to the ecosystem might be limited, because the majority of fruits will be extracted from the forest, thus decreasing the quantity of palm food for animal populations and reducing the dispersal of palm seeds. Indeed, adult trees with no fruit will do little to increase the carrying capacity of game populations.

Agroforestry Production of Palms   Local people will not fell palm trees when they occur in small private plots, but will collect fruits for both household consumption and market sale. Many communities in the Peruvian Amazon realize the damage inflicted by cutting palm trees and have begun agroforestry systems that incorporate these species. Interestingly, many palms grow only 2–5 m tall when planted in open systems without competition for sunlight. Thus, these cultivated palms do not require cutting or special climbing equipment. The production of cultivated palm fruits will enable inhabitants to have a renewable supply of palm fruits for market sale and subsistence consumption and will leave wild palm fruits as food for game animals. The major disadvantage of this strategy is the delay between planting and fruit production, which is about 6–8 years. During cultivation, harvesting of wild palms will continue; however, once agroforestry plots become productive, the number of wild harvested palms should decrease.

COMBINING AGROFORESTRY AND HARVEST TECHNIQUES   This would appear to be the best strategy for palm trees, game populations and local people. By combining agroforestry and harvesting methods, it will be possible for local people to harvest wild palm fruits while trees in agroforestry plots mature. Once palm trees in agroforestry plots become productive, harvesting of wild fruits can cease.

Integrated Management of the Pacaya-Samiria National Reserve

Local people of the Peruvian Amazon usually take a true concern in managing palms and wildlife, and are interested in determining the best ways to secure economic gain from extraction using techniques that will not compromise future availability of these resources. This requires strong links between community representatives, extension workers, and biologists. There is a true need to coordinate natural resource management by rural communities with biological information and good intentions of extension workers. Poor management will upset both rural people and natural areas by not restricting overharvesting of species currently vulnerable to local extinction, and possibly underusing species that are currently sustainably harvested.

Managing palm fruits and wildlife will have long-term advantages over cutting of wild trees and overhunting of game. For one, inhabitants will have a renewable supply of palm fruits and game meat for market sale and subsistence consumption. In addition, palm fruits in natural habitats will be mostly left for animal food, which, in turn, should strengthen game populations.

Managing these resources will require increased labor in the short term. But this labor cost will result in increased revenues as agroforestry plots become productive and as populations of game become more abundant. Indeed, by combining the management of wildlife and palm fruits, a true system of sustainable use can be developed that is likely to increase economic incomes of local people over the long term.

Acknowledgments

We wish to thank Tula Fang, James Penn, Carlos Rengifo, Dennis McCaffrey, Eduardo Durand, Eloy Pezo, and Luis Benitez for their kind assistance. Logistical and financial support was provided by the Tropical Conservation & Development Program of the University of Florida, The Nature Conservancy, the Fundación Peruana para la Conservación de la Naturaleza, Instituto Veterinario de Investigaciones Tropicales y de Altura, INRENA–Ministerio de Agricultura, the Universidad Nacional de la Amazonia Peruana, and the Amazon Conservation Fund.

Literature Cited

Alvard, M. 1995. Shotguns and subsistence hunting in the Neotropics. Oryx 29: 58–66.

Anderson, A. B. 1990. Alternatives to deforestation: Steps toward sustainable use of the Amazon rain forest. Columbia University Press, New York.

Ayres, J. M. 1993. As matas de várzea do Mamirauá. MCT-CNPq, Brasilia.

Bayley, P. B. & M. Petrere Jr. 1989. Amazon fisheries: Assessment methods, current status and management options. Pages 385–398 *in* D. P. Dodge, ed., Proceedings of the International Large River Symposium. Canadian Special Publications on Fisheries and Aquatic Science 106.

Bodmer, R. E. 1989. Frugivory in Amazonian Artiodactyla: Evidence for the evolution of the ruminant stomach. Journal of Zoology 219: 457–467.

————. 1990a. Fruit patch size and frugivory in the lowland tapir. Journal of Zoology 222: 121–128.

————. 1990b. Responses of ungulates to seasonal inundations in the Amazon floodplain. Journal of Tropical Ecology 6: 191–201.

————. 1991. Strategies of seed dispersal and seed predation in Amazonian ungulates. Biotropica 23: 255–261.

————. 1995a. Susceptibility of mammals to overhunting in Amazonia. Pages 292–295 *in* J. Bissonette & P. Krausman, eds., Integrating people and wildlife for a sustainable future. Wildlife Society, Bethesda, MD.

————. 1995b. Managing Amazonian wildlife: Biological correlates of game choice by detribalized hunters. Ecological Applications 5(4): 872–877.

————, **T. G. Fang, L. Moya I. & R. Gill.** 1994a. Managing wildlife to conserve Amazonian forests: Population biology and economic considerations of game hunting. Biological Conservation 67: 29–35.

————, **P. Puertas, L. Moya & T. Fang.** 1994b. Estado de las poblaciones del tapir en la Amazonía peruana: En el camino de la extinción. Boletín de Lima 88: 33–42.

Browder, J. O. 1992. The limits of extractivism: Tropical forest strategies beyond extractive reserves. BioScience 42: 174–182.

Caughley, G. 1977. Analysis of vertebrate populations. John Wiley, New York.

Ehrlich, P. R. & G. C. Daily. 1993. Science and the management of natural resources. Ecological Applications 3: 558–560.

Fragoso, J. M. 1994. Large mammals and the community dynamics of an Amazonian rain forest. Ph.D. dissertation, University of Florida.

Holling, C. S. 1993. Investing in research for sustainability. Ecological Applications 3: 552–555.

Kahn, F. 1988. Ecology of economically important palms in Peruvian Amazonia. Pages 42–49 *in* M. Balick, The palm—Tree of life: Biology, utilization, and conservation. Advances in Economic Botany 6. New York Botanical Garden, Bronx.

Levin, S. A. 1993. Science and sustainability. Ecological Applications 3: 545–546.

Padoch, C. 1987. The economic importance and marketing of forest and fallow products in the Iquitos region. Pages 74–89 *in* W. M. Denevan & C. Padoch, eds., Swidden-fallow agroforestry in the Peruvian Amazon. Advances in Economic Botany 5. New York Botanical Garden, Bronx.

————. 1988. People of the floodplain and forest. Pages 127–141 *in* J. S. Denslow & C. Padoch, eds., People of the tropical forest. University of California Press, Berkeley.

Peters, C. M., A. H. Gentry & R. O. Mendelsohn. 1989. Valuation of an Amazonian rainforest. Nature 339: 655–656.

Phillips, O. 1993. The potential for harvesting fruits in tropical rainforests: New data from Amazonian Peru. Biodiversity and Conservation 2: 18–38.

Redford, K. H. & J. G. Robinson. 1991. Subsistence and commercial uses of wildlife in Latin America. Pages 6–23 *in* J. G. Robinson & K. H. Redford, eds., Neotropical wildlife use and conservation. University of Chicago Press, Chicago.

Robinson, J. G. & K. H. Redford. 1991. Sustainable harvest of neotropical forest mammals. Pages 415–429 *in* J. G. Robinson & K. H. Redford, eds., Neotropical wildlife use and conservation. University of Chicago Press, Chicago.

———— & ————. 1994. Measuring the sustainability of hunting in tropical forests. Oryx 28: 249–256.

Vasquez, R. & A. H. Gentry. 1989. Use and misuse of forest-harvested fruits in the Iquitos area. Conservation Biology 3: 350–361.

The Pacaya-Samiria Project: Enhancing Conservation and Improving Livelihoods in Amazonian Peru

Eduardo Durand and Dennis McCaffrey

The Pacaya-Samiria National Reserve and Its Area of Influence

Parts of what is today the Pacaya-Samiria National Reserve were first set aside for conservation by the Peruvian government in 1940 and 1944 (COREPASA, 1986). Since then, a series of decrees has enlarged the reserve and solidified its protected status. The most recent decree, under which the reserve is currently administered, is Decreto Supremo No. 016-82-AG, promulgated in 1982.

Pacaya-Samiria National Reserve is located in northeastern Peru, in the Department of Loreto, between the Marañon and Ucayali Rivers. The juncture of these two rivers marks the extreme northeast corner of the reserve. The reserve takes its name from two secondary rivers in its boundaries: The Pacaya flows into the Ucayali and the Samiria flows into the Marañon.

Pacaya-Samiria is Peru's largest protected area, comprising roughly 2,150,770 ha, according to a 1993 measurement (CDC, 1994). This area is subject to continual change as the borders of the rivers that bound the reserve shift over time. In general, the Ucayali is moving gradually eastward, whereas shifts in the Marañon do not show a particular trend in direction.

The Pacaya-Samiria Project covers the entire reserve as well as an area of influence surrounding it (see Fig. 1). This area of influence is defined as a strip, 10 km wide, bordering the entire reserve except for its southwestern edge. In the area of the Puinahua Canal, an arm of the Ucayali that forms much of the eastern boundary of the reserve, the area of influence extends across the canal to the main stem of the Ucayali, and thence 10 km further east.

The Pacaya-Samiria Project area lies east of the Andes, in what is known as the "Ucamara depression." This depression has filled with sediments in recent geological periods (Rodriguez et al., 1995). Today the area is characterized by large and small rivers with low gradients, numerous oxbow lakes, permanent swamps, seasonally flooded forests, and emergent river levees. Some hills occur in the extreme southwestern

233

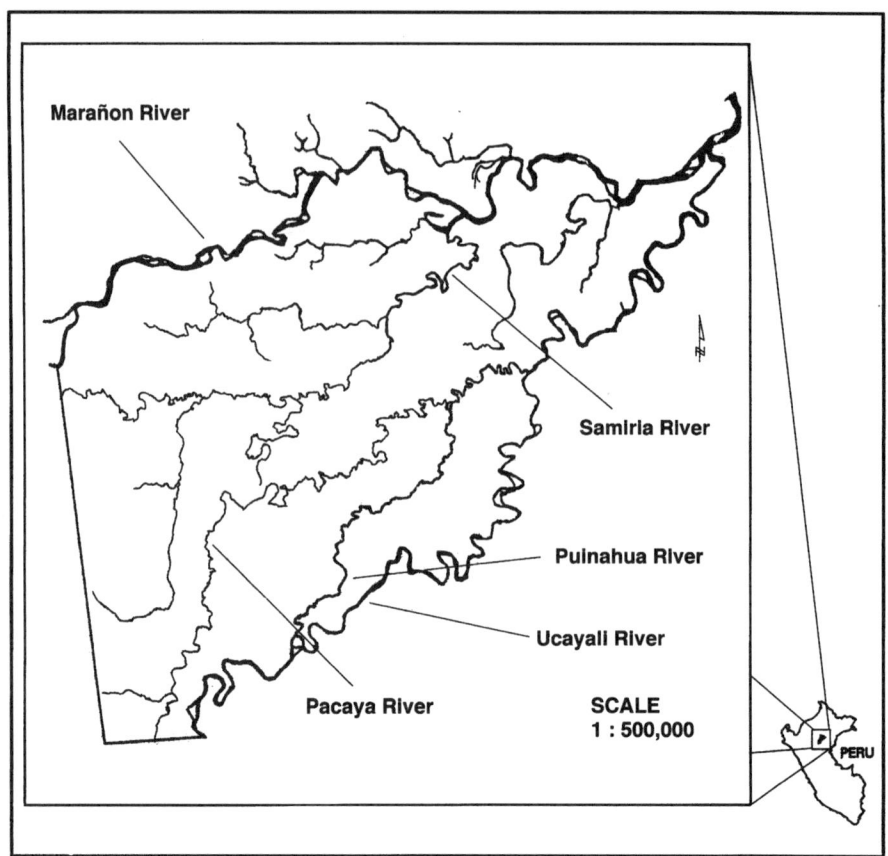

Figure 1. Pacaya-Samiria Natural Reserve.

part of the reserve, north of the Marañon River, and east of the Ucayali River, and range in elevation from 110 m to 200 m.

The dominant physical features of the area are its four major rivers: the Marañon, Ucayali, Pacaya, and Samiria. Water levels in all of these rivers fluctuate seasonally, with the low-water period occurring between July and November and the high-water period between December and June. In a typical year, the difference between maximum high water level and minimum low water level is 8–10 m. Both abnormally high water and abnormally low water occur fairly frequently. Notably high water caused floods in 1982–1983, 1986, and 1992–1993 (Rodriguez et al., 1995). In the reserve, the annual flood pulse of the rivers is smoothed somewhat by the effect of extensive flooding and flow of water into swamps, especially on the Pacaya side.

Pacaya-Samiria is classified as tropical moist forest. The climate of the region is hot, humid, and rainy, with annual precipitation of 2000–3000 mm. Mean monthly minimum and maximum temperatures are 20°C and 33°C, respectively. A relatively dry season occurs from June through September and relatively lower temperatures between April and June (Rodriguez et al., 1995)

A rapid ecological assessment of the Pacaya-Samiria Project area was conducted in 1993 (CDC, 1994). The assessment classified the vegetation in the area into five types, with their respective percentages of cover, as shown in Table I. The Table also includes areas of permanent water and areas in which the existing vegetation results from human activity.

The Pacaya-Samiria Reserve was established for the principal purpose of protecting and managing aquatic fauna (COREPASA, 1986). Aquatic mammals include the Amazonian manatee (*Trichechus inunguis*), classified by the IUCN as a vulnerable species; the giant river otter (*Pteronura brasiliensis*), rare throughout the region and considered vulnerable; the pink river dolphin (*Inia geoffrensis*), also classified as vulnerable (IUCN, 1990); and the gray river dolphin (*Sotalia fluvialitis*). Both river dolphins are common in the reserve.

Aquatic reptiles include three species of caiman and several species of turtle, notably the Amazon River turtle (*Podocnemis expansa*), the largest turtle in the Amazon basin, considered an endangered species (IUCN, 1990) due to excessive exploitation of eggs and mature females.

Approximately 259 species of fish occur in Pacaya-Samiria (Rodriguez et al., 1995). Of these, the paiche (*Arapaima gigas*) is noteworthy. It is the largest fish in the Amazon, reaching an adult size of nearly 3 m. The earliest efforts to establish Pacaya-Samiria as a protected area in 1940 were for the purpose of conserving the paiche (COREPASA, 1986).

Diversity of terrestrial fauna in Pacaya-Samiria is restricted by the frequent occurrence of floods (R. E. Bodmer, pers. comm.). Several terrestrial mammals in the area, like the tapir (*Tapirus terrestris*) and the capybara (*Hydrochoerus hydrochaeris*), are well adapted to watery conditions. Others, like the white-lipped peccary (*Tayassu pecari*), are capable of migrating to avoid floods (Bodmer et al., this volume).

Avifauna is well represented in the area, with 330 species and 58 families identified

Table I. Vegetation types in the Pacaya-Samiria Project Area

Type	Percent of cover
Riverine forests	14.1
Permanent swamps (palmaceous, herbaceous, and arboraceous)	36.6
Flooded forests	32.6
Gallery forests	1.4
Upland forests	4.6
Human activity	5.3
Permanent water (rivers and lakes)	5.4

Source: Rodriguez et al., 1995: 46–47.

during the rapid ecological assessment (CDC, 1994). Arboreal mammals are prominent, with 13 species of primates present (Bayley et al., 1992) and 62 species of bats reported from the vicinity of Genaro Herrera on the Ucayali River (Ascorra et al., 1993). This is the largest diversity of bats ever recorded from any site in the world.

Social and Economic Significance

Approximately 77,000 people live in 173 villages and towns in the Pacaya-Samiria area, mostly on the edges of the major rivers (IIAP, 1994). Most of the villages (89%) have fewer than 500 inhabitants. A significant fraction of the regional population, however, does live in the area's three larger towns: Requena, on the Ucayali; Nauta, near the mouth of the Marañon; and Lagunas, on the Huallaga near its juncture with the Marañon. Together these three towns comprise about 59% of the area's population. Each of these large towns is an important regional center and gateway to Iquitos and to other regions of Peru (Rodriguez et al., 1995).

In 1993 a study was conducted of the productive capacity of the land and natural resources in the Pacaya-Samiria area (IIAP, 1993). Results of the study indicated that most of the area, both inside the reserve and along its boundaries, had some potential for forest production but very little potential for agricultural production. Most of the land suitable for agriculture occurred along the edges of rivers or in upland areas outside the reserve. Table II summarizes the assessment of land-use capability stated in the study.

Even though the Pacaya-Samiria area has relatively low agricultural capacity, agriculture is an important activity in the area. A very large number of annual and perennial crops are grown. A partial list of the most important crops would include rice, maize, beans, yuca, sugar cane, plantain, lemon, jute, papaya, peanuts, and orange (IIAP, 1994). Roughly 3% of the Pacaya-Samiria area is under cultivation, yet this represents approximately 25% of the total area under cultivation for the entire department of Loreto. Much of the agricultural product is consumed locally, but a significant fraction—61% from the Marañon valley and 27% from the Ucayali—enters regional markets (Rodriguez et al., 1995). Much of the land under cultivation floods every year, and thus agriculture is a seasonal activity.

Fishing is a very important economic activity throughout the Pacaya-Samiria area. Fish consumption is high, providing, on average, about 40% of protein in the local diet.

Table II. Land use capability in the Pacaya-Samiria
Project Area

Capability class	Percent of cover
Agriculture	8.7
Permanent crops	4.6
Forestry	59.8
Protection	26.9

Source: Rodriguez et al., 1995: 30–31.

At least 28 different species of fish are commercially important (IIAP, 1994). Pacaya-Samiria produces most of the fish consumed in Loreto, providing more than 80% of the fresh fish consumed in Iquitos. The Ucayali and its tributaries produce the greatest volume of fish for market, regularly providing 50–90% of the annual volume of fish available in Iquitos (Rodriguez et al., 1995).

The Reserve also serves as a major area of extraction of other wild commercial products. The most important of these are turtle eggs and turtle meat, wild game meat, aguaje palm fruits, and heart of palm. All of these, as well as lesser products, are also used directly by local people (Rodriguez et al., 1995).

All of the major commercial uses of land and extraction of resources fluctuate seasonally in conjunction with the annual rise and fall of the rivers. Agriculture and fishing are most intense during low water, whereas hunting and most kinds of extraction are most intense during high water.

Basic Characteristics of the Employment and Natural Resource Sustainability Project

The Employment and Natural Resource Sustainability Project, or Pacaya-Samiria Project, began in October 1991 and is ongoing. Funding for the project is provided under a cooperative agreement between the U.S. Agency for International Development (USAID/Peru) and The Nature Conservancy (TNC). The agreement enables the investment of US$5 million in the project: $3.6 million as a grant from USAID and $1.4 million as the counterpart contribution from TNC.

The Nature Conservancy carries out the project in close collaboration with the Fundación Peruana para la Conservacion de la Naturaleza, or Pro Naturaleza, which is the local implementing nongovernmental organization (NGO). Both organizations work closely with the Peruvian government, especially the regional office of Instituto de Recursos Naturales (INRENA) in Iquitos, which has jurisdiction over the Pacaya-Samiria Reserve. The goal of the project is to create for the Pacaya-Samiria National Reserve and surrounding area a balance between, on one hand, conservation of natural resources and protection of ecosystems and, on the other hand, an environmentally sustainable economic growth, on the other (USAID, 1991). Four principal results are expected from the project:

1. emplacement of a management system in the reserve to protect its natural resources and provide for their sustainable extraction
2. organization of communities in and around the reserve and their participation in its protection and management
3. development of income-generating activities compatible with the protection and management of the reserve
4. adoption by local communities of these income-generating activities.

The rationale for the Pacaya-Samiria Project is that it will foster economic development in Loreto, one of the least developed parts of Peru, by improving conservation of one of the region's most important sources of natural resources, the Pacaya-Samiria National Reserve. The department of Loreto, now officially denominated as the Region of Amazonas, was considered in the past to be an economically and politically marginal region of Peru, despite its sizable population, economic activities, and un- or ill-exploited natural resources (Durand et al., 1991). The recent crises in the country have been acutely reflected in this region. These include economic recession, expressed in the drastic reduction of tourism, commerce, and employment; fiscal limitations on public expenditures and lack of private investments; and generalized poverty in urban and rural areas.

The situation has changed positively since the project began field operations in 1992. The national economy has improved, inflation has been brought under control, and other factors such as terrorism and cholera no longer affect tourism. By late 1994, people both in cities and in rural areas had become noticeably more confident about their future. Prospects for development had improved significantly as compared with conditions at the beginning of the project. But it will take much more time for the situation in rural areas to improve to the same degree that it has in the main cities in the country, as even Iquitos is still experiencing the lingering effects of the crises.

One main objective of the project is to improve the way the population uses resources in the reserve and surrounding area (McCaffrey, 1991). The project's development program will attempt to gradually increase family incomes by a combination of improved extraction practices, improved processing, storage, and transport of products, and increased agricultural and livestock productivity. This should result in less pressure on the reserve while improving the livelihoods of local people (The Nature Conservancy, 1992).

Project Strategy

When the project began in October 1991, it was faced with the task of quickly formulating a strategy for implementation. The project organizers determined that the first step was to analyze existing knowledge about the reserve and the project area. The next step, coincident with starting field operations, was to augment this knowledge with in-depth studies commissioned by the project.

In late 1991 the project began this informational process by contracting an environmental impact study of the project (Bayley et al., 1992). Beginning in mid-1992, the project contracted three in-depth studies: a rapid ecological assessment (CDC, 1994), an analysis of land and natural resource use and capacity (IIAP, 1993), and a socioeconomic analysis (IIAP, 1994).

One of the issues explored by the environmental impact study was how to devise and apply a strategy for the project that would enable it to meet its dual purposes of conservation and economic development. The environmental impact study considered the question of project strategy from two competing perspectives. The first point of view concentrates on extraction of renewable resources from within the reserve by bordering communities. The second concerns the promotion of alternative generators of income, such as agriculture and fisheries programs, outside the reserve.

The first perspective is potentially advantageous because local communities would have an interest in conserving the reserve's resources; but its potential to generate income is inferior to that of the second perspective. The second perspective, apart from not providing an incentive to protect the reserve, runs the risk of attracting too many people to the buffer zone, thereby increasing pressure on the reserve.

Based on the assumption that the project would partially adopt both perspectives, the environmental impact study analyzed the level of income generation that could be obtained from each of three alternative strategies. The three strategies would combine the two perspectives in differing degrees, as follows:

- Strategy A: Concentrate primarily on using resources from inside the reserve.
- Strategy B: Concentrate approximately equally on using resources from inside the reserve and on developing resources outside the reserve.
- Strategy C: Concentrate primarily on developing resources outside the reserve.

Figures 2 and 3 show predictions of the results that could occur from the application of the three alternative strategies. Figure 2 relates the three alternative strategies to the two perspectives. It predicts that income-generating capacity will increase as one moves from strategy A to B to C.

Figure 3, which contains the crux of the analysis, predicts the degree of future resource protection that would result from the application of the three strategies. It also predicts future protection from an assumed strategy of no action.

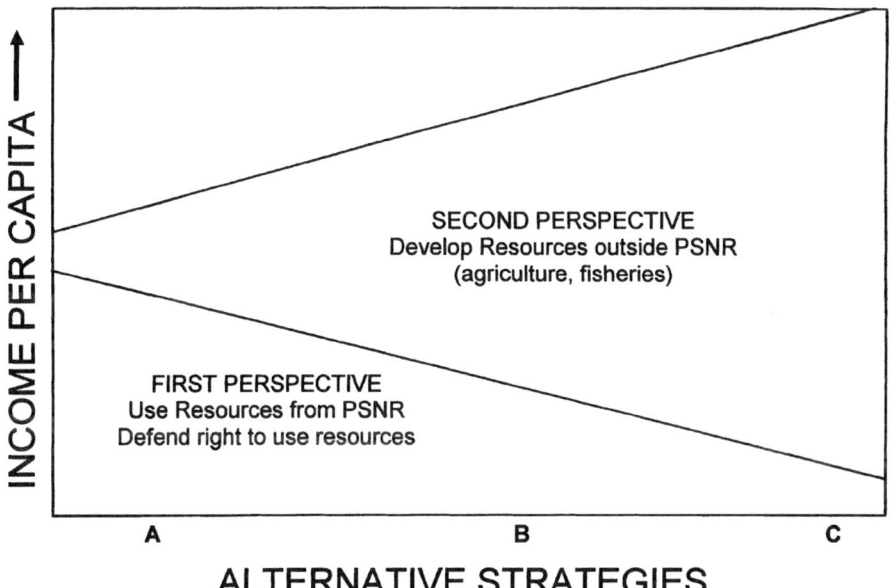

ALTERNATIVE STRATEGIES

Figure 2. Predicted changes in income generation under three alternative strategies. **Strategy A:** Concentrate primarily on using resources from inside the reserve. **Strategy B:** Concentrate approximately equally on using resources from inside the reserve and on developing resources outside the reserve. **Strategy C:** Concentrate primarily on developing resources outside the reserve. (From Bayley et al., 1992: 4.)

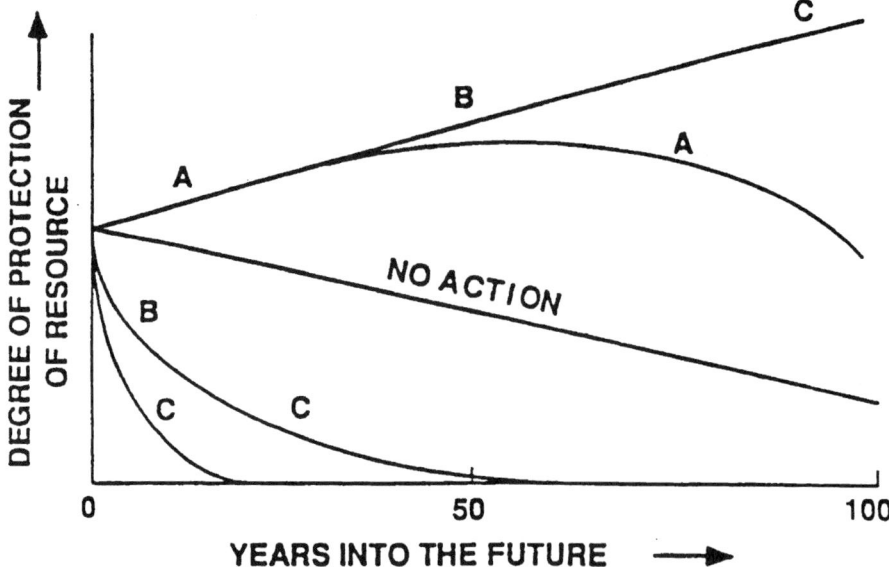

Figure 3. Predicted changes in level of resource protection under three alternative strategies. **Strategy A:** Concentrate primarily on using resources from inside the reserve. **Strategy B:** Concentrate approximately equally on using resources from inside the reserve and on developing resources outside the reserve. **Strategy C:** Concentrate primarily on developing resources outside the reserve. (From Bayley et al., 1992: 5.)

Under the no-action strategy (i.e., intervening neither in protection nor in development), existing trends of resource use in the reserve area would continue into the future indefinitely. This would lead to gradual but persistent reduction of protection in the reserve. In summation, the environmental impact study recommended a long-term program of conservation and development which would emphasize protection in the early years and shift gradually to an emphasis on development. The project adopted these recommendations and applied them in setting up the structure for the project and in defining and carrying out its actions.

Project Structure

The main office for the project is located in Iquitos, Peru. The Nature Conservancy has one full-time person based in Iquitos and provides support to the project from its home office in Washington, DC. The rest of the permanent staff in Iquitos, consisting of the project director and logistical and administrative support staff, are employed by Pro Naturaleza, which provides assistance from its headquarters in Lima. Both The Nature Conservancy and Pro Naturaleza maintain regular contact with USAID in Lima and with the regional government of Loreto and INRENA in Iquitos.

Most project staff are located in the field—approximately half in the Marañon-Samiria watershed and half in the Ucayali-Pacaya watershed. Fieldwork is supervised by two coordinators, one for each watershed. Infrastructure in the field consists of 13 control posts within the reserve and four Community Centers for Conservation and Development, or CECODES, located in four towns around the edges of the reserve. Approximately 40 reserve conservationists perform the function of patrolling the reserve. Twenty-five of the reserve conservationists are employees of the project and 15 are employees of INRENA. Twelve promoters work at the CECODES, which also serve as offices for the two coordinators. All field personnel live in the towns where the CECODES are located or at the control posts.

The project has installed a radio communications network that connects each control post and each CECODES and also links to Iquitos and to Lima. Energy to operate the radio network is supplied by solar panels installed by the project. The project provides boats and motors for the operation of the control posts and the CECODES and for transport between the field and Iquitos.

Major Project Components and Activities

The project is organized into two programs, one oriented toward the project's goal of conservation and the other toward the goal of economic development. All project activities fall within one or the other of these two programs. The project tries to carry out the combined activities under its program for management of the reserve in articulation with the activities under its program for sustainable use of natural resources. Involvement of government, especially regional and local government, and involvement of local communities are central to this articulated approach. Also central to the approach is the belief that conservation and development, in general and especially in Peruvian Amazonia, are inextricably linked.

This approach represents the adoption and application by the project of evolving thinking on the relationship between conservation and development for the humid tropics in general (Cluesner-Godt et al., 1992) and for the future of Amazonia (Dourojeanni, 1990; Comisión Amazónica de Desarrollo y Medio Ambiente, 1992). It is also consistent with the evolving thinking of conservationists that parks and reserves cannot be protected in isolation but must be incorporated into the economic development of communities and nations (IUCN, 1993; Munasinghe & McNeely, 1994). USAID is incorporating this approach into its programs as a valid strategy for economic development (USAID & WRI, 1993).

PROGRAM FOR MANAGEMENT OF PACAYA-SAMIRIA NATIONAL RESERVE   The project has made an important investment in reinforcing and supporting regional government agencies in charge of the protection and management of the reserve. This has consisted, in part, of building or refurbishing the 13 guard posts and administrative bases located at critical points in the reserve. The project has also trained and equipped the 40 reserve conservationists who work under government officials to patrol the reserve and engage in resource management and conservation activities in general. The project further supports these measures by covering the costs of operating boats, motors, radios, and other equipment, plus much of the costs of fuel and supplies needed for protection operations.

Working closely with regional and local government, the project has begun a process of long-range planning for the reserve. This process includes convocation by the regional government of meetings of local officials from the communities surrounding the reserve. The meetings serve to educate the local officials about the reserve and provide opportunities for them to express their views on how it should be managed.

The project is working toward the goal of developing and applying a system of zoning to the reserve. The three studies the project has carried out (CDC, 1994; IIAP, 1993, 1994) would provide the principal base of information for the zoning. The zoning plan would build on this information to obtain support from government agencies and local communities for the definition of zones and their location on the grounds of the reserve. Application of the zoning plan would require continuing involvement by regional government and local communities.

The project promotes scientific research into the basic biology of species in the reserve and into reserve ecology. It has financed the publication of 41 reports on research conducted in the reserve between 1979 and 1994 which previously had been unavailable to the public (CDC, 1995). The project also partially finances new research related to reserve management (Bodmer et al., this volume). The purpose of this research is to increase scientific understanding of the reserve so as to incorporate it into reserve management and economic development of communities.

PROGRAM FOR SUSTAINABLE USE OF NATURAL RESOURCES   The program for the sustainable use of natural resources operates from the CECODES constructed by the project. The promoters employed by the project work directly with the communities around the CECODES to carry out the various project activities in the program.

Step one in putting the program of sustainable use of natural resources into operation was to select a methodology for working with the communities. The methodology selected was participatory rural appraisal, as defined and developed by Grupo de Estudios Ambientales (GEA), Mexico City, Mexico. The project invited two practitioners of participatory rural appraisal from GEA to conduct a training workshop in the methodology in Iquitos in September 1993. All project personnel engaged in the natural resources program attended the workshop. Following the workshop, project personnel began applying participatory rural appraisal in various communities in which the project was working. As of September 1995, the project had conducted participatory rural appraisal in eight communities.

The participatory rural appraisal work with the communities enables the project to shape its activities in each community according to the community's expectations, preferences, and particular circumstances. All of the communities engage in a mix of extractive activities from the reserve combined with agriculture and local commerce. This is consistent with the economies of riverine communities in Peruvian Amazonia over many decades (Coomes, 1995). The exact mix varies from community to community. In general, the communities along the Marañon derive more of their living from agriculture than from fishing, whereas the converse is true for communities along the Ucayali. Agriculture and fishing combined account for over 90% of economic activity throughout the project area. Hunting, extraction of turtle eggs, and gathering of plant products from the reserve figure less importantly in the communal economies. Nonetheless, these are important commercial activities for some families, and contribute to the subsistence of most households.

The project supports applied research into household economies in the project area and into particular kinds of extraction and agricultural production (Bodmer et al., 1994). Anticipated results of this research, review of experiences of other projects in the region, and the experience of project staff have led to definition of three categories of income-generating activities the project supports: alternative uses and management of resources, improved technology for extraction and production, and improved postproduction technology.

In the first category, the project is attempting to improve the balance between extracting the fruits of aguaje palm (*Mauritia flexuosa*) for direct use and sale and leaving the fruits in the forest for consumption by game animals (Bodmer et al., this volume). The project is working with communities to establish plantings of aguaje in the communities and to use tree-climbing gear, rather than cutting down trees, to harvest fruit in the wild. Another activity in this category is rearing hatchlings of taricaya (*Podocnemis unifilis*) in captivity and then releasing them into the wild.

The second category focuses on agriculture and fishing. It includes introduction of improved seed and improved planting and cultivation practices for the principal commercial crops of the region: maize, rice, and peanuts. It also entails introducing improved stock of chickens and pigs and improved practices for feeding, health, and protection of the animals. Project activities geared toward improvements in fishing focus on technology both for catching fish and for processing fish after catch (e.g., brine salting and solar drying).

For improving postproduction technology the project is developing ceramic vessels that can be made locally for storing fish and grain. It is also working with producers, transporters, and wholesalers to increase efficiency in transporting and marketing. The object is to increase prices to producers by increasing the volume and quality of product reaching the market.

Some General Observations on the Project

The project is encountering many of the constraints and obstacles that affect development projects all over the world. The two most significant of these, as seen through the eyes of the project, are (1) the short time allotted to such projects versus the longer time needed for social and ecological change and (2) the challenge of generating income while using natural resources in a sustainable manner.

First, the project faces the typical time constraint for projects supported by international assistance in which the investment period is only four to five years. These very short timelines are often extended once a project is underway. But they influence initial planning and goal setting, and they force a rapid pace of project expenditures and actions that exceeds the optimal rate for change in social behavior or ecological conditions.

Second, increasing income through the sustainable use of natural resources is a fairly new approach for development projects. It requires refined tools that are not presently available for every situation and case. Every new case, both as a whole and for each resource affected, demands a lot of attention and applied research to balance ecological, cultural, and economic costs and benefits. At the same time, however, the expectations of the people in the project area have to be satisfied from the very beginning of the project, in order to insure credibility and cooperation over the whole course of the project.

These conflicts pose a tremendous challenge for the Pacaya-Samiria Project. They underscore the need for permanent monitoring and evaluation of project actions, and the surrounding physical and social environment, in order to ensure that the strategy being followed is working.

Alternative Economic Activities in the Context of Geographic Isolation

One of the most critical aspects of the project arises from its isolated location. Protected areas are often located far from urban areas and densely populated rural areas, whereas traditional production areas are usually close to population centers. Thus, in order to compete successfully, people who live near protected areas must rely on trading and marketing products that are unique to their area, thereby obtaining comparative advantage. But many potential products from protected areas originate from protected or endangered species.

The project involves promotion of agricultural and fisheries programs based on technological improvements that are likely to be adopted also in other areas of the region that are closer to potential markets. Introduction of new practices for producing commercial crops or managing wildlife could bring initial success. On the other hand, rapid imitation of these new practices in other areas with geographical advantage in relation to markets could reduce their value in the project area.

Persistence of Paternalistic Patterns in Development Projects

The Pacaya-Samiria Project is subject to the general tendency in underdeveloped countries of being perceived as a paternalistic vehicle to bring welfare in the forms of infrastructure, social services, and financial aid for production activities. To overcome this perception, the project emphasizes participation and local commitment in its relations with the people. But tradition and prior experiences with other development projects have left a persistent legacy of expectations that is hard to avoid. The applied participatory methodology for working with communities is gaining momentum and acceptance. Even so, the inertia of attitudes of paternalism still pervades day-to-day relations with the project. This applies both to the population and to the project promoters who, because they are culturally close to the population, also become unconsciously affected by this bias.

Cultural Gaps in Community Promotion and Project Implementation

Related to the constraint of paternalistic attitudes is the issue of cultural gaps in the chain that links the theoretical design of the project with the implementation of its activities. Even when anthropological studies are conducted, these tend to relate intrinsically to the population as a subject to work with, not as equal partners in development. Such studies usually disregard the influence of social and cultural traits within the project staff and in its relation with the people in the field. Points of view and approaches devised by scholars, when transmitted to local professional people and, from them, on to people in the field, have to bridge many cultural and intellectual differences and contexts. There is high risk that the message or objective will be lost or misinterpreted, if understood at all. The same can happen in a "bottom-up" scheme, in which upper-level project managers fail to understand the signals coming from the technical level in the field.

Though seldom considered in project formulation, this problem can be disruptive in cross-cultural contexts, as in the case of the Pacaya-Samiria Project and others in similar protected areas. It seems necessary to gear more-specific studies toward solving problems of internal communication and gradual translation of abstract concepts to actions, not only in the interaction with the target population but also within the very organization of the project.

Literature Cited

Ascorra, C. F., D. L. Gorchov & F. Cornejo. 1993. The bats from Jenaro Herrera, Loreto, Peru. Mammalia 57(4): 533–552.

Bayley, P. B., P. Vasquez R., F. Ghersi P., P. Soini & M. Pineda P. 1992. Environmental review of the Pacaya-Samiria National Reserve in Peru and assessment of project (527-0341). Environmental assessment contract, completed for The Nature Conservancy, Arlington, VA.

Bodmer, R. E., J. W. Penn & E. Durand. 1995. Costos y beneficios del establecimiento de una extracción de recursos más sostenible en la Amazonía occidental. Pages 215–222 *in* G. A. B. da Fonseca et al., eds., Abordagems interdisciplinares para a conservação de biodiversidade e dinâmica da uso da terra no Novo Mundo. Belo Horizonte, Conservation International do Brasil, Universidade Federal de Minas Gerais, and University of Florida.

Centro de Datos para la Conservación (CDC). 1994. Evaluación ecológica de la Reserva Nacional Pacaya-Samiria. Universidad Nacional Agraria La Molina, Lima.

———. 1995. Reporte Pacaya-Samiria; Investigaciones en la Estación Biológica Cahuana 1979–1994. Universidad Nacional Agraria La Molina, Lima.

Clusener-Godt, M., I. Sachs & J. I. Uitto. 1992. Final report: Conference on environmentally sound socio-economic development in the humid tropics. UNAMAZ, MAB-UNESCO, UNU, TWAS; Manaus, Brazil.

Comisión Amazónica de Desarrollo y Medio Ambiente. 1992. Amazonía sin mitos. IDB, UNDP, TCA, Brasilia.

Coomes, O. T. 1995. A century of rain forest use in western Amazonia: Lessons for extraction-based conservation of tropical forest resources. Forest and Conservation History 39(3): 108–120.

Comite Local de Desarrollo de la Reserva Nacional Pacaya-Samiria (COREPASA). 1986. Plan maestro de la Reserva Nacional Pacaya-Samiria. Loreto, Perú.

Durand, E. L. H., A. D. C. Camino, F. T. Torres & P. Fernandez-Davila. 1991. Economic, social and institutional analyses. Report on the Employment and Natural Resources Sustainability Project (527-0341). U.S. AID, Lima.

Dourojeanni, M. J. 1990. Amazonía: Que hacer? Centro de Estudios Teológicos de la Amazonía, Iquitos.

Instituto de Investigaciones de la Amazonia Peruana (IIAP). 1993. Informe final del estudio: Evaluación del uso y capacidad de la tierra y de los recursos naturales de la Reserva Nacional Pacaya-Samiria. IIAP, Iquitos.

————. 1994. Estudio socio-económico de las poblaciones vecinas a la Reserva Nacional Pacaya-Samiria. IIAP, Iquitos.

International Union for the Conservation of Nature (IUCN). 1990. 1990 IUCN red list of threatened animals. IUCN, Gland, Switzerland.

————. 1993. Parks for life: Report of the IVth World Congress on National Parks and Protected Areas. IUCN, Gland, Switzerland.

McCaffrey, D. 1991. Employment and Natural Resource Sustainability Project; No. 527-0341; Technical Analysis Report. Under contract to USAID/Peru, Lima.

Munasinghe, M. & J. McNeely, eds. 1994. Protected area economics and policy: Linking conservation and sustainable development. The World Bank, Washington, DC.

The Nature Conservancy. 1992. The Pacaya-Samiria Bioreserve; Draft strategic plan, revised. The Nature Conservancy, Arlington, VA.

Rodriguez A., F., M. Rodriguez A. & P. G. Vasquez R. 1995. Realidad y perspectivas; La Reserva Nacional Pacaya-Samiria. Pro Naturaleza, Lima.

United States Agency for International Development (USAID). 1991. Project paper-like document: Employment and Natural Resource Sustainability (527-0341). USAID, Lima.

———— **& World Resources Institute (World Resources Institute).** 1993. Green guidance for Latin America and the Caribbean; Integrating environmental concerns in A.I.D. programming. USAID, Washington, DC.

Equity, Sustainable Development, and Biodiversity Preservation: Some Questions about Ecological Partnerships in the Brazilian Amazon

Deborah de Magalhães Lima

The mobilization of local populations to defend natural resources that are essential for their livelihood, and the growing number of nongovernmental organizations that work for the preservation of the environment, are two social movements found in Amazonia today. These two movements frequently form alliances, as in the case of the rubber tappers' movement and the movement for the preservation of lake resources. Hall (1994) named these partnerships "socio-environmental movements" and pointed out that the strength of local mobilization actually depends on the alliance it establishes with organizations that work for the productive conservation of the Amazon. In some cases the political pressure exerted by these movements has been successful and has gained government support to legalize their proposals. One of the legal mechanisms used has been the creation of new categories of conservation units or the revision of already established ones. The founding of several conservation units in Amazonia is based on alliances of this type; these include the Mamirauá Reserve, the 11 federal and state extractive reserves, the Jaú National Park, and the National Forest of the Tapajós.

The National System of Conservation Units (SNUC) is the legislation that regulates the definition of categories of protected areas in Brazil. It divides protected areas into two main types: those for "direct use," formed by categories of conservation units that allow different forms of sustainable management (national and state forests, areas of environmental protection, and extractive reserves); and conservation units for "indirect use," which comprise totally preserved areas (national parks, biological reserves, ecological stations, and private reserves of natural heritage). There are 112 state and federal conservation units in the Brazilian Amazon, covering 420,000 km^2, or 8.7% of the region officially recognized as the Legal Amazon (Rylands, 1995). The total area covered by direct-use conservation units is 245,910 km^2 (4.9% of the Legal Amazon), an area equivalent in size to the United Kingdom. These data give an idea of the dimension of the area where local populations and conservationists are or will be working in partnership. There are, moreover, several sustainable development projects being implemented in areas of the Amazon that do not belong to conservation units.

The majority of projects that involve ecological partnerships were begun at the end

247

of the 1980s and the beginning of the 1990s, when changes in the socioeconomic scenario of the Amazon and the development of new theoretical concepts in conservation biology contributed to the formation of a setting favorable for their implementation. The main factor was the general acknowledgment of the negative social and ecological consequences of large-scale capitalist enterprises in the Amazon. These enterprises led to rapid deforestation and local extinction of natural resources; they endangered the region's biodiversity and threatened the survival of local populations. The construction of highways and dams, fiscal incentives for the occupation of land by capitalist enterprises, cattle ranching, urban growth, and increased commercial fishing and pressure on várzea fish stocks—these are the things that directly affected local people's lives and attracted the attention of the scientific community and the media to the consequences of the deforestation of the Amazon. At the same time, the decline of the traditional system of *aviamento*, a patron–client relationship based on debt bondage, opened a space for the political engagement of local populations previously constrained by patronage (Aramburu, 1994). Local people were then able to mobilize to defend their rights over the areas that they occupy. The alliance between conservationists and local populations has also benefitted from the existence of financial policies of "First World" institutions that, following new strategies for conservation, favor projects that integrate conservation and local people.

An important issue that requires a thorough discussion is the concept of "traditional populations," present in environmental thought and recently included in the ongoing revision of Brazil's National System of Conservation Units (SNUC). In response to the demands of the socio-environmental movement, a new version of the legislation, currently in debate, proposes the creation of new categories of conservation unit that include human groups, labeled in the proposal as "traditional populations." Another important issue in need of discussion is what criteria are used to identify certain social groups as "traditional populations" and to grant these groups—and deny to others—the right to remain in the conservation units. In many cases, the ecological partnership incurs the risk of becoming involved in local conflicts and being manipulated by social groups that compete for territories and for exclusive rights over important economic resources. (For instance, see Araújo's (1994) discussion of the conflict between communities of Lago Grande, near Monte Alegre, Pará. The dispute resulted from legislation that closed an area of the lake to commercial fishers. The communities included in the preserved area do not allow the entrance of members of communities that are located on the same lake but were left outside the protected area.)

The generalized use of the concept of traditional populations tends to simplify the diversity of social situations. More problematically, it implies that small-scale family production is typical of the group. Social groups that maintain this form of production are favored by the environmental movement exactly because of their greater disposition to accept models of sustainable use rather than turn to capitalist enterprises. Without adequate reflection, however, the expectations of the sustainable-use model can work against the autonomy of these populations in deciding their futures in the face of modern aspirations of consumption and definitions of well-being.

The ecological partnerships have actually been based on social inequality. Poor populations are accepted as residents of conservation units without there being a clear definition of the social goals to be achieved. The only norm defined for development, understood as the improvement of living conditions, is restrictive: that human activities must not jeopardize the objective of biodiversity preservation. From the conservationists' point of view, the permanence of human populations is itself a benefit offered to local people. Under the alternative, "pre-partnership" legislation, any local population would have been relocated upon the gazetting of a conservation unit. The very presence of human groups in conservation units is seen as a risky concession, exchanged for the political acceptance of the conservation unit and the adoption of norms of sustainable use that serve to buffer the boundary zones of preserved areas with their unrestricted surroundings. For the local people, on the other hand, their inclusion involves a series of sacrifices that are not equally shared with the rest of society. It is true that no project being now implemented uses coercion. On the contrary, open dialogue and community empowerment are accepted procedures. Even so, we cannot fail to acknowledge the fact that the economic behavior expected of these populations, justifiable in terms of an ideal model of human adaptation to the environment, is not demanded of other social sectors.

The implementation of integrated projects entails several changes in the social organization of the local populations, particularly in the social relations of production that define their access to and use of natural resources. As the resources found in areas destined for human use tend to be communally owned, it is possible that the population involved may exhibit a rather equitable internal social order, founded on the absence of exclusive rights over the principal means of production, land and water. The more or less egalitarian social organization, however, will be surrounded by a more diverse social order, upon which the local population depends and which offers the urban zone, not the rural, its most valuable benefits. It is necessary, therefore, to think also about the social boundaries of this model of biodiversity conservation.

Several issues concerning the concept, use, and expectations of traditional populations have to be addressed: the definition of such groups, the economic expectations in terms of enterprise development, and, lastly, the demographic effects of anticipated improved basic living conditions. The aim of this paper is to discuss the involvement of local populations in conservation units based on an analysis of the experience of the Mamirauá Reserve, a conservation unit developed in 1996, in partnership with communities engaged in the movement for the preservation of lakes.

The Experience of Mamirauá

The Mamirauá Reserve is a conservation unit of 1.2 million hectares located between the Solimões, Japurá, and Auati-Paraná Rivers (see Ayres et al., this volume). Originally, Mamirauá was decreed an ecological station (Estação Ecológica do Lago Mamirauá). By law, this is a category of conservation unit of the "indirect use" group, which does not allow the presence of people within its boundaries or the use of resources for purposes other than scientific research. It was thus a category incompatible with the model being implemented, one in which local people were being integrated into the management of the unit. The fact that Mamirauá is a reserve declared by the state of Amazonas helped the project to continue in its "irregular" status. The state's governor at the time it was established was opposed to strict conservation projects and lent the project his political support. If the reserve were linked directly to the federal government, the project would have faced greater difficulties because of the controversy over human presence in protected areas. The problem of the illegal status of the model adopted was resolved in 1996 with the change to a new category of protected area—"sustainable development reserve," defined by legislation enacted by the state of Amazonas. The state enacted this innovative legislation on protected areas because the legal models presented in the current national system of conservation units could not deal adequately with the reality of Mamirauá.

The new category "sustainable development reserve," now getting its first test in Mamirauá, is characterized essentially by the conjunction of three objectives: preservation of natural heritage, research on biodiversity, and promotion of sustainable development in order to combat poverty. The legislation foresees a gradual implementation of the reserve, with each phase ending with the approval of a management plan. Thus, the first phase of implementation is being concluded with the publication of the first management plan that refers to the focal area of the reserve (Mamirauá: Plano de Manejo, 1996; Ayres et al., this volume).

The appearance of projects known in the literature as "integrated conservation and development programs" (ICDPs) and "community-based conservation" (CBC) began several discussions on the implications of the integration of human populations to the objectives of biodiversity preservation (Redclift, 1989; Robinson, 1993; Western & Wright, 1994). All ongoing projects employ the concept of sustainability, especially after the United Nations Conference on the Environment and Development, held in Rio de Janeiro in 1992. With the controversy generated by associating the concept of conservation with the notion of development (sustainable development implying constant growth, a contradiction in terms that does not convey the intended meaning), the alternative concept of "sustainable use of natural resources" has been preferred. Nonetheless, the objective of promoting development, in the sense of improving the living conditions of the poor people who inhabit the majority of the areas where these projects are implemented, has been pursued.

The integration of two concepts—not always well defined—is involved: sustainability and development, the latter as a condition to eliminate poverty, which is considered an agent of environmental degradation. The concepts of sustainability and development denote processes, whereas the condition of poverty emerges as the subject upon which these processes would act. Indeed, contemporary literature on conservation presents the involvement of local people as a globally accepted strategy. In Brazil, however, resistance against the application of this model is still found within some official environmental institutions, especially federal ones, that defend the orthodox notion of integral preservation, excluding any human interference.

Integrated projects employ the concept of sustainability (or maintenance over time) in two ways: first, to denote the sustainability of biological and evolutionary processes that are based on noninterference by humans in specific areas designated for total biodiversity preservation; and second, to refer to the sustainability of the resources exploited by the human population. These projects present several strategies to integrate human populations: total integration, without zoning; integration with zoning of core areas of preservation and definition of buffer zones where human activities are allowed; or, as developed in Mamirauá, patches of areas with different categories of use that, in a sense, represent multiple zoning of core preservation areas with occasional overlaps. Ideally, the local people should ensure the two principles of sustainability. By adopting management rules, local people promote sustainable use, and by assuming the role of guards, they guarantee the sustainability of evolutionary processes and maintenance of biodiversity in the untouched areas.

This assumption implies to the local people, a restriction of their free use of space. This sacrifice, demanded of them in integrated projects, is only accepted if it is accompanied by the concession of some benefits in exchange. In general, this sacrifice is justified by assuring the local populations that their mode of living will be maintained—that is, securing that the natural resources essential for their survival will not become extinct. Therefore, a third proposal of sustainability is defined, that of continuity of the human population. This factor implies the maintenance of the population's reproduction, a process that includes both social and ecological variables. The guarantee of maintenance of this social reproduction requires a continuous program of research and monitoring that integrates natural and social sciences, as will be discussed below.

The presence of people in conservation units provokes a number of questions, the majority still without answers. The more direct questions relate to the development of basic research on the biology of species with economic value, directed toward the definition of criteria of sustainability and management rules. There are more-complex questions that require strategic decisions, specific to each conservation unit. For instance, in Mamirauá the question of human settlement zones faces a specific problem of várzea, namely, the instability of that environment. The solution found by local people to survive in an area of intense geomorphologic modification is to change the place of settlement whenever needed (Lima Ayres & Alencar, 1994); hence the definition of human settlement zones must be considered temporary and be periodically revised.

Another important aspect is demographic density. In Mamirauá, for example, current population density is 0.6 inhabitants per square kilometer. Although total population figures are stable, this population is highly mobile; that is, there are both individuals and whole families constantly leaving or entering the area. The stability in number of inhabitants is maintained by a high emigration rate that compensates for the annual rate of population growth, which, at around 4.1%, would lead to a doubling of population every 18 years. Whether or not this situation is compatible with the goal of biodiversity conservation is a question that cannot be answered at once. In any case, to ensure the level of population density regarded as ideal will demand highly evolved and complex discussions with the local people in order to develop mechanisms to sustain it. This is only a sample of the questions that future research needs to address.

Integrated projects aim for an improvement in the standard of living of the local population, as compensation for their share of sacrifice and in order to promote their acceptance of the preservation project. The political and ethical reasons for incorporating human populations, and not only endangered species, in the efforts to preserve and maintain adequate conditions for reproduction are also pertinent. But this extension of the concept of nature in need of preservation is still feeble. The inclusion of human populations in conservation units still presents many ambiguities. Beside the question of understanding and securing the social processes of human reproduction is the issue of defining the aimed-for standard of living for this population. What exactly is intended, and in what areas do we have the right to intervene?

The Mamirauá Project continues to carry out extension work in the areas of health, sanitation, and environmental education, limiting project intervention to areas outside of the obligations of the state. Community health education, training courses for health agents and school teachers, and a pilot project on sanitation are examples of such activities carried out by the project. The promotion of community participation also contributes to the improvement of living conditions, in the sense that it respects the people's claims over the territory they inhabit and promotes their awareness of their civil rights. There are also extension activities in agriculture and forestry to improve production, and a plan to implement ecotourism. It is acknowledged that current means of social reproduction are not satisfactory in relation to modern standards of well-being. Indicators such as a high rate of infant mortality, low rates of school enrollment and adult literacy, and poor sanitation testify to inadequate conditions of reproduction.

One way economists measure welfare is to look at consumption. In entirely capitalist societies, consumption can be measured by levels of expenditure and savings as percentages of income. Compared to developed capitalist economies, the economic condition of the populations found in conservation units presents distinctive features common to modern peasantries. Their consumption originates both from their own production and from the purchase of manufactured goods acquired from the sale of products extracted or cultivated in the reserves.

A change in income not only is a partial indicator of living conditions but also involves subjective criteria of level of satisfaction of consumption needs. Given the priority of biodiversity conservation—also considered an indicator of living conditions—it is likely that protected-area populations will have norms and limitations dictated to them, concerning exploitation of threatened species. This may inhibit growth in income and consequently affect the consumption of manufactured goods. The type of social reproduction typical of Amazonia's rural population, however, appears to facilitate the definition of limits to the volume of production. As household producers, their production capacity reflects mainly the number of family members capable of work. Their reproduction is linked to a simple circulation of commodities. This simple circulation has a limited potential for growth, whereas the merchant-capitalist has direct growth possibilities.

While rural populations are characterized today by a peasant type of social reproduction, if the management systems designed by scientists show positive results and increase the densities of managed resources, they might change their peasant features and adopt new economic strategies, as petty capitalists or entrepreneurs. This possible modification can have two effects: Tighter control and regulation of resource uses may be required to prevent overexploitation, or, if the enterprise becomes effectively organized, such controls may facilitate the promotion of sustainable management based on spontaneous interest of "enlightened" entrepreneurs to plan the management of natural resources.

The Movement for the Preservation of Lakes

The várzea of the middle Solimões is formed by hundreds of water bodies that are regionally called "lakes" but are in fact various types of channels that are enlarged in spots. During the low-water season these areas become isolated, and in the high-water season they become interlinked and the fish disperse. These lakes are a matter of dispute between river dwellers and professional fishers, especially during the low-water season when fishing is more productive due to the concentration of fish in the lakes. The history of the movement for the preservation of lakes effectively began in the 1980s, when the Catholic Church, through its Land Pastoral Commission (Comissão Pastoral da Terra, or CPT), supported its organization (cf. McGrath et al., this volume). The conflict between local residents and fishers, however, dates back to the beginning of the twentieth century, and its beginning coincides with the fall of the aviamento system.

Because of the annual floods, the várzea of the middle Solimões is not amenable to private property legislation. Nevertheless, in the settlements that specialized in fishing (feitorias), patrons controlled the mouth of the lakes, where they traded manufactured goods for pirarucu (Arapaima gigas fish) and turtles. When the patrons left, the pattern of human occupancy of the area changed. Settlements became located almost exclusively in the margins of the larger rivers where itinerant traders called "regatões" travel, and the use of the lakes became open to the residents of neighboring settlements. The current system of informal territorial control was then formed. Each settlement, comprising an average of nine households, all linked through kinship ties, occupies an area that corresponds roughly to the resident population's demand for fishing lakes and for higher lands suitable for agriculture.

During the early 1980s the Movement for Grass-Roots Education (Movimento de Educação de Base, or MEB) began to train settlement leaders and, by means of an association between social extension and religious teaching (the organization had links with the Catholic Church), formed a system of political representation based on elected posts of president, vice-president, secretary, and treasurer. The settlements that adopted this system, previously called "villages" or "hamlets," were then labeled "communities" (comunidades). In the middle Solimões region, therefore, the term "community" has the meaning of a particular type of political organization linked to the Catholic Church. Subsequently, several government and nongovernmental institutions recognized the usefulness of this organization, and today it no longer suggests a close association with a Church movement as it did in its early days. Nevertheless, community organization is stronger in those communities that maintained their links with the Catholic Church and that still participate in training courses and meetings promoted by the Diocese of Tefé, the main town on the middle Solimões.

The growth of Amazonian towns from the 1960s onward led to an increase in the demand for fish. Professional fishers, using equipment that allows higher volumes of production, such as gill nets, began to exploit, with advantage, the same lakes that river folk continued to fish using artisanal equipment. The aggravation of the conflict between professional fishers and local people propelled the CPT of Amazonas state to organize several meetings between the two parties to discuss the issues involved. Locals based their complaints not only on the inequity of the competition, but mainly on the overexploitation of the lakes by the fishers, thus reducing the stocks of the most important commercial species of fish. Since fish is their main source of both protein and cash, local residents proclaimed their case was a fight for survival. They adopted defensive measures known as *empates* (standoffs) to guard their lakes following the example of the resistance of the rubber tappers against the destruction of the forest.

In 1986, the Diocese of Tefé designed a system for preserving lakes in order to assure local people's means of survival. In several meetings organized by the diocese, community leaders were encouraged to reserve two lakes: one, called a "procreation lake," would be totally protected; the other, called a "maintenance lake," would be for subsistence fishing. Because the seasonal rise of water level recurrently connects most lakes, the closing of procreation lakes guarantees a regular supply of fish for the communities' maintenance lakes. Although the communities formed fishing committees that were in charge of guarding the lakes, problems among community members began to occur. Some members did not follow the proposal and negotiated the entrance of professional fishers into the lakes in exchange for promises of money or material goods such as engines; these promises were not always fulfilled. On other occasions, the community itself decided to fish the preserved lakes when forced to choose between preservation and survival.

The movement for the preservation of lakes succeeded in winning the support of a number of town councils in the middle Solimões, and lists of procreation and maintenance lakes were included in municipal legislation. The movement did not, however, gain any legal support from the Brazilian Federal Institution for the Environment (IBAMA) to ban outsiders from their lakes or ensure the closing of certain lakes. Professional fishers, represented by the Tefé Fishers' Union, pleaded that municipal and federal legislation are incompatible, and argued for their constitutional right of free movement. Without effective legal support, the movement currently can count only on the moral backing of nongovernmental institutions such as the Church and on the perseverance of community leaders who believe in the positive results of lake management.

In the area of the Diocese of Tefé, comprising nine counties of the State of Amazonas (Tefé, Japurá, Alvarães, Maraã, Fonte Boa, Uarini, Jutai, Carauari, and Itamarati), there are today 143 preservation lakes and 167 maintenance lakes that directly involve 2136 families (Ternus, 1996). In 1992, the Movement for Grass-Roots Education (MEB) formally joined the group of institutions that support the lake preservation movement by creating the Group of Preservation and Development (GPD) that comprises 35 communities from three counties (Tefé, Alvarães, and Maraã). Because of the movement's connection to the Catholic Church, Pentecostal communities have not joined in. A criticism of the Catholic model made by some Pentecostals is that the movement promotes poverty rather than development, since it does not include a category of lakes for commercialization. Recently, however, the Catholic leadership dealing with the issue (the CPT), revised its original proposal and suggested, among other things, the following six items: (1) general support for conservation efforts, (2) a request for scientific support from research institutions to develop appropriate management systems, (3) the study of economic alternatives, (4) the definition of criteria for judicious commercialization of fish, (5) the institution of a rotational system for preservation and maintenance lakes, and (6) the creation of an association of community conservationists to strengthen the movement (CPT-Amazonas, 1996).

Among the several nongovernmental institutions acting in partnership with local communities, the CPT is probably the one that presents its ideological position most clearly:

> The activities of the CPT (must) be understood as the action of the community directed toward the complete liberation and development of the human being and of nature; the CPT, in solidarity with the resistance struggles of the lower class of the hinterlands, seeks to reinstate the term "subsistence" as the essence of the construction of a new society. . . . The CPT understands the ecological question as the need to fight for the preservation and liberation of the human being integrated into and with a duty toward Creation (Genesis, 20). Creation, as a continuous act of God, entails the universal right to life and places the human being as co-creator or re-creator of nature. The permanent struggle against chaos (social inequity and environmental destruction), in search of integral liberation, recovers this re-creationist dimension of the human being. This position presumes a profound change in the current perception of progress as profit and accumulation. [CPT–A Grande Região Noroeste, 1996]

Inspired by Liberation Theology, the CPT is also the institution that presents the most radical proposal in relation to the transformation of society, considered the fundamental requirement to reach the goal of conservation of nature. Presenting the most utopian project, it is probably the institution that presents the greatest difficulties in sharing its ideals with the population with which it works.

The Participation of Local People in Mamiraua

Beginning in 1991, the involvement of the population in the establishment of the Mamiraua Reserve was facilitated by the precedence of the movement for the preservation of lakes. The first step taken was to consult the resident population that was not aware of the creation of the reserve, to know if they wanted to participate in the setting up of the protected area. Project activities proceeded only because the response of the majority of the local people was positive, since the reserve responded to their need for legal support for the lake preservation movement.

The experience of involving the population of Mamiraua illustrates the difficulty of developing a new model of conservation unit based on a proposal that did not originate locally. The transformation of a vertically organized project into a horizontal one, with the participation of local people in the management of resources and in the elaboration of the management plan, was a long process. The Mamiraua Project took five years to obtain the support of almost all of the 60 communities of an experimental focal area that were directly affected by the creation of the reserve. (Besides the resident population, the activities of the project involve adjacent communities that were identified as users of resources within the reserve and that depend on this use for their subsistence. There are 5277 individuals directly affected: 1668 that live in 23 settlements located inside the reserve and 3609 users from 37 adjacent communities.)

The establishment of a system to promote community participation was facilitated not only by the precedence of the above-mentioned lake preservation efforts but also by the fact that the settlements already had a system of community organization developed since the late 1960s by the MEB, including a pattern of democratic discussions of their problems. In order to help community leaders in deciding on conservation areas, the project divided communities of residents and reserve users into nine sectors, each comprising an average of six neighboring communities. Each reserve sector has a coordinator that organizes meetings among leaders, and all community leaders meet annually in general assemblies.

Despite the fact that community representatives were active participants in the choice of preservation lakes in the focal area of the Mamirauá Reserve, the decisions made in formal meetings regarding the choice of lakes were disputed, and several changes took place. These conflicts over the authority to choose lakes relate to internal political problems, associated with the social organization of settlements. Settlements are formed by domestic groups linked by kinship ties, and disputes stem from issues of rank between different kindreds, usually based on precedence of occupation. The formal political organization implanted by the MEB and reproduced by the Mamirauá Project is superimposed on this primary organization. In some cases, the formal leadership is more legitimate in its function of articulating the community with external institutions than in dealing with internal affairs. Moreover, there are often divergent opinions among household heads with different economic specializations, mainly between those who obtain the major part of their income from agriculture and those who specialize in fishing.

These differences affected the choices of lakes to preserve and the commitment to effectively protect them. Of the 616 lakes registered inside the focal area of the reserve, 200 were classified by the communities in one of the three zoning categories defined for lakes: preservation, subsistence, and commercialization. The disputes over some lakes, both within and between neighboring communities, reflect different concepts of use and opposing economic interests. In the area of Mamirauá, two existing conflicts— one between the Fishers' Union and the communities, the other between the Indian community of Porto Praia and other local groups—reveal the difficulty of acting impartially in search of a solution compatible not with particular interests but with the objectives of the conservation unit.

Research on fish offered for sale in the Tefé market indicated that 10–20% of fish sold comes from the reserve. Based on these data, the reserve residents' second general assembly decided to allow commercial fishing in a specific area of the reserve. It was agreed that the Fishers' Union would negotiate with the Jara005 sector, the reserve sector that includes the highest number of fishing lakes (see Fig. 1). Through negotiation, the choice of lakes to be exploited by professional fishers as well as the type of fishing equipment to be employed would be decided. Until today, however, no agreement has been reached, because the union insists on fishing in the same lakes the communities use. The union also disputes the basis of differentiation between them and community

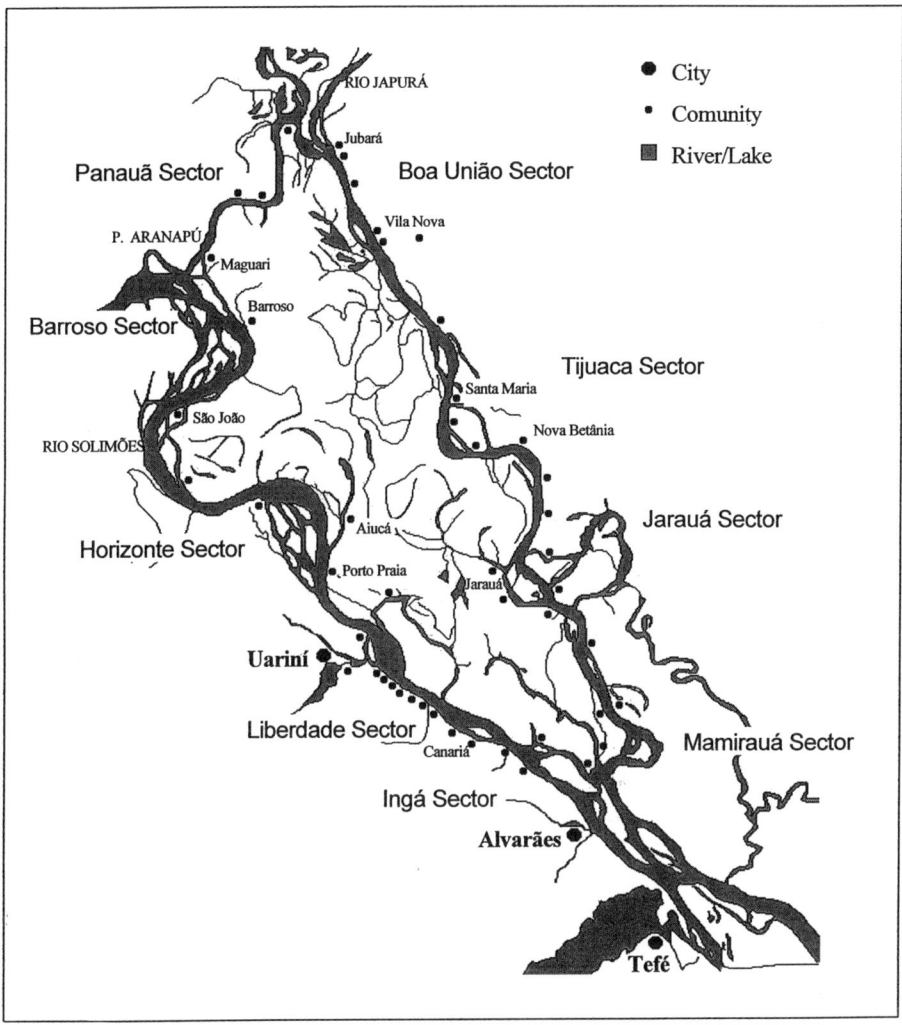

Figure I. The sectors of Mamirauá, the communities, the main towns of the region, and some of the 616 lakes that have been registered thus far.

residents, arguing that if the latter fish for sale they must also be considered commercial fishers, and should therefore be officially registered as professionals in order to be allowed to sell fish. Feeling marginalized by the popularity gained by the movement for preservation of lakes, union leaders have chosen not to attend meetings set up to discuss the matter, thus halting the negotiations.

Another case that shows how the socio-environmental movement can become involved with local disputes over territories and the right to exclusive use of resources is the dispute between the indigenous community of Porto Praia, whose inhabitants are descendants of Ticuna Indians, and neighboring communities of the Liberdade sector (see Fig. 1). The categorization of two lakes, Urucuri and Baú, was disputed by these communities even before the creation of the reserve. After several meetings proposed by members of the project, the thirteen communities that form the Liberdade sector decided to use a number of lakes and reserve the lake Urucuri for preservation. The Porto Praia community does not respect this decision, and frequently invades the preservation areas of the sector to fish, hunt, and log. The conflict was aggravated in 1994 by an official proposal for the creation of an Indian reserve for Porto Praia issued by the Brazilian Federal Indian Bureau (FUNAI). This proposal gave rise to much disagreement, not only because the Mamirawhat Reserve was already in place at the time the Indian reserve was proposed, but also because the proposed Indian reserve, comprising 10% of the focal area, would affect communities in the Jarauá and Horizonte sectors. After the proposal was announced, Porto Praia denied neighboring communities access to what they considered their territory, despite the fact that the Indian reserve has not been officially declared.

The complaints of local residents regarding the Indian reserve are based on the fact that many of the residents have lived in the area since the early years of this century, whereas the residents of Porto Praia arrived from the upstream town of Fonte Boa in 1972 and now "want to put up a sign post and take over the area," as one community leader said. The assertion of their ethnic identity gives Porto Praia residents the privilege of winning a territorial dispute that began in the 1980s. It is a well-known fact in the region that residents of Porto Praia have not participated in the conservation movement. The demarcation of a territory and the assertion of their ethnic identity are being used to exempt this particular community from the control of the social movement for preservation. The Mamiraua Project has officially questioned the creation of the Indian reserve and requested the revision of the original proposal (Reis, 1995). Inhabitants of two other Indian reserves are users of the reserve, and the project has established friendly and productive relationships with these other groups. As in other cases, the strategy adopted by the project for the definition of reserve users and for mediation of conflicts is based on the criteria of the extent of dependency on reserve resources, antiquity of settlement and resource use, and, chiefly, commitment to the goal of biodiversity conservation.

Identity and Political Participation

In order to understand the specific ways "traditional peoples" relate to environmental groups, it is important to know the social context in which they live and their local history. In general, besides their own identity, local people have an image, often stereotypical, attributed to them by the social groups with whom they interact in the regional society. In Mamiraua, the local population presents two categories of reference. One, that of *caboclos*, is used by the local society to identify in general terms all rural peoples along Amazonian rivers; and the other, that of the poor, is the broadest category that lower-class rural people use to classify themselves. Some facts about the local history of the population occupying Mamiraua and their notions of identity are necessary for understanding their intentions and expectations in allying themselves with the Mamiraua Project.

The term "caboclo," from the Tupi word *caa-boc*, means "that which comes from the woods," and was originally used by tribes of the coast to denote Indian groups of the interior. The meaning of "otherness" was maintained both in the initial use of the term by the Portuguese colonizers and in the present use. In the first use it denoted settled Indians as well as the mixed-breed, and today it is used by the urban population to refer to the rural population. The image of the "typical caboclo" is that of a riverine inhabitant who uses a canoe as his or her means of transportation and has a vast knowledge of the natural environment. The cultural representation of a typical rural Amazonian dweller is not restricted to this simple portrait but includes also moral evaluations of the riverine inhabitant. Indolence and laziness of the caboclo are also part of the stereotype that offers a moral interpretation of their poverty.

As the caboclo population grew in number and in importance in relation to the Amazonian labor force during the second half of the nineteenth century, when racist ideas dominated the elite's social thinking, the social position of the caboclos was explained as a consequence of the deleterious effects of racial admixture. This search for the caboclos' own attributes as the cause of their low social position continues today. Just as the role of colonial politics in the constitution of a peasant class subordinated to the colonial elite was ignored in the last century, today the laziness attributed by the stereotype substitutes a proper understanding of the unfavorable conditions for the reproduction of Amazonian peasantry in the context of the expansion of merchant-capitalism in rural Amazonia (Lima Ayres, 1992).

The picture of a caboclo, however, does not correspond to a social identity, and the term is usually rejected by the people it refers to or it is transferred to other classes and social categories considered inferior to the speaker's. The rural people's own construction of identity does not give them a notion of belonging to a common social group marked by a clear social differentiation as the notion of caboclo can imply. In their own speech the broadest term of identity used is that of the "poor," followed recently by the identity of "ribeirinho" (river folks), which was introduced during the work of the Catholic Church. Less extensive categories that effectively distinguish groups within the rural population are linked to their economic activity ("agriculturists," "fishers"), the environment they live in ("vargeiros," [floodplain dwellers] and terra-firmeiros [terra firme dwellers]) and their religion ("Catholics," "Protestants"). The identity of "Indians" is the most excluding and has the strongest political connotations. In the region, however, it is considered an artificial identity because of its use for political objectives and especially because there are no clear cultural distinctions between those who identify themselves as Indians and the ribeirinhos.

The loose identity of the poor is the self-representation that most influences the way Mamirauá residents relate to social categories standing in higher political and economic positions. In a certain way, they indirectly embody the stereotype, because their structurally underprivileged position allows them to negate any responsibility for their destiny and to position themselves as "naturally" entitled to assistance. While the stereotype attributes the cause of their poverty to the natural indolence of their "race," their own interpretation is that since they are not responsible for their social condition they are needy people who deserve to be assisted. This negative self-image, manipulated by patrons and politicians, especially during elections, is one of low self-esteem and adds to the existing difficulties they face in improving their standard of living.

As a relative concept, the notion of poverty is always defined in relation to a superior or better condition, thus the importance of understanding what is being referred to by inferiority and need. In historical terms and in the sense of representing a social class, poverty is associated with the system of aviamento and patronage. The notion of class, associated with domination by merchants who ruled the region during the first half of this century, is already built into the local culture. This is evident by local sayings, such as "There can be rich only if there are poor to buy." Although this perception of the origin of class division still persists, the meaning of poverty has been expanded by the process of modernization and the narrowing of relations between urban and rural environments.

The identity of poor shared by the Mamirauá people does not mean that the region does not offer the basic means of subsistence. Although instruments of work are simple, basic needs are met by fishing, agriculture, hunting, gathering, and timber extraction. The condition of poverty refers mainly to the fact that the income obtained from labor invested in production destined to the market is too low and allows only for the simple reproduction of the domestic groups (as in other situations where the domestic form of production is inserted into a market economy). In the economic sense, therefore, it conveys the meaning of the limit of what can be obtained from family labor. Among Mamirauá residents, for example, the mean annual income of a domestic group of seven members is US$900. This is obtained from the sale of an average production of 500 kg of fish, 20 m^3 of timber, and 200 kg of manioc flour. The major part of the money is spent in basic supplies. Sugar, coffee, soap, cooking oil, powdered milk, and salt are the main items of standard monthly supplies costing US$50.

In its more extensive and recent use, the concept of "poor" refers to the fact that the Amazonian rural population does not have direct access to the basic institutions of the modern world such as education, health, and the market itself. This sense of social exclusion is clearly reflected in politics. The local and state governments' lack of political interest in the rural area contributes to increased urban migration, where consumer needs and social assistance can be obtained. Therefore, although economic conditions allow basic subsistence needs to be met, the rural area does not offer full conditions for social reproduction and depends on urban centers for its existence. This partiality, discussed in the anthropological literature of the 1950s, based on the characterization of modern peasant societies as "part societies" (Redfield, 1953; Kroeber, 1948), is also related to the notion of poverty, in the sense of political, economic, and cultural dependency and inferiority of the rural zone in relation to urban centers.

Extension workers in Mamirauá involved with health, education, and community participation matters assert that they face many difficulties in their work because they have to deal directly with the locals' negative self-identity as poor. Project personnel have found that villagers have exceedingly high expectations of extension projects, hoping that they will find an immediate solution for their problems. They constantly request material aid and react with resistance when extension proposals demand their active engagement and commitment. Moreover, poverty imposes overwhelming expectations on agreements reached during general assemblies and other official meetings, because the immediacy of needs determines when an agreement should be broken and how long it can be sustained.

The partnership between conservationists and community residents involves specific interests that may or may not converge. For this reason, it is important to negotiate agreements with care, taking into account the particular nature of each party's interests. We should not forget that the interest of the so-called traditional populations for conservation is based on a specific economic benefit: their own subsistence.

In order to have the norms of sustained use respected, it is necessary to show the local population how these measures will benefit them economically. Changes in economic behavior will take time, and they require a specific program of extension work. The role of the community organization will also have to be expanded, for it will be the communities' responsibility to manage economic production collectively in relation to the control of zones of use. At the moment, the economic production of the communities is anarchic in the sense that resources are not regulated collectively by community members. The promotion of a system of community management of resources may change the opportunistic rationality that characterizes the economic behavior of Amazonian populations and which hinders the implementation of measures to regulate resource use.

The Mamirauá experience showed that there are important cultural differences not only between project staff and local people but also among researchers of the natural and social sciences. An interdisciplinary approach imposes an additional burden. Differences in academic backgrounds tend to show up in different concepts of the role, expectations, and rights of local people. These different perspectives are reflected, for instance, in budget decisions and in ways of dealing with local people. This is yet another factor that slows the process of implementing a new type of protected area, since it calls for a convergent analysis of social and biological data with no trained specialists, or much previous experience, in such an enterprise. It is necessary, during the process, to learn to understand our different languages and to see how different issues appear from different viewpoints.

Conclusion

Projects that integrate conservation and social development are ambitious intellectual efforts to create a model of sustainable use of natural resources in partnership with local people. These integrated projects aim to reconcile the shifting demands of the modern world with the carrying capacity of ecosystems and the preservation of biodiversity. In addition, these projects attempt to maintain a democratic partnership whereby the process is subject to the evaluation of the people involved. Authoritarian expressions of existing social inequality between the two parties, especially in terms of the authority of scientific knowledge, are avoided, and the people's knowledge of their environment is not depreciated but is accorded respect.

The situations of rural people in the Amazon differ greatly. Forms of kinship, identity, access to land and water, systems of inheritance of land and other property, rules of usufruct of communal resources—all are examples of features that distinguish social categories and types of settlement. Rubber tappers and their *"colocações"*—the *"colônias"* of northeastern migrants in the lower Amazon, the recent Transamazon Highway migrant settlements called *"ramais"* or *"travessões,"* the groups dependent on cracking nuts in babaçu palm fields, the descendants of slaves and *"quilombos"*—are only a few examples of the diversity of social identities and forms of settlement. This diversity of social settings demands that conservation projects, rather than simply employing a single model of local involvement, first understand the local forms of social reproduction and then develop models of participation, management, and conservation appropriate to each situation.

The involvement of local people in protected areas cannot follow a rigid scheme. Each project must be built through a process of continuous interaction with the particular population involved, adjusting local demands and habits to the needs of creating a system of sustainable use of the environment combined with biodiversity preservation, that may together lead to an improvement in the population's standard of living. Ongoing experiences also show that the processs of implementing a protected area in partnership with local people does not end. The evolution of society, and changes in demography, as well as in population density and access to natural resources caused by the protected area itself, makes it necessary to periodically readjust the management plan. Modifications need to be based on long-term monitoring of natural and social indicators and on a continued dialogue with the population.

The concept of sustainable use and the involvement of local populations in biodiversity preservation efforts are clearly positive developments in worldwide conservation. The need for this approach is reflected in the growing proportion of international development and conservation funding destined for projects run in partnership with local inhabitants. It is all the more important, therefore, that the complexities of such partnerships, as revealed by pioneering experiences like those of the Mamirauá Project, be carefully evaluated.

Acknowledgments

A Portuguese version of this paper was presented at the III BRASA Conference, held in Cambridge, UK, in September 1996. The Overseas Development Administration (ODA–UK) financed travel expenses. Thanks are due to participants of the conference session "Rural Social Movements in the Amazon," especially its chairman, Jean Hebétte, and Roberto Araújo, for lively discussions. Marta Mirazón Lahr kindly provided editorial assistance with the English version.

Literature Cited

Araújo, R. 1994. Manejo ecológico, manejos políticos: Observações preliminares sobre conflitos sociais numa área do Baixo Amazonas. Pages 301–308 *in* M. A. D'Incao & I. M. da Silveira, eds., Amazônia e a crise da modernização. Museu Paraense Emílio Goeldi, Belém.

Comissão Pastoral da Terra – A Grande Região Noroeste. 1996. Documento sobre a Grande Região Noroeste para Encontro de Março de 1996. Manuscript.

Comissão Pastoral da Terra – Amazonas. 1996. Report of the 12th Meeting of Ribeirinhos. Manuscript.

Hall, A. 1994. Social movements for productive conservation in Brazilian Amazon. Paper presented at the 48th International Congress of Americanists, Stockholm.

Kroeber, A. 1948. Anthropology. Harcourt Brace, New York.

Lima Ayres, D., E. Moura & J. M. Ayres. 1995. Mamirauá: Ribeirinhos e a preservação da biodiversidade da várzea amazônica. Pages 90–115 *in* G. Fonseca et al., eds., Abordagens interdisciplinares para a conservação da biodiversidade e dinâmica do uso da terra no Novo Mundo. Belo Horizonte, Conservation International do Brasil, Universidade Federal de Minas Gerais, and University of Florida.

————. 1992. The social category caboclo: History, social organisation, identity and outsider's social classification of the rural population of an Amazonian region (the middle Solimões). Ph.D. thesis, University of Cambridge, UK.

———— & E. Alencar. 1994. Histórico da ocupação humana e mobilidade geográfica de assentamentos na área da Estação Ecológica Mamirauá. Pages 353–384 in Anais do IX Encontro Nacional de Estudos Populacionais. Vol. 2. Associação Brasileira de Estudos Populacionais, Caxambu, Minas Gerais.

Mamirauá: Plano de Manejo. 1996. Brasília: Sociedade Civil Mamirauá, and National Council of Scientific and Technological Development; Manaus: Institute for Environmental Protection of Amazonas State.

Redclift, M. 1989. Sustainable development: Exploring the contradictions. Routledge, London.

Redfield, R. 1953. The primitive world and its transformations. Cornell University Press, Ithaca, NY.

Reis, M. 1995. Área Indígena Porto Praia—Uma avaliação da pertinência de sua criação em área preservada da Estação Ecológica Mamirauá. Manuscript.

Robinson, J. 1993. The limits to caring: Sustainable living and the loss of biodiversity. Conservation Biology 7(1): 20–28.

Rylands, A. 1995. Áreas protegidas na Amazônia Brasileira. Pages 183–198 in G. Fonseca et al., eds., Abordagems interdisciplinares para a conservação da biodiversidade e dinâmica do uso da terra no Novo Mundo. Belo Horizonte, Conservation International do Brasil, Universidade Federal de Minas Gerais, and University of Florida.

Ternus, A. I. 1996. Projeto da Comissão Pastoral da Terra—Amazonas, 1996–1998. Manuscript.

Western, D. & R. M. Wright. 1994. The background to community-based conservation. Pages 1–14 in D. Western & R. M. Wright, eds., Natural connections: Case studies in community-based conservation. Island Press, Washington, DC.

Section 4:
Soils and River Dynamics

Introduction

Daniel Nepstad

Despite its continental size (or perhaps because of its large size) Amazonia has been the target of a large number of oversimplifications coming from the scientific community. For example, the Amazonian literature tells us that the upland terra firme forests are perched on thin, nutrient-poor soils, that they are "fragile" and easily destroyed by human activities, and, hence, that they are unsuitable for permanent agriculture. In contrast, the seasonally flooded várzea forests along Amazonia's immense whitewater rivers are naturally fertile because of annual inputs of nutrient-rich sediments, they are subjected to frequent natural disturbances and are therefore resilient to human activities, and, hence, they are suitable for permanent agriculture.

New research on Amazonian forests and soils is modifying this dichotomous view of terra firme and várzea. At least one-third of terra firme forests appears to depend on the water stored in deep soils to maintain green, physiologically active leaf canopies during annual dry seasons of three to five months' duration (Nepstad et al., 1994; Negreiros et al., in press). These forests have access to far more nutrients than was previously believed, because their deep mycorrhizal root systems allow them to absorb soil nutrients to depths of several meters (Nepstad, 1989; D. Markewitz & E. Davidson, unpubl. data). Hence, terra firme forests rapidly regrow following numerous cycles of slash-and-burn agriculture (Vieira et al., 1996; Saldarriaga et al., 1988) and following several years of pasture grazing and burning (Uhl et al., 1988; Buschbacher et al., 1988; Nepstad et al., 1991, 1996). Evidence from charcoal dates, river sediment deposition patterns, and the diversification of languages and ceramic design reveals that vast areas of the terra firme forests of Amazonia were subjected to severe droughts and fire well before European invasion of Amazonia (Meggers, 1994; Sanford et al., 1985). Large areas of Amazonian terra firme forest are neither fragile nor "pristine."

The notion that terra firme forests are unsuitable for permanent agriculture has also come into question. Cattle pastures on land cleared of terra firme forest served primarily as a mechanism for capturing the escalating value of land in the 1970s and 1980s (Hecht et al., 1988; Browder, 1986). However, in the 1990s a new generation of cattle pastures can sustain profitable cattle production for ten years or more with small inputs of fertilizer (Mattos & Uhl, 1994). Agroforestry systems such as those practiced by Japanese-Brazilian farmers in Pará state (Uhl & Subler, 1988) and colonist farmers in Acre state further demonstrate the potential for permanent agriculture following clearing of terra firme forest when problems of local organization, land tenure, marketing, and credit are surmounted.

This volume provides a collection of articles that represent a substantial advance in developing a more sophisticated understanding of várzea ecosystems. In this fourth section, four papers describe new findings in the dynamics of river banks and river islands (Kalliok et al.), the interaction between soil properties and appropriate uses of várzea ecosystems (Ohly & Junk), vegetation development following abandonment of agricultural fields (Pinedo-Vasquez), and general trends in sediment chemistry along the main channel of the Amazon River (Zarin). These papers form the basis for a modified view of várzea ecosystems that goes beyond the notion of a fertile, disturbance-adapted ecosystem that is highly suitable for agriculture.

The Fragility of Várzea Ecosystems

Fragile ecosystems are those that are slow to recover following disturbances and those that lose their ability to fully recover following disturbances of relatively low intensity or short duration. In this sense, várzea ecosystems have been viewed as resilient ecosystems because they occur within a natural regime of high disturbance through annual or tidal flooding, sedimentation, and erosion. Várzea ecosystems are in a state of frequent renewal as water channels swing across the floodplain and new vegetation colonizes recently formed sedimentary deposits; várzea ecosystems should therefore recover rapidly following the disturbances associated with agricultural activities.

However, the trajectory of ecosystem recovery in the várzea depends on the type of agriculture that preceded field abandonment and the type of landform on which the field is located. Both plant succession and sedimentation patterns varied in the Napo-Amazon floodplain as a function of landform (silt bars vs. backslopes of levees) and agriculture (rice vs. jute), as Pinedo-Vasquez discusses here. In one of the first studies of várzea plant succession following agriculture, his paper reveals that, unlike the terra firme forests of Amazonia, ecosystem recovery following the abandonment of agricultural fields in the várzea includes the influence of the recovering vegetation on patterns of sedimentation and erosion. Hence, the very ground on which these ecosystems develop depends on the presence of the ecosystem. The highly dynamic nature of várzea soils, and their dependence on vegetation, indicates that some várzea ecosystems are very fragile, and will undergo long-term alteration through agriculture. Similar analyses are needed for other land-uses of Amazonia's many várzea types.

Not all of the movement of river and stream channels is a function of the vegetation that occupies várzea soils. River channels are "capricious," and their movements are difficult to predict, as is demonstrated by Kalliola et al. in their paper here. Hence, when large scales of time and space are employed, the changes in vegetation and sedimentation patterns documented by Pinedo-Vasquez may be erased as major river channels sweep across the human-altered várzea landscape.

The Fertility of Várzea Ecosystems

Várzea ecosystems are considered "fertile" because of the annual or tidal input of riverborne sediments. Hence, nutrients removed from várzea ecosystems through the harvest of agricultural products, or through the cutting and burning of várzea vegetation, are less likely to limit the long-term productivity of these sites when compared to terra firme soils, where nutrient inputs are primarily through rainfall. The chemical composition of the sediments deposited in the várzea soils of the Amazon's great whitewater rivers is not uniform, however. The quality of river-borne sediments for plant nutrition declines as one moves down the main channel of the Amazon River, indicating that várzea ecosystems that are closest to the Andes should be more fertile than those which are close to the Amazon estuary. Zarin explains in his paper here that the most likely reasons for this decline in the concentration of nutrient cations in river-borne sediments include (1) the dilution of the Andean waters with drainage water from nutrient-poor geological formations, (2) the weathering of sediment cations during the journey downstream and the capture of these cations by várzea vegetation, and (3) biological uptake of these cations.

Várzea Fragility, Fertility, and Appropriate Land Uses

Appropriate land uses for the Amazonian várzea vary greatly from one type of várzea to the next, but some general principles emerge from the four papers in this section. Although várzea ecosystems are very resilient in the face of flooding and channel migration, they are important determinants of soil formation and are, in this sense, more fragile than terra firme ecosystems on highly consolidated soils that are resistant to erosion. Different forms of agriculture foster different trajectories of plant succession and soil formation in the várzea, and these trajectories should be included in evaluations of the sustainability of different forms of land use.

But the interaction between land use and soil formation may be superseded by large-scale changes in river channels that have little to do with the vegetation that occupies the river banks. Erosional and land-forming processes of Amazonian rivers create uncertainty about the future of várzea land that must be included in evaluations of land-use sustainability.

In their paper here, Ohly and Junk suggest that both the analysis of existing forms of várzea land use and the design of future uses should integrate knowledge of spatio-temporal variability in the physical and chemical environment with the social and economic characteristics of várzea regions. Rapid turnover of land through erosion and land formation, and annual and/or tidal flooding, greatly restrict the types of land uses that are viable in the várzea. Labor shortages and poorly developed infrastructure for transport and marketing further constrain the forms of land use that are appropriate for the várzea.

Literature Cited

Browder, J. 1988. The social costs of rain forest destruction: A critique and economic analysis of the "hamburger debate." Interciencia 13: 115–120.

Buschbacher, R., C. Uhl, & E. A. S. Serrão. 1988. Abandoned pastures in eastern Amazonia II. Nutrient stocks in the soil and vegetation. Journal of Ecology 76: 682–699.

Hecht, S. B., R. B. Norgaard & G. Possio. 1988. The economics of cattle ranching in eastern Amazonia. Interciencia 13: 233–240.

Mattos, M. M. & C. Uhl. 1994. Economic and ecological perspectives on ranching in the eastern Amazon. World Development 22(2): 145–158.

Meggers, B. J. 1994. Archeological evidence for the impact of mega-Niño events of Amazonia during the past two millennia. Climate Change 28: 321–338.

Negreiros, G. H., D. C. Nepstad & E. A. Davidson. In press. Profundidade mínima de enraizamento das florestas na Amazônia brasileira. *In* C. Gascon & P. Moutinho, eds., Floresta Amazônica: Dinâmica, regeneração e manejo. Smithsonian Institution, Washington, DC.

Nepstad, D. C. 1989. Forest regrowth in abandoned pastures of eastern Amazonia: Limitations to tree seedling survival and growth. Ph.D. dissertation, Yale University, New Haven.

————, **C. R. de Carvalho, E. A. Davidson, P. Jipp, P. Lefebve, G. H. Negreiros, E. D. da Silva, T. Stone, S. Trumbore & S. Vieira.** 1994. The role of deep roots in the hydrological and carbon cycles of Amazonian forests and pastures. Nature 372: 666–669.

————, **C. Uhl & E. A. S. Serrão.** 1991. Recuperation of a degraded Amazonian landscape: Forest recovery and agricultural restoration. Ambio 20(6): 248–255.

————, ————, **C. Pereira & J. M. C. da Silva.** 1996. A comparative study of tree seedling establishment in abandoned pasture and mature forest of eastern Amazônia. Oikos 76: 25–39.

Saldarriaga, J. G., D. C. West, M. L. Tharp & C. Uhl. 1988. Long-term chronosequence of forest succession in the upper Río Negro of Columbia and Venezuela. Journal of Ecology 76: 938–958.

Sanford, R. L., Jr., J. Saldarriaga, K. Clark, C. Uhl & R. Herrera. 1985. Amazon rain-forest fires. Science 227: 53–55.

Uhl, C. & S. Subler. 1988. Asian farmers: Stewards of Amazonia. Garden 1988: 16–31.

————, **R. Buschbacher & E. A. S. Serrão.** 1988. Abandoned pastures in eastern Amazônia I. Patterns of plant succession. Journal of Ecology 76: 663–681.

Vieira, I. C., R. Salomão, N. Rosa, D. Nepstad & J. Roma. 1996. Renascimento da floresta no rastro da agricultura. Ciencia Hoje 20(119): 38–45.

Fluvial Dynamics and Sustainable Development in Upper Rio Amazonas, Peru

Risto Kalliola, Päivi Jokinen, and Eeva Tuukki

Introduction

Amazonia is a region of conflicting demands of conservation and development. Strict environmentalists tend to glorify the biological richness of this tropical lowland area and argue for its protection. On the other hand, one should not forget that the Amazon basin is inhabited by people and that this population needs to find economically sound ways to use the natural resources available, preferably in a sustained manner (Eden, 1990; Torres, 1993).

Recent development reflects a reluctance of planners to introduce modern development into seasonally inundated alluvial lands. This trend is exemplified in the vicinity of the city of Iquitos, in northern Peru, where new roads are being paved in the unflooded terra firme, where soils are less likely to support permanent agriculture than nutrient-rich soils of inundated areas. This type of development accelerates deforestation, thereby reducing the ecological value of the area.

In contrast, the economy of the Iquitos region is traditionally associated with the use of alluvial plains. Substantial settlement is present along the river margins, and the rural people (*ribereños*) use the riparian zone for a multitude of income-generating purposes (Padoch, 1988). One common practice is the consistent cropping of rice and other cultivated plants on fluvial bars during the low-water season. Further, a contrasting trend of resource use, in terms of sustainability, is the extractive logging of valuable timber trees, which are now absent from the most easily accessible areas (Vásquez & Gentry, 1989).

Ecologically, the floodplain conditions of turbid whitewater rivers such as the Amazon mainstream are strikingly different from other types of environments in Amazonia (Sioli, 1984; Puhakka et al., 1992). These rivers carry sediment of primarily Andean origin along very unstable courses. As a result, the natural vegetation of the floodplains grows in an environment that is at the same time productive and harsh (Encarnación, 1985; Junk, 1989; Kalliola et al., 1993). For example, newly deposited sediment is rich in extractable mineral nutrients, a productivity reflected in the high growth rate and vegetative strength of the early successional communities that occupied these areas.

Both the production potential and the naturally disturbed nature of these alluvial areas suggest that they could support market-oriented land management and production. However, sustained wetland exploitation strategies should be founded on an appropriate understanding of the fluvial geomorphology, hydrology, and stability of these areas. Because the majority of fluvial traits tend to vary both within floodplains and between rivers, no model can readily form the basis for such land use. Basic surveys, therefore, are needed to pinpoint significant conditional variations and assist in the acceptable compromise between ecological and developmental interests. With this in mind, this paper will examine the fluvial dynamics of the Amazon mainstream area, with the aim of discovering fresh insights within the discussion of sustainable development in Amazonia.

Study Area

This study covers the upper reach of the Amazon river, from the confluence of the rivers Ucayali and Marañon to the downstream end of Isla de Iquitos (Fig. 1). Within this reach of approximately 140 km, the Amazon breaks into several channels where islands of varying sizes cover a total area of approximately 210 km^2. At its widest, the central river channel occupies a belt 12 km wide. But these multichannel reaches are separated by relatively narrow single channels. For purposes of clarity we will indicate different locations along the river as distances from the confluence of Ucayali and Marañon (e.g., Iquitos is located at km 118).

The annual precipitation in the study area is approximately 3000 mm (Peñaherrera, 1986). Rainfall is consistent during the year, although July and August are somewhat drier than the other months. Seasonal rainfall within the catchment area causes pronounced fluctuations in water levels and an overflow of the Amazon mainstream, which presents an annual elevation range of 6–7 m.

Vegetationally, the study area represents typical Amazonian várzea, the floodplain vegetation of a whitewater river. Features of fluvial disturbance and riparian succession are common sights throughout the study area, as pioneer plants easily occupy newly exposed fluvial bars. The remaining natural vegetation comprises older successional stands and mature floodplain forests (for details of riparian succession, see Encarnación et al., 1990; Lamotte, 1990; Kalliola et al., 1991). Human impact is most pronounced in the immediate vicinity of the river, especially on the heavily disturbed islands near Iquitos. Scattered settlements occupy stable shores in other parts of the study reach.

Methods

Channel migration was surveyed using an extensive set of remote sensing imagery (semicontrolled aerial photos and mosaics, radar, and satellite imagery) representing the Amazon River in different years (Table I). Only the northern study reach was photographed in 1948, 1962, and 1965, whereas analysis from the rest of the study area begins after 1972. The course of the river in 1993 was drawn on the basis of a 1990 satellite image, with the help of 1993 field observations. Five chosen dates formed the basis for a sequence of four consecutive time intervals of nearly equal lengths: 1948–1962 (14 years), 1962–1972 (10 years), 1972–1983 (11 years), and 1983–1993 (10 years). Other images were used only to verify the details of change in certain complicated situations.

To minimize inaccuracy, all studied images were converted to a scale of 1:150,000,

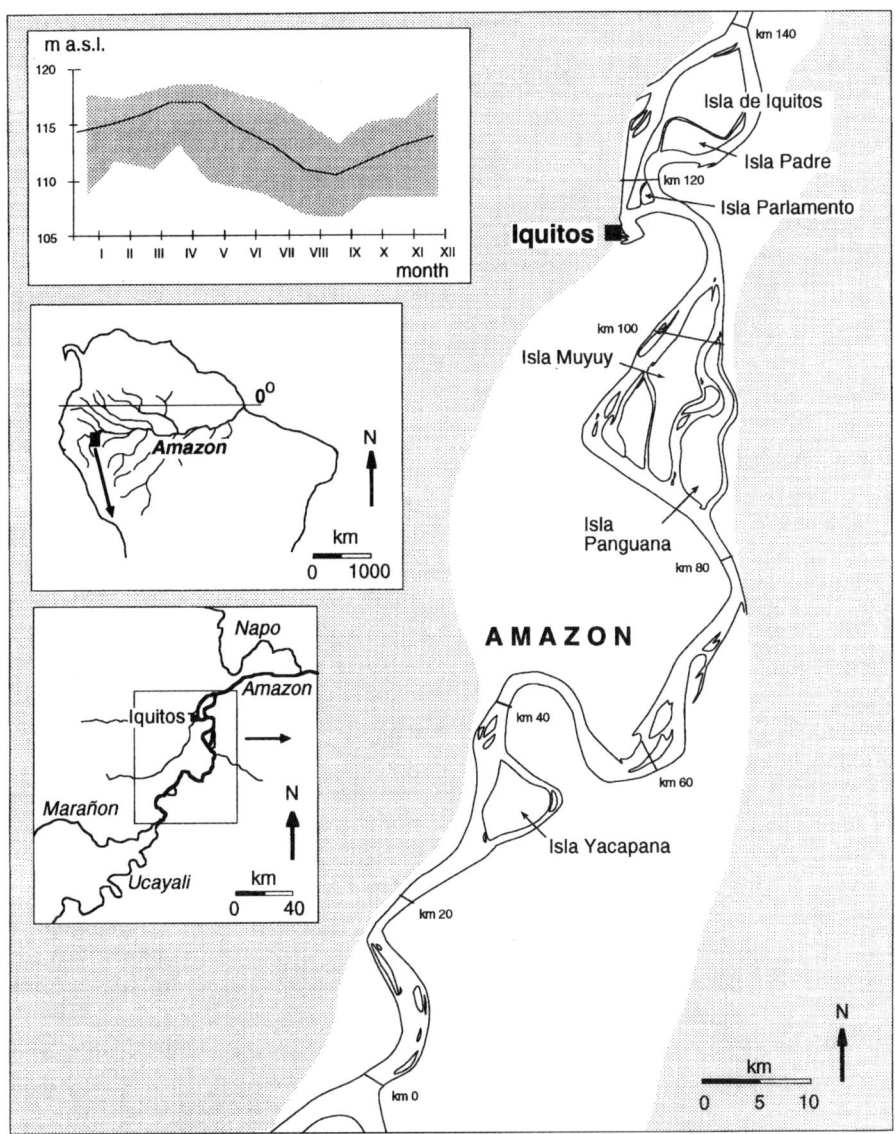

Figure 1. Location map of the Amazon mainstream and its alluvial plain, 1993. **Inset:** Water level fluctuations in the Amazon River near Iquitos, based on data from 1968–1992. The range between the minimum and maximum values is shaded and the average is represented by a solid line. (Data source: SENAMHI.)

Table I. List of the remote sensing material used in the change detection analysis

Year	Date	Water level[a]	Original scale	Material	Sheet/project
1948	2 Nov	113	1:140,000	Aerial photo mosaic	SAN: 3094
1948	2 Nov	113	1:30,000	Vertical aerial photo	SAN: 3094
1962	24 Nov	113	1:110,000	Aerial photo mosaic	SAN: 92-62-A
1965	28 Jan	No data	1:122,000	Aerial photo mosaic	SAN: 132-64-A
1972	16 Jun–1 Oct	116–113	1:150,000	Aerial photo mosaic	SAN: 214-72-A
1972	16 Jun–1 Oct	116–113	1:20,000	Vertical aerial photo	SAN: 214-215-72
1972	Unknown	No data	1:125,000	Radar (SLAR)	13-1-D
1983	19 Sep	111	1:250,000	Landsat MSS	IFG 1984
1985	16 Jul	112	1:250,000	Landsat TM	006-063
1989	24 Aug	111	1:250,000	Landsat TM	006-063
1990	May–Jun	Ca. 114	1:100,000	Radar (SAR)	DMA 1991

[a] Meters above sea level on the date of data registration.

both photographically and with the help of a photocopying machine. For each set of images a fitting operation was executed against the Landsat MSS planimetric map from 1983 (IFG, 1984), using distinguishable features, such as permanent lakes, as markers in the overlaying operation. Thereafter, river courses were drawn on transparencies from each set of imagery, and these drawings were used in detecting change between the monitored years.

The migration pattern of the active river channel was surveyed with a line transect method. The active river channel comprises one or more passageways for water which are divided by channel islands. We defined channel axis as the central line of this belt, and drew the lines separately for each corresponding year. Change detection was made through pairwise comparisons of the channel axes. The magnitude and direction of channel axis shift was determined along line transects drawn at intervals of 2 km perpendicular to the previous channel axis. Altogether, 129 transects were studied in this manner and represented two consecutive time periods (Fig. 2).

While the above cartographic operations were executed manually, errors in the overlaying operation may have introduced inaccuracy in the detection of particularly small change units. The distinction of erosion was always easier than that of sedimentary forms, and unequal water levels may have influenced these results. The general accuracy of the change maps was controlled with a ground truth survey performed from May to August 1993, when two small-plane flights were made over most of the study reach and detailed inventories of vegetation were carried out. The results of these studies will be included in later publications.

Results

The active channel of the Amazon River shows an alternating pattern of stretches of change and those of relative of stability (Fig. 2). The latter areas tend to be straight and

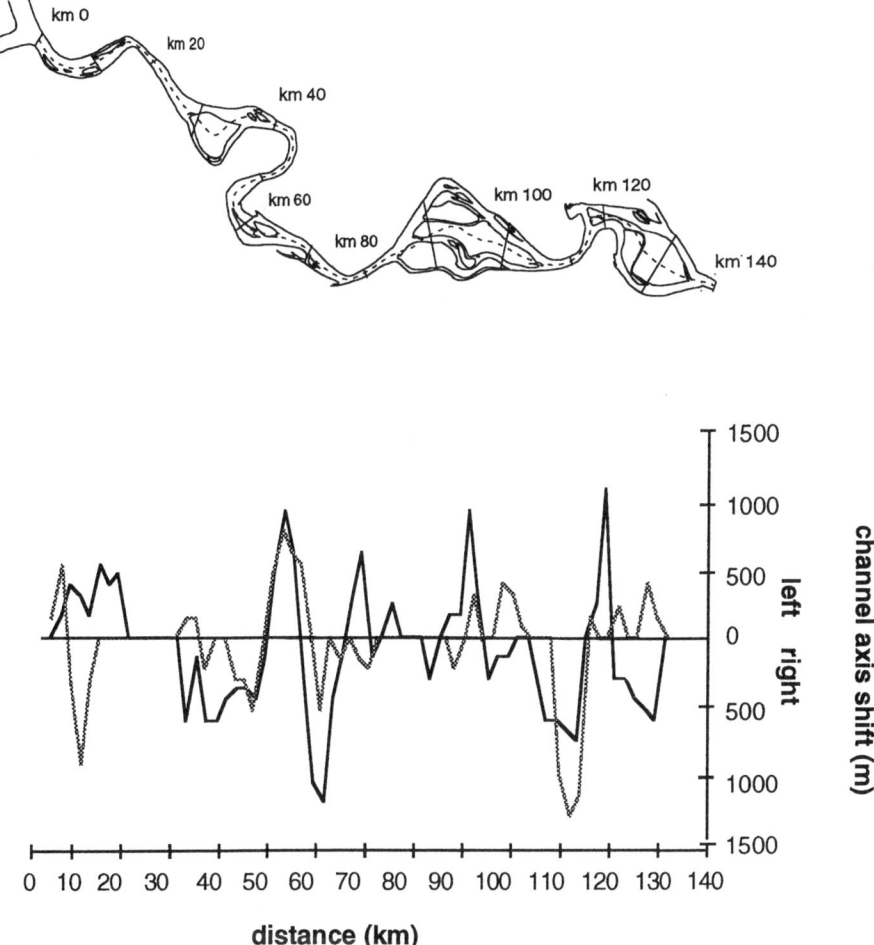

Figure 2. Top: Channel migration of the Amazon mainstream, 1972–1993, with river course and channel axis shown for orientation. **Bottom:** Channel axis shift refers to the distance from the new channel axis to that of the older date. Black line = change from 1983 to 1993; gray line = change from 1972 to 1983.

narrow single channels, although not all such river sections are stable. There is no inclination to unidirectional migration of the river. At km 55 and km 117, the active channel constantly migrated to the left, whereas the opposite trend occurred at km 60 and km 110. At km 10 and km 130, migrations occurred back and forth. The multi-channeled river sections from km 90 to km 105 and from km 115 to km 135 changed significantly all through the monitored period, whereas in many other areas the change rates were distinctively unequal between the different time periods.

Due to the unequal distribution of units of erosion and those of deposition along the length of the river, the channel islands of these multichanneled river sections have undergone many alterations in size and shape (Fig. 3). The individual change units are variable in form and size. For example, one distinctive change unit type is a long but narrow section that follows the shore of the river, indicating constant building or planing of the riverbank. However, the change type of a given shore area itself does not remain constant over time but changes considerably.

Certain river areas appear especially susceptible to repeated changes. The western margin of the multichanneled section, from km 90 to km 100 (near Isla Muyuy), has recently been subjected to numerous alterations, during which a reworking of small islands has taken place. In contrast, the eastern side-channel, near Isla Panguana, has maintained its form and shape for nearly half a century. As for the central parts of this channel reach, both erosion and deposition have increased toward the latter part of the monitored period. Though these river margins are occupied by large trees, we have to assume that the type of shore vegetation does not significantly influence erosion rates.

The islands near the city of Iquitos (km 115 to km 140) have changed in a comparable manner. The Amazon River bifurcates at the upstream end of Isla de Iquitos by a continually changing wedge. Considerable erosion of this zone is one source of change, while the gradual migration of an island (Isla Parlamento) from the apex of an adjacent meander (1962) to join this wedge is equally noteworthy. A less common cause of change is the unification of Isla Padre with Isla de Iquitos. Intuitively, one would hardly anticipate this merge from an examination of the situation as it was in 1948, when a wide active river branch separated those two islands.

One important consequence of the above migration processes is that the river margins consist of land areas of different ages and types of origin. Small islands are particularly young and unstable. Terrain more than 45 years old was found only on the interior of large islands, where it emerges along with younger surface types. Abandoned channel sections, swales, and other swampy depressions often mark the borders between different terrain types. Contrasting habitat classes occur in a variety of combinations, and all islands present their unique arrangement of these patches. Any type of alluvial land is susceptible to future river erosion. River islands thus differ considerably from the sequential build-up of regular meandering loops, such as the one between the two above-discussed multichanneled reaches.

Discussion

Capricious Amazon

The ever-changing nature of the Amazon River is one of the most widely cited facts about Amazonia. As early as 1761, Jesuit missionaries noted the abrupt channel migrations of the Amazon mainstream near Iquitos and compared this river to an animal ("bestia es este gran río" [Uriarte, 1986]). Scientific analyses of fluvial dynamics emerged 200 years later as Sternberg (1960) monitored migrations of the Solimões-Amazon mainstream in Brazil, and Irion et al. (1983) and Meade et al. (1985) made detailed studies of river sediments and their deposition. Mertes (1985) used radar images and navigation charts along with hydraulic data to distinguish five distinct reaches of the Amazon alluvial plain in Brazil.

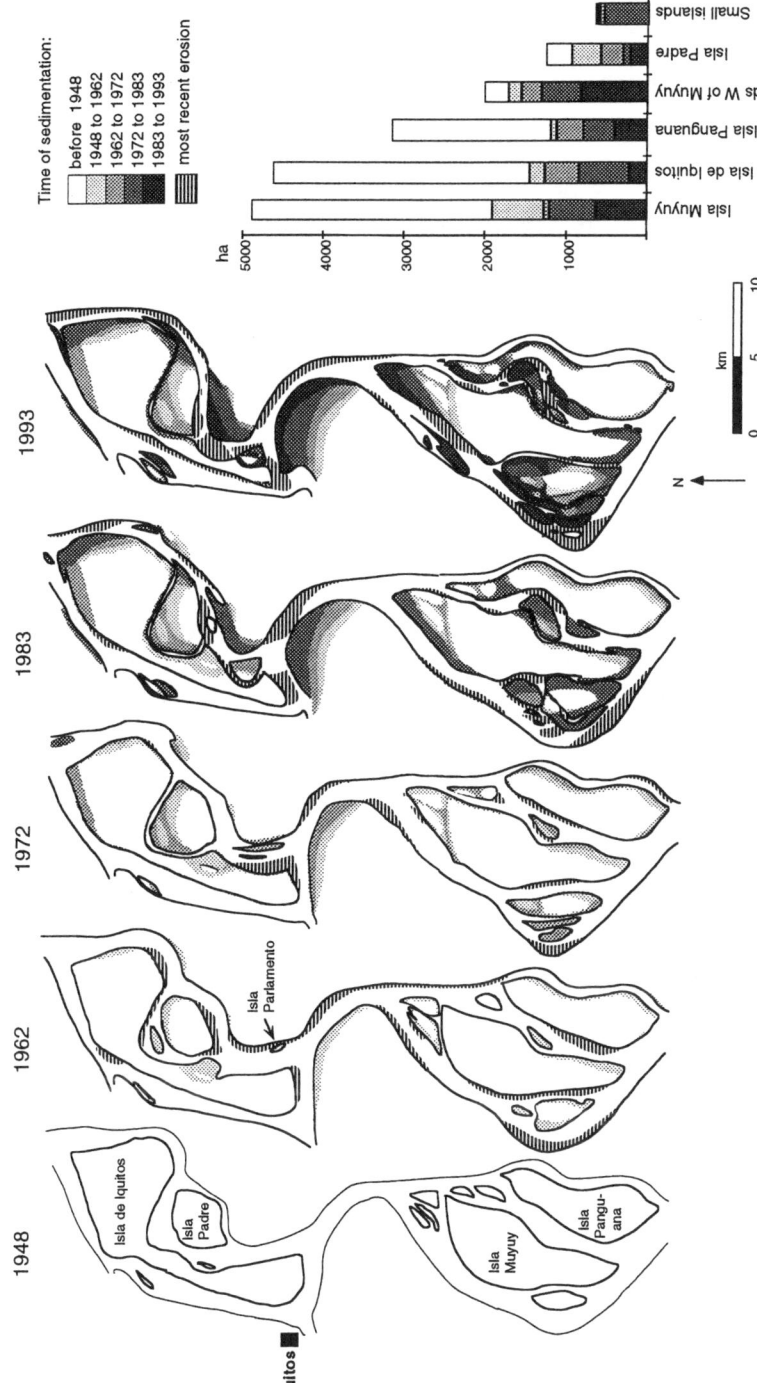

Figure 3. Migration pattern and bank evolution of the Amazon mainstream between km 80 and km 140, 1948–1993.

In Peru, studies have been done on the migration of the Amazon, especially in relation to the development of the region's urban centers (García, 1987; García & Bernex de Falen, 1994). Kalliola et al. (1992) used Landsat MSS images to show that the migration patterns of the anastomosing Marañon and Amazon Rivers differ considerably from those of the Ucayali. The most distinctive difference was that anastomosed channels exhibit patchlike distribution of change units in contrast to the sequential arrangement of such patches along the meandering Ucayali.

This study adds to the present knowledge by showing distinctive variations in the tendency for channel change both spatially (along the length of the same river) and temporally (periods of change/stability at same sites between different years). The most stable reaches tend to be single-channeled sections of relatively narrow streams. Material removed from the upper regions flows through these passages and then is deposited further downriver, where the channel is wider and also very unstable. Indeed, one of the lessons of the present study is to avoid any anticipation of changes at the local level, as our data showed no clear tendencies that would enable effective prediction. On the contrary, Amazonia seems to make a fool of many simple intuitive interpretations of its future behavior. Yet the above results aid in developing ideas for future work with more prediction-oriented goal-setting. One important insight proposes that the anastomosed Amazon is a continuum, where channel sequences may interact in a pulselike manner and no river section should be seen as a separate unit. Moreover, erosion should not be seen as the dominant force of channel migration, as sediment deposition within the channel also promotes changes that can be difficult to anticipate.

Furthermore, the geomorphology of the present alluvial plain adds to observations of Amazon behavior. At its widest, the belt reaches a width of about 40 km, indicating an obvious instability. In addition to the above types of processes, it is likely that even complete river sections may at times be depleted as a result of channel shortening caused by avulsion. Such changes may be relatively fast and locally disastrous.

Urban Planning

The city of Iquitos is the principal commercial center and foreign trade river port in eastern Peru. Ocean-going freighters frequently come up the Amazon to Iquitos, despite navigational problems caused by channel changes and underdeveloped port facilities. The latter circumstance results in part from the constant need to move the location of the port with river channel fluctuations: A comparison of an air photograph taken in 1948 with the current situation shows that during the latest half of this century, the main port of Iquitos has been moved some 4 km downriver.

Other aspects of the urban environment are greatly affected by landslides and the extensive deposition of fluvial sediments. The changing appearance of Malecon, a park-like street by the Amazon River, illustrates the influences of these contrasting phenomena. For example, whereas in the 1960s bank erosion resulted in the loss of urban terrain, by the late 1980s the very same shores turned depositional. These newly formed fluvial bars were first used for seasonal cropping. A more recent cultural consequence, however, has been the dispersion of poor colonist settlements from the district of Belém among the areas of most distinguished appearance in Iquitos. The formation of poor settlements in inundated terrain, in the midst of more exclusive urban sectors that occupy unflooded land, is the root of a sharp social division.

The above examples serve to show how channel changes constitute a constant cause of concern for urban planning. In the early 1960s, a North American consulting company was hired to make a survey of shore erosion because some important areas of the city were threatened by advancing landslides. It was concluded that the problem "should be treated as one of slope stability rather than erosion resulting from river action" (Anonymous, 1965). This opinion was concretized by the construction of extensive shore-protecting structures in 1972 (García, 1987), which ultimately dissolved with hardly any noticeable results. Now, about 20 years later, suggestions for a new engineering effort are again emerging, especially after an incident in April 1994, when the Amazon cut through Isla de Iquitos and landslides destroyed riverine industrial plants. One group of opinion suggests continuous dredging, while others dream of making the Amazon run through an artificial channel.

The present study points out that no project can be justified without an understanding of the dynamic relationship between the upper and lower river sections. Also, the sedimentation pattern in a multichanneled river, where the discharge and sediment load constantly change both within and between years, is complex. We should keep in mind that the Amazon is one of the most unstable rivers in the entire world, so it will be by no means easy to control its migrations. Therefore, it is recommended that future investments which aim to settle the problems caused by the river's instability should favor cultural adaptation, rather than forceful efforts to make the Amazon simply obey our wish.

Floodplain Exploitation

Sustainable production systems should be simultaneously ecologically acceptable and economically profitable. Because whitewater floodplains are both fertile and naturally unstable, they are likely to withstand anthropogenic disturbance. This makes the question of ecological fragility less critical than usual in tropical lowlands. On the other hand, investments for floodplain exploitation are at an increased risk that the business created will not pay well, if at all: River erosion may suddenly destroy an agro-silvicultural area that was once established with great attention. Who wants to risk losing their investment into the river?

Opportunistic exploitation strategies may help to overcome the problem of economic uncertainty. It is a reality that any floodplain area can experience erosion or related changes, and one should accept this high risk and, accordingly, heed the axiom "Don't put all your eggs in one basket." We will exemplify this approach with the case of channel islands, which are, undoubtedly, among the most challenging areas to exploit.

Our results showed consistent changes in the sizes and shapes of river islands, due to which these islands consist of patches of unevenly aged land. On a local level, the environmental conditions reflect unique developmental histories, and also depend on factors such as the actual vicinity of the closest river channel. Over the years, these conditions will change as a consequence of channel migration or because flood depth and duration, or the rate of sedimentation, will differ from time to time. Small islands are repeatedly washed by fluvial disturbance, and only accommodate pioneer vegetation. In contrast, the major part of larger islands consists of areas that have existed for half a century or more, and preserve mature floodplain forests with scattered swamps. Still, the shapes of even these islands are ever-changing in nature.

The above conditions should be taken as the baseline for all land use on river islands. The most stable areas can sustain longstanding activities such as agroforestry or timber production, provided that the species lost to overexploitation will be reintroduced and appropriately managed. In turn the most unstable sites can support short-lived cropping. Also, hunting, extraction, and related nonagricultural activities may contribute substantially to sustenance and income generation. Indeed, it seems possible to find ecologically and economically appropriate production methodologies for a number of different habitat types. The diversity of environmental conditions should simply be met with a diversity of extensive resource use practices (Fig. 4).

Although diversified resource use characterizes the traditional culture of the Peruvian ribereños (e.g., Hiraoka, 1995), our vision suggests that floodplain areas could support much more effective yet still sustained management practices. The major difference is that integrated management models, provided that they are based on a sufficient understanding of local ecological conditions, could support important industries on a permanent basis. This vision requires that land use will be based on those management and exploitation strategies that focus attention on assuring constant production instead of peaklike overexploitation.

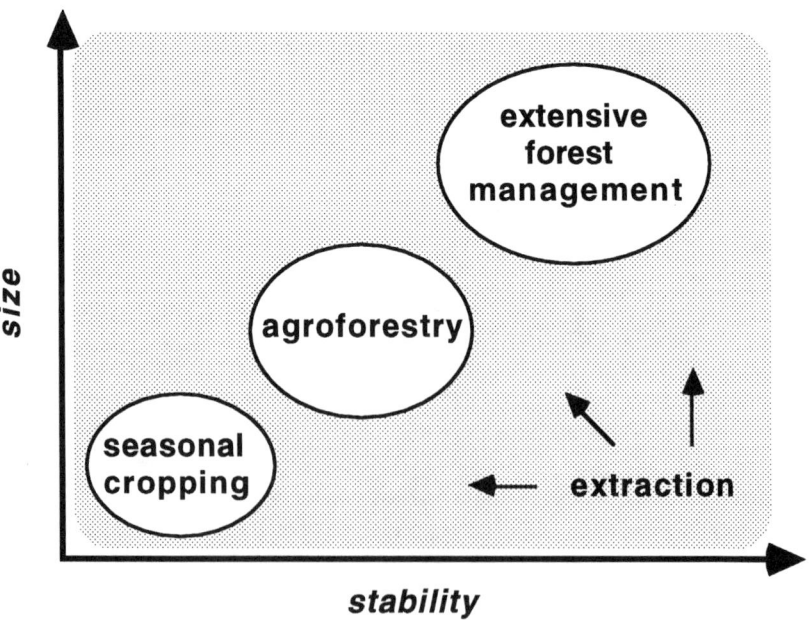

Figure 4. Scheme for the adjustment of integrated exploitation strategies for different site types in alluvial plains.

One example of a natural plant species with an obvious potential is the giant reed-grass *Gynerium sagittatum* (caña brava), which is abundant at practically all the newly deposited fluvial bars of the Amazon River. Each culm of this species produces several one-meter-long decorative panicles, which are used locally for ornamental purposes. The fact that these products are of high quality and can be harvested in overwhelming quantities suggests that this market could be enlarged worldwide. If the sustainability of the harvest were to be certified, the "Amazon giant reed" could be sold as one of the products that serves as an alternative model to the destructive migratory farming in fragile upland areas.

Acknowledgments

We thank our colleagues in the Amazon Project of the University of Turku, and A. Sarmiento, A. Layche, J. Arimoya, INRENA (Instituto Nacional de Recursos Naturales), and UNAP (Universidad Nacional de la Amazonía Peruana) for collaboration; and SENAMHI (Servicio Nacional de Hydrografía y Navegación) and SAN (Servicio Aerofotográfico Nacional) for acquisition of data. The study was supported by the Academy of Finland (SA 30228), the European Union (STD3 TS3*CT94-0314), the Society of the University of Turku, and the society of T. and J. Wallden.

Literature Cited

Anonymous. 1965. Iquitos riverbank protection. Second part of phase II of studies for riverbank protection along the Amazon River at Iquitos, Peru. Fondo Nacional de Desarrollo Económico, Lima; and McCreary Koretsky Engineers, San Francisco. Unpublished report.

Eden, M. J. 1990. Ecology and land management in Amazonia. Belhaven Press, London.

Encarnación, F. 1985. Introducción a la flora y vegetación de la Amazonia peruana: Estado actual de los estudios, medio natural y ensayo de una clave de determinación de las formaciones vegetales en la llanura amazónica. Candollea 40: 237–252.

———, **R. Aquino & J. Moro.** 1990. Flora y vegetación de la Isla Iquitos y Padre Isla (Loreto, Peru): Su relación con el manejo semiextensivo de *Sanguino mystax*, *Saimiri sciureus* y *Aotus*. *In* N. E. Castro-Rodríguez, ed., La primatologia en el Peru. DGFF, IVITA, OPS/OMS, Lima.

García, P. J. 1987. El río que se aleja. CETA, Iquitos.

——— **& N. Bernex de Falen.** 1994. El río que se aleja. Ed. 2. CETA, Instituto de Investigaciones de la Amazonia Peruana, Iquitos.

Hiraoka, M. 1995. Aquatic and land fauna management among floodplain ribereños of the Peruvian Amazon. *In* T. Nishizawa & J. Uitto, eds., The fragile forests of Latin America. United Nations University Press, Tokyo.

IFG (Institute for Applied Geosciences [Finland]). 1984. Mapa planimétrico de imágenes de satélite 1: 250000. IFG, Neu Isenburg.

Irion, G., J. Adis, W. Junk & F. Wunderlich. 1983. Sedimentological studies of the "Ilha de Marchantaria" in the Solimões/Amazon River near Manaus. Amazoniana 8: 1–18.

Junk, W. J. 1989. Flood tolerance and tree distribution in central Amazonian floodplains. Pages 47–64 *in* L. B. Holm-Nielsen et al., eds., Tropical forests: Botanical dynamics, speciation, and diversity. Academic Press, Cornwall, UK.

Kalliola, R., J. Salo, M. Puhakka & M. Rajasilta. 1991. New site formation and colonizing vegetation in primary succession on the western Amazon floodplains. Journal of Ecology 79: 877–901.

———, ———, **T. Häme, M. E. Räsänen, M. Puhakka, M. Rajasilta & R. J. Neller.** 1992. Upper Amazon channel migration: Implications for vegetation perturbation and succession using bitemporal MSS images. Naturwissenschaften 79: 75–79.

————, A. Linna, M. Puhakka, J. Salo & M. Räsänen. 1993. Mineral nutrients from fluvial sediments in the Peruvian Amazon. Catena 20: 333–349.

Lamotte, S. 1990. Fluvial dynamics and succession in the lower Ucayali River basin, Peruvian Amazonia. Forest Ecology and Management 33/34: 141–156.

Meade, R. H., T. Dunne, J. E. Richey, U. de M. Santos & E. Salati. 1985. Storage and remobilization of suspended sediment in the lower Amazon River of Brazil. Science 228: 488–490.

Mertes, L. A. K. 1985. Floodplain development and sediment transport in the Solimões-Amazon River, Brazil. M.S. thesis, University of Washington, Seattle.

Padoch, C. 1988. People of the floodplain forest. Pages 127–140 in J. S. Denslow & C. Padoch, eds., People of the tropical rainforest. University of California Press, Berkeley.

Peñaherrera del A., C. 1986. Geografia fisica del Perú. Gran geografia del Perú 1, vol. 1. Manfer-Juan Mejia Baca, Barcelona.

Puhakka, M., R. Kalliola, M. Rajasilta & J. Salo. 1992. River types, site evolution and successional vegetational pattern in Peruvian Amazonia. Journal of Biogeography 19: 651–665.

Sioli, H. 1984. The Amazon and its main affluents: Hydrography, morphology of the river courses, and river types. Pages 127–165 in H. Sioli, ed., The Amazon: Limnology and landscape ecology of a mighty tropical river and its basin. W. Junk, Dordrecht.

Sternberg, H. O. 1960. Radiocarbon dating as applied to a problem of Amazonian morphology. Pages 7–12 in Comtes Rendus 18 Congrès International de Géographie, vol. 2. Comité National du Brésil, UGI, Rio de Janeiro.

Torres Vásquez, J. 1993. Manejo forestal, un camino hacia la conservación de los bosques en la selva baja. Pages 221–234 in R. Kalliola et al., eds., Amazonia peruana—Vegetación húmeda tropical en el llano subandino. Proyecto Amazonia, Universidad de Turku, and Oficina Nacional de Evaluación de Recursos Naturales, Jyväskylä.

Uriarte, P. M. J. 1986. Diario de un misionero de Maynas. IIAP & CETA, Iquitos.

Vásquez, R. & A. H. Gentry. 1989. Use and misuse of forest-harvested fruits in the Iquitos area. Conservation Biology 3: 350–361.

Multiple Use of Central Amazon Floodplains: Reconciling Ecological Conditions, Requirements for Environmental Protection, and Socioeconomic Needs

Jörg. J. Ohly and Wolfgang J. Junk

Introduction

High population growth rates, particularly in so-called Third World countries, create a constant need to increase food production. However, there are natural and economic limits to a continual expansion in agriculture, and land once considered marginal is needed for cultivation. At the height of the "Green Revolution," irrigation, fertilization, and increased yields due to advances in plant breeding were seen as progress. Today, almost half of the irrigated lands are so badly affected by salinization that their continued use may no longer be feasible. That is why there is renewed interest in cultivating the marginal, or at any rate problematic, lands, namely the wetlands.

One of the principal statements of the workshop "Wetland Soils" (held in Los Baños, the Philippines, in 1985) was that "the greatest potential for expansion of food-producing lands may be in wetland soils" (IRRI, 1985). The most important and intensively used tropical and subtropical wetland areas are found in South Asia and Southeast Asia. The wetlands of Africa, especially the larger areas in Central Africa, have "a limited suitability for conversion to productive use" (IRRI, 1985). The South American wetlands along the coastal zones of Venezuela, the Guianas, and Suriname are already being extensively cultivated for sugar cane, rice, and other field crops. Parts of the Pantanal in Mato Grosso, Brazil, are used for large-scale ranching. But the wetlands of the Amazon basin are still largely untouched, with the exception of areas in the vicinity of towns and minor settlements along the rivers. The Los Baños workshop reported that "tropical South America probably has more wetland with high potential for development than any other continent."

In the past, many authors familiar with the situation of the Amazon floodplains recommended balanced development that took into account the peculiar environmental conditions (Junk, 1989; Sternberg, 1966; Zimmermann, 1987). One of the most enthusiastic was Camargo (1958), who developed an ideal model of complementary use of the flooded alluvial soils and the adjacent unflooded terra firme. The model was described and discussed in detail by Petrick (1978). Klinge et al. (1981), warned that the várzea should not be looked at as a "future paradise." According to the "environmental ranking" proposed by Goodland (1980), the floodplain should only be used extensively (e.g., for keeping water buffalo) due to its short production period and high risk of floods.

The present paper refers specifically to the region around the city of Manaus, with its particular hydrological and socioeconomic conditions. Any conclusions or recommendations must therefore be adjusted before being applied to other regions.

Characteristics of the Floodplain

There are different types of Amazonian floodplains: Those deriving from blackwater and clearwater rivers are locally known as *igapó* floodplains, and those formed by sediment-rich whitewater rivers (i.e., the Amazon-Solimões River, the Purús River, and the Madeira River) are known as the *várzea* (Sioli, 1984).

According to Mertes (1985), the regularly inundated surface area of the Amazon várzea covers approximately 50,000 km^2. This area is classified as the "modern floodplain." About 10,000 km^2 of permanent lakes are included in this estimate. According to other estimates based on the geomorphological and geological maps of RADAM BRASIL, the floodplains of the Amazon and its large tributaries, including river courses and lakes, extend over about 300,000 km^2, of which roughly 200,000 km^2 belong to the várzea category and 100,000 km^2 to the igapó (Junk, 1993). The width of the várzea floodplain varies along the course of the rivers; it is about 16 km wide near Itacoatiara and 50 km wide near Parintins.

Following Walter and Breckle's (1984) ecological classification of climatic zones, the central Amazon region has a humid equatorial climate, with variations in temperature and humidity during the day more pronounced than are differences between seasons. Mean annual rainfall in central Amazonia is 2100 mm. A drier season (locally called summer) lasts from June to November; the rainy season, from December to May. More important than average precipitation is the general occurrence of a few, heavy, and brief rainfall events of 50 mm or more (Ohly, 1987). The monthly distribution of rainfall is unfavorable for agriculture, especially for arable farming, and the heavy rainfall events cause erosion and can damage crops.

Soil Fertility

In the Amazonian context, várzea soils are relatively fertile due to the periodic flooding that annually deposits new sediments (see also Zarin, this volume). Weischet (1977) classified them as "soils of exceptional fertility in the tropics," a statement that reflects the poor qualities of the majority of other soils rather than the favorable conditions of these alluvial soils. Irion (1978) also stressed the fertility of várzea soils in comparison to terra firme soils. Sombroek (1966), however, pointed out that physical and chemical parameters of várzea soils in the Amazon basin vary considerably. Junk (1989) described the small-scale patchy distribution of habitats in the várzea due to differences in edaphic conditions. Because of such variations, general recommendations for agricultural use of the várzea are not useful.

According to the commonly used Brazilian system of soil taxonomy, várzea soils in the central Amazon region are classified as *"gley pouco húmico"* (slightly humic gley) (Cochrane & Sanchez, 1982). The heavy soils often have a high clay content, of 20% or more; the silt fraction is within the 70–80% range. The rather narrow levees are exceptions, with an average content of fine sand (around 15%). The clay fraction is dominated by montmorillonite and illite (Irion, 1978). Montmorillonite has a high capacity for water retention, swelling, shrinking, and absorption. The result is that tillage of the arable soils is problematic and most crops need daily irrigation during the drier season (September–November) or during pronounced dry spells (Gutjahr, 1995).

The chemical characteristics of these soils are generally more favorable. Central Amazonian várzea topsoils show a high base saturation, high levels of exchangeable calcium, and medium levels of exchangeable magnesium, phosphorus, and apparent cation-exchange capacity. The levels of exchangeable potassium are low, and those of nitrogen, sodium, and organic matter are very low. The carbon:nitrogen ratio ranges from 5 to 6 and is rather reduced (Ohly, 1987). According to Alfaia and Falcão (1993), levels of micronutrients (zinc, copper, manganese, iron) vary within a wide range but are high, considering the demand of most crops. Very low concentrations of exchangeable aluminum were found, compared to the generally high levels found in major terra firme soils. The soils are generally low in pH (4.3–5.5). When only natural fertility is considered, the suitability of these soils for most of the cash crops that are now being grown on várzeas is limited (Landon, 1991). For high-yielding varieties of corn, beans, rice, and vegetables, the low nitrogen concentration is a major constraint. Farmers rarely apply urea, because the relative costs and benefits are unknown to them. Further applied research is needed if we are to achieve a better understanding of element flows and water balance in várzea soils.

The Flood Regime

According to the "flood pulse concept" (Junk et al., 1989), the area of a river-floodplain system, which oscillates between a terrestrial and an aquatic phase, is the aquatic terrestrial transition zone (ATTZ). Most forms of utilization of floodplains by man concentrate on this zone. Existence, productivity, and interactions of the major biota in the ATTZ depend on the flood pulse.

Figure 1 shows the average curve of the annual fluctuation of the Rio Negro at Manaus harbor, reflecting as well the situation of the nearby Amazon-Solimões River. Floods are at their maximum levels in June–July. They then fall rapidly, reaching their minimum in October–November, and then rise again gradually. Although variations between the years can be considerable, over medium- and long-term periods, they are predictable. Elderly people remember only the disastrous flood of 1953. The high floods in 1989, 1993, and 1994 destroyed most of the papaya and banana plantations, certain crops like manioc and sweet potatoes, and many home gardens and houses. Yet these floods did not surprise the riverine folks.

The flood pulse strongly influences many forms of human use of the alluvial plains, including the practice of arable and livestock farming, wood extraction, and fisheries. The terrestrial phase, which permits agricultural activities, lasts approximately from Au-

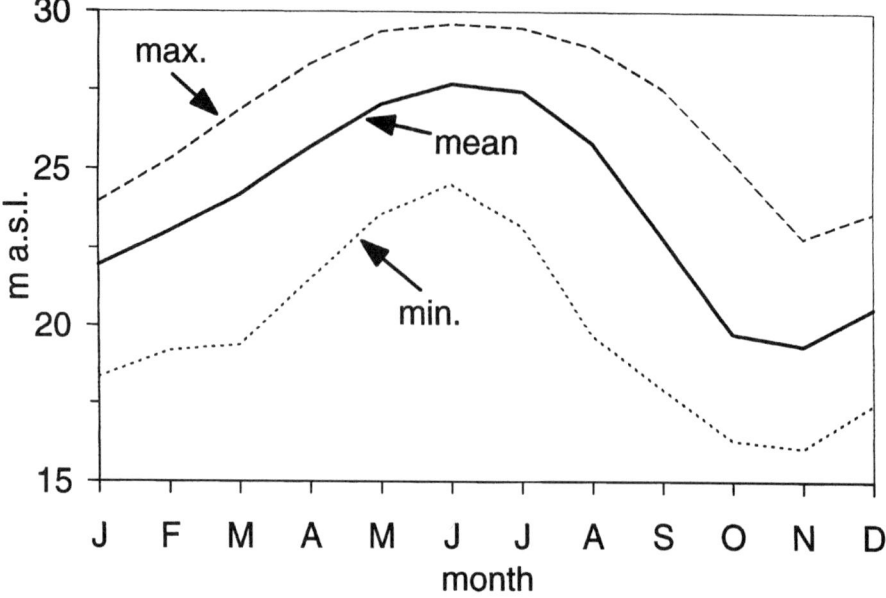

Figure 1. Mean monthly water level fluctuations and absolute monthly minima and maxima of the Rio Negro at Manaus. (Data from PORTOBRAS, Manaus harbor, 1974–1993.)

gust to March. Timber felled in this period can only be removed during the high-water phase when the floating trunks are tied together in large rafts. Although fishing is possible all year round, fish become scarce in the high-water period (when they enter the floodplain forest for feeding and shelter).

The river system has always been and still is the most important means of transportation in the central region of the Amazon basin. Currently the only paved roads in the region link Manaus with two nearby towns: Itacoatiara (288 km) on the left bank of the Amazon River, and Manacapurú (ca. 80 km) on the left bank of the Solimões River. Unpaved roads are impassable or in very poor condition during the rainy season. The road (2294 km) connecting Manaus with Cuiabá (Mato Grosso) via Porto Velho (Rondônia) has been unfit for travel between Manaus and Porto Velho (with the exception of some stretches) since the mid-1980s. The partly paved dirt road (667 km) connecting Manaus with Boa Vista (Roraima) is generally closed for transport for weeks during the rainy season. Perishable goods (i.e., fresh vegetables) are flown into the region during the flood period when the várzea is out of production. Less perishable goods (i.e., onions, potatoes, or deep frozen meat) are transported by trucks shipped on large rafts from Belém in Pará (1713 km).

Houses, sheds, and home gardens can be found on levees which in central Amazonia have an elevation of 27–29 m above sea level. Land is used for grazing and crops down to 23–24 m. That means that the lowest area of a farm is, on average, free of flooding for five or six months of the year (Junk, 1989). Growing time is indeed limited, since pastures need up to eight weeks for the sward to build up (Ohly, 1987), and farmers must prepare their fields. The mean growing period for crops is eight months for the levees and four months or less for the low-lying areas. Depending on the topographic conditions of the site, farms occupy a 20–50 m stretch along the river or channel. Large cattle ranches are sometimes 1000 m wide. All farms need access to the river or channel for water supply (drinking, washing, irrigation) and transport. The shape of landholdings is usually rectangular, and length depends on available manpower. Small farms that depend entirely on family labor may be only 50–200 m long. Large cattle farms reach further inland and can have an extension of several hundred hectares. Due to devastating erosion along some reaches of the mainstream and major channels of the Amazon (paranás), many farms were forced to move further inland several times in the last couple of decades, and many lost 50% or more of the original land. Other farms have disappeared altogether (Sternberg, 1956).

Present Floodplain Populations

Apart from reports of travelers (e.g., Acuña, 1891) and a few archaeological findings, very little information is available on the precolonial population density of the floodplains. Estimates made by Denevan (1976) and Roosevelt (1991) indicate that previously there were significantly higher population densities, both in the floodplains and in the terra firme, than there are today (see also Roosevelt, this volume). Although these estimates might have a considerable margin of error, there is no doubt that comparatively complex Amerindian cultures had developed in the várzea (Meggers, 1984). But all disappeared and have been replaced by what some modern authors call "caboclo culture." This term is not well-defined and is rather confusing (caboclo means "coming from the forest" in Língua Geral). Several studies exist regarding the origin of the present riverine population (e.g., Sternberg, 1956; Grenand & Grenand, 1993), but it is obvious that current settlers, farmers, or fishermen of the Manaus-area floodplains are descendants of poor migrants from Brazil's famine-stricken northeast who entered the region since the 1870s. Although there is constant migration to the region (mainly to urban centers), two major periods when labor was in great demand can be distinguished: the "golden era of the rubber boom" in the second half of the nineteenth century which ended abruptly during the World War I, and the short revival of this boom during the World War II (when the Allies ran short of raw material due to Japanese occupation of the major production centers in South Asia and Southeast Asia). Many of the migration workers who tapped rubber stayed in Amazonia. Some traditions (dances, traditional songs), habits, and terms used in daily life (e.g., regional names for weeds) derive from the "Nordeste" of Brazil.

The state of Amazonas covers 1.57 million km^2. About 28.5% of its approximately 2.01 million inhabitants live in rural areas (IBGE, 1991). In terra firme areas the population density is below 0.5/km^2, whereas districts along the mainstream of the Amazon, from Manaus to Óbidos in the east of the state, have 1–10 persons/km^2. Some várzea districts in the western part of the state bordering Peru, however, have population densities of 0.5–1/km^2. Only the stretch along the Solimões from Tefé to the district Manacapurú has a low population density as low as that of the terra firme. There are three very obvious reasons why people choose to live in floodplain areas: the fertility of várzea soils, easy access to fish as a cheap and valuable source of animal protein, and the availability of the river as a means of transport to and from markets on or near the mainstream Amazon. Such is the case of the city of Manaus, located on the left bank of the Negro River.

Manaus, the capital of the state, falls outside the typical population distributions in the two major ecosystems. The city is home to about 60% of the state's overall population. The free-trade zone established in the early 1970s is the main driving force for urban development and a major source of income in the state (Kohlhepp, 1984). Migration to this city from rural areas within and outside the state is a constant concern for developers.

Is the Present Land Use Adapted to the Ecosystem?

There are four major systems of várzea land use: cultivation of crops, animal husbandry, timber extraction, and fishing. Agroforestry systems are not dealt with separately here because they are economically insignificant today. Highly diversified home gardens supply farm families with fruit, fuel wood, and medicinal plants (Lima, 1994). Many of the remaining rubber and cocoa plantations have been transformed into pastures (Bahri, 1993). The surplus of some products of home gardens (e.g., mangos, bananas) is sold in local markets, but this is not a major contribution to the floodplain economy. Rubber tapping still takes place, to a certain extent, but local and world markets demand a high quality of latex, and prices are low.

Agriculture

Many crops planted today on floodplains are poorly adapted to the climatic conditions (e.g., tomatoes, lettuce, cauliflower). Crops in use (e.g., modern varieties of maize) have a production potential that cannot be achieved by traditional low-input agriculture with no fertilizer application. These "exotic" varieties are often very susceptible to pests and insect attacks (Lourd, 1993; Gutjahr, 1995) and have to compete with very vigorous native vegetation. Weeding is usually done manually. Herbicides are too expensive, and as a result of the relatively short terrestrial phase, low-lying areas are very risky for arable farming (see Fig. 2).

Some orchard or plantation crops can withstand short and shallow inundation (e.g., mango, coconut), whereas others have to be replanted if flooded (e.g., papaya, passion fruit, bananas) and should be grown only on the levees (restingas). Cassava tubers, a staple food, must be harvested before flooding. Although there are cassava varieties in use with a short growing period, floodplain-grown cassava cannot compete in quality and yield with cassava grown on terra firme (Gutjahr, 1995).

Poultry and swine are kept only for household consumption. Cattle and, to a lesser extent, water buffalo (Ohly, 1986) are raised almost exclusively for meat production on large and extensively managed farm units. During the flood period, herds have to be

Figure 2. Flood threatens a late-planted lettuce field near Manaus. Beginning of the high-water period 1993. (Photograph by J. J. Ohly.)

transferred to terra firme pastures for two to four months. Wilhelmy (1969) used the term "tropical transhumance" for this production system. Usually only small dairy herd units remain in the várzea and are kept and fed in "*marombas*" (sheds on wooden piles or rafts) with aquatic grasses such as *Echinochloa polystachya* (Fig. 3).

The economics of producing and marketing annual food crops (Gutjahr, 1995), orchard crops, plantation crops, and livestock in the area (Hund, 1995) have been investigated in detail. Table I shows the productivity of land and labor for some common production systems.

If production systems are ranked according to their labor productivity, intensive vegetable and horticultural production systems are superior. They also show high levels of land productivity. These data suggest that a farmer can achieve a high income on a relatively small area using such intensive methods. Although productivity per man-day is the highest, these production systems are rather labor intensive, and hired labor is scarce. Therefore, only small units of these production systems can be found (see Fig. 4). Size depends almost entirely on family labor.

Among animal production systems, water buffalo husbandry shows the highest productivity per unit of labor. Large herds can be managed by only one or two men. Generally no milk or cheese is produced except for the consumption of families living on the farm. Land productivity is significantly lower than that of cattle production systems. Water buffalo roam freely on native pastures and are well-adapted to the semi-aquatic conditions of the ecosystem (Ohly, 1987). These data show that the keeping of water buffalo is highly competitive where costs of land are relatively low (which is the

Figure 3. A small herd of dairy cows is fed in a maromba-type shed with the aquatic grass *Echinochloa polystachya* (Canarana). Careiro Island, Paraná do Cambixe. Highwater period 1993. (Photograph by J. J. Ohly.)

case in districts distant from major markets). Other animal production systems such as beef cattle, dairy cattle, and specialized cheese production have a significantly lower labor productivity and a slightly higher land productivity. Their disadvantage when compared to raising water buffalo are their comparatively higher maintenance costs for fencing, establishment and maintenance of artificial pastures (*Brachiaria radicans* and *B. mutica*), labor, and additional pastures on the terra firme. The demand for beef has remained high and prices are stable, owing to the fact that about 85% of the beef consumed in the state of Amazonas is imported from other Brazilian regions. High costs of forest clearing make beef production unprofitable in areas below 25–26 m elevation (Hund, 1995).

Vegetables, horticultural crops, and permanent or perennial crops (i.e., orchard and plantation crops) are generally competitive (Table I) with animal production systems and some other field crops (e.g., maize and fiber crops). Cacao was an important tree crop on floodplains in the past, but today the market value is low. The expansion of these production systems is limited by several factors, including the present capacity of the local market, insufficient infrastructure for transportation of perishable goods such as most vegetables, and inadequate processing capacity for fruits, juices, or frozen pulp, as well as the scarcity of labor in rural areas.

Field crops generally show low productivity per unit of labor, but in comparison to animal production systems their productivity per unit of land is relatively high. Growing crops is an alternative to animal production if farms are small and sufficient labor is

Table I. Productivity of labor and land of some agricultural production systems in the central Amazon region

Production system	Labor productivity (US$/man-day)	Land productivity (US$/ha)
Vegetables/horticultural crops (intensive production)		
Lettuce	16.8	2990
Tomato	13.5	1920
Sweet peppers	11.6	1260
Cucumber (*C. sativus*)	11.5	1220
Cabbage (var. *capitata*)	10.3	940
Vegetables/horticultural crops (extensive production)		
Watermelon	16	430
Lettuce	14.9	2160
Sweet peppers	10.9	740
Tomato	10.7	1130
Cabbage (var. *capitata*)	5.7	550
Cucumber (*C. sativus*)	3.3	350
Animal production		
Water buffalo (meat)	11.3	33
Cattle (milk)	7	60
Cattle (cheese)	6.3	58
Cattle (beef)	5.9	41
Orchard/plantation crops		
Papaya	8.5	1896
Passion fruit	8	1846
Banana	8	916
Mango	7.5	685
Coconut	5.6	675
Cacao	3.1	70
Field crops		
Cassava (sweet)	7.9	109
Sweet potato	4.2	162
Cassava (bitter)	3.6	81
Maize (grain)	2.9	70
Malva	2.5	45
Jute	1.8	48

Source: Adapted from Gutjahr, 1995; Hund, 1995.

Figure 4. Aerial view of the southern coast of Marchantaria Island showing the patchy distribution of smallholder farms 2000 to 3000 m² in size. Low water period 1992. (Photograph by J. J. Ohly.)

available. Sweet potatoes are produced on a commercial scale in Amazonas only on floodplains; sweet potatoes are one of the few locally produced crops that satisfy 100% of local demand, including the Manaus market. Cassava has been included in this list, as it is produced primarily as a subsistence crop on the várzea. Today the growing of maize, malva, and jute is not economically sound when the costs of hiring labor are considered; labor productivity in these crops is close to or below daily wages. Fiber production continues in the várzea because it is a cash crop that is not perishable. Maize, like rice, has never been a typical várzea crop in the Central Amazon Region, although it is rather successfully grown in the floodplains of the lower Amazon (Lima, 1956).

Wood Extraction

Timber extraction has been an important economic activity for over a hundred years. Traditionally, most raw timber comes from the alluvial plains. The main reason for this is that the cost of transport by the "river highways" to the processing centers is low.

The highly flood-adapted várzea forest occupies around 5.5 million ha and has approximately 90 m^3/ha^{-1} of standing timber. Only a few tree species, about 25–30, are exploited commercially. Of these, 7–10 species account for a significant portion of the wood locally processed. They are all classed as softwood and have a relatively low market value. Locally exploited timber is often of low quality, and the utilization rates of trunks range between 54% in sawmills and 56% in the plywood industry (M. Klenke, pers. comm.). Almost 80% of the processed wood is consumed within Amazonas for low-cost housing, building joinery, and, to a lesser extent, furniture. A substantial part of the timber used in construction is used for temporary site structures and concrete framework. Most of the remaining 20% is exported to expanding markets in central Brazil. International trade plays a minor role in the wood industry of Amazonas, although it is important in the lower Amazon region, including the state of Pará.

Higushi et al. (1994) give a detailed description of the operations of wood procurement, harvesting, and transport. Logging is done selectively, with few stems removed per area; it is carried out by hand or with a power saw. The logging season is the low-water period. When the floodwaters reach the timber felled along the margins of the rivers and channels, large rafts are assembled and pushed to the processing centers. The availability of raw timber is very seasonal. In years of low floods (e.g., 1991 and 1992), logs could not be moved and wood became scarce and expensive. Although legal norms and regulations for the exploitation of native forests exist, commercial-scale management or even reforestation is not practiced. Over the last few years this lack of management has resulted in the overexploitation of species such as samaúma (*Ceiba pentandra*).

A real obstacle for the várzea ecosystem and the timber industry is the fact that until today there has been no industrial forest management. Land or felling permits on government land are inexpensive, timber has been abundant in the past, government control of protected wood species is inefficient, and research has not provided foresters and industry with specific management techniques and plans.

Although reliable statistics are not available, the wood industry does offer a source of income for seasonal hired labor in many regions where other sources of income do not exist.

Fisheries

Várzea fisheries are explored at length in section 1 of this volume. Here we will mention that Bayley and Petrere (1989) estimate the fishery potential of Amazonian inland waters at about 1 million tons per year, of which 20–25% is actually used. A major part of the fish derives from fertile floodplains of whitewater rivers which show a fish biomass about five times larger than that of blackwater rivers (Saint-Paul, 1994). These data indicate the great importance of the floodplains in general and the várzea specifically for the supply of animal protein to the Amazonian population.

Large parts of the rural population and low-income groups in major cities or towns cannot afford other sources of animal protein. Table II shows the results of data collected in Manaus (Comissão de Desenvolvimento do Estado do Amazonas, 1988). Only people earning the minimum wage (around US$75 per month) or more were interviewed. This cut-off excluded a considerable part of the urban population. The situation in rural areas is very different because beef, pork, mutton, or sausages are rarely available or are too expensive when compared to fish. When income rises, other sources of animal protein can be substituted for fish. Beef is favored and, as expected, the consumption

Table II. Consumption of meat, meat products, and fish in Manaus (kg per capita, annually), by income group

	Income group (multiples of the minimum salary US$75/month)		
	I (per capita kg/yr [%])	4 (per capita kg/yr [%])	>9 (per capita kg/yr [%])
Beef	10.20 (25.4)	14.40 (30.5)	51.00 (60.2)
Poultry	5.88 (14.7)	10.56 (22.4)	18.12 (21.4)
Other	1.08 (2.7)	2.16 (4.6)	12.36 (14.6)
Fish	22.92 (57.2)	20.04 (42.5)	3.24 (3.8)
Total	40.08 (100)	47.16 (100)	84.72 (100)

Source: Adapted from Comissão de Desenvolvimento do Estado do Amazonas, 1988.

of all meat and meat products increases significantly. Apart from a few fish species of high commercial value, fish, together with rice, beans, and cassava meal, is the primary food source for the poor.

In rural areas almost every family engages in subsistence fishing. Any surplus also provides an important additional income. Depending on the technique, tradition, preferences, and season, a wide spectrum of species are fished (Petrere, 1978a, 1978b).

Fish yields are highly competitive per unit of area when compared with other animal production systems. Hund (1995) based his calculations on estimates of a potential total catch of 47.1 kg/ha (Bayley, 1986) and the average market price for fish in Manaus. He concluded that beef production is only more profitable than fishing along the highest parts (28–29 m elevation) of the floodplains. The demand for fish and beef alike will increase while the prices for these products will continue to be stable. But increasing demand for fish will force fishermen to move to remote areas, and this will consequently raise production costs and possibly cause an expansion of animal husbandry.

The establishment of pastures for cattle and water buffalo has a direct negative effect on the fish populations. If the várzea forest is removed on a large scale, fish will lose feeding areas and shelter during the flood period (Goulding, 1980; Saint-Paul, 1994). There are reports of substantial decreases in the fish populations in intensively used ranching regions such as Parintins. A constant conflict between subsistence and commercial fisheries has also existed over the last 40 years because both types of fishermen compete for the same fishing grounds (Junk & Honda, 1976).

Commercial fisheries concentrate on a relatively small number of species of high market value that can be purchased by consumers with higher incomes and that are offered in restaurants and tourist hotels; these include the tambaqui (*Colossoma macropomum*), pirarucu (*Arapaima gigas*), and tucunaré (*Cichla ocellaris*). Commercial fishermen also seek some schooling species that can be captured in large numbers for low-income groups, such as jaraquí (*Semaprochilodus* spp.) and the curimatá (*Prochilodus nigricans*).

Brazilian legislation restricts fishing of some species during certain periods, sets the minimum size of some species, and regulates the methods of fishing that are permitted. It is at present impossible, however, to effectively control commercial fishing.

The Benefits and Limitations of the Central Amazon Várzea for Human Use: A Summary

The benefits include the following:

- The várzea provides the only large nutrient-rich area in the central Amazon region.
- The land is "fertilized" annually by periodic floods.
- It is today the only area with cheap means of transport.
- Most food consumed is locally produced.
- It furnishes subsistence and permanent residence for a major part of the rural population.

There are, however, arguments against and constraints on the expansion of the land use of várzeas:

- Production is highly seasonal.
- The risk of floods restricts access to credit and necessary investments.
- In many sections of the river, the várzea comprises rather small structural units that do not permit the large-scale production units that are a prerequisite for modern conventional agricultural land-use systems.
- Present land-use systems often are poorly adapted to the peculiarities of the floodplains.
- There is a lack of knowledge about the actual carrying capacity of the area over medium or long periods of time under present production systems.
- There is widespread uncertainty both about legal rights to land ownership and concerning use restrictions.

At present, Brazilian federal policies do not regulate the use of most Amazonian floodplains. Earlier plans to include these várzeas in a countrywide wetland development project (PROVARZEAS) have not been realized. Under the environmental law—the "Código Florestal" of 1988—the margins of major rivers seem to be excluded from any land use and further development. However, Vieira (1992), considering Brazilian environmental law and other regulations concerning floodplains, wetlands, and water rights, concluded that, under current Brazilian legislation, a rational exploitation of Amazonian floodplains is both permissible and possible.

The specific socioeconomic and natural conditions of central Amazonia demand a rational and well-thought-out form of exploitation of the floodplains, addressing economic needs without ignoring the peculiarity of the complex ecosystem. The history of the extractive economy over the last few centuries, the failure of the "conquest" of the vast terra firme in the 1970s and early 1980s (Operation Amazonia), and the establishment of the Free Trade Zone at Manaus (Zona Franca), which led to heavy migration from the hinterland to the capital, demonstrate that an integrated development of the region as a whole has not taken place in recent times. Various economic events of the past (e.g., the rubber boom) or in the present (e.g., industries created in the Free Trade Zone) created a dependency on very distant regions of Brazil or elsewhere. In this respect there is little difference between the situation of the eighteenth or nineteenth century and that of today (Agassiz & Agassiz, 1969; Le Cointe, 1922; Wallace, 1853). From the agricultural point of view, it is hard to understand why the gap between the supply and demand for food could not be reduced if one takes into consideration the potential of the fertile floodplains and the extremely low population density.

Utilization versus Environmental Protection

Use of the várzea includes activities carried out during the terrestrial phase (mainly arable farming, animal husbandry, and logging) and activities of the aquatic phase (mainly fishing and transport of timber). The impact of various utilization systems varies, leading, in turn, to increasing conflicts between the different users and between use and conservation values as well.

Agricultural activities are concentrated on higher elevations, mainly the levees, to maximize the production period for crops and domestic animals. These areas are covered originally by a highly flood-adapted floodplain forest that has multiple functions in the ecosystem—for instance, as stabilizer of sediments, as a brake for water currents and fire, as habitat for many plant and animal species, as a timber source, and as an important food source for many fish species with economic value. Large-scale destruction of the floodplain forest has negative impacts on the ecosystem, reduces habitat and species diversity, and diminishes stability of the floodplain. Extensive destruction also has major socioeconomic effects, especially on timber extraction and fisheries and on the populations that depend on these activities. On the other hand, the maintenance of the floodplain forest restricts agricultural activities, thus reducing rural development and capacity of many to remain in rural areas. There are no good arguments to prohibit any utilization of fertile soils in the Amazon, especially because this would increase pressure on the much more fragile terra firme ecosystems.

It is very likely, therefore, that the utilization of the várzea will increase in the future. An optimal utilization of all resources will be possible only when all production systems are carried out in a more balanced form than they are at present. A concept of multiple use of the várzea is ecologically, economically, and socially desirable. Such a concept requires an agro-ecological zonation that differentiates the use of the várzea along the flood gradient—that is, according to the length of the terrestrial and aquatic phases and by distances to markets. In very general terms it would include the following:

First, arable farming and milk production should be concentrated on levees in the vicinity of the markets to allow fast and relatively inexpensive market access for the perishable products. Some animal husbandry can serve as a side business for small farmers, who use the animals for traction and as a sort of insurance for specific expenses. Clear-cutting of the forest should be done only to an elevation of about 24–25 m, corresponding to an average five- to six-month terrestrial phase in the central Amazon region. Stretches of forest or plantations of trees for fruit or timber production, perpendicular to the river current, should be maintained on the levees to act as brake for fire and water current and to maintain biological diversity. Because of the small-scale pattern of habitats in the várzea and great demand for labor, small holdings should be promoted.

Areas below 24–25 m should not be deforested on a large scale but could be used for timber extraction, including the enrichment of the forest by planting valuable local flood-resistant tree species. This would help maintain the stability of the floodplain as well as its biological diversity, and would favor the growth of frugivorous fish. In addition, this would compensate partially for losses of agricultural production during high floods.

Native pastures found at elevations below 21–22 m, corresponding to an average of less than three to four months of terrestrial phase, could be used periodically as additional grazing areas for cattle from terra firme or várzea ranches.

Large-scale cattle or buffalo husbandry for meat should be concentrated in areas adjacent to the nonflooding terra firme, to which the animals could be easily transferred during periods of higher floods. Animal husbandry could be concentrated, too, in areas distant to markets, since live animals are traditionally transported to slaughter. Meat production systems should not compete for várzea land in areas close to markets; these areas should be reserved for the production of perishable goods.

Specific areas should be selected and maintained as nature reserves and totally protected or used with restrictions. Floodplains of blackwater rivers should be used exclusively for fishing, owing to the low natural fertility of the soils and the extremely slow growth rate of igapó forest.

The type of management outlined above would allow for multiple and sustainable use of all resources and products of the floodplain without the destruction of habitat and species diversity. It would fix a considerable part of the population in the várzea. Cost-benefit data to support the economic aspects of this model are available.

Acknowledgments

Most of the ideas expressed in this paper are the result of research conducted in the central Amazon region by colleagues and the authors, within the framework of the scientific cooperation program between the National Research Institute of the Amazon (INPA) and the Max-Planck-Institute for Limnology. Part of this cooperation includes the "Multiple Land Use on Central Amazon Floodplains" project, sponsored by the Federal Ministry for Education, Science, Research, and Technology (BMBF) of Germany (Project SHIFT ENV 14, grant no. 0339372A), the National Research Council (CNPq) of Brazil, and the Brazilian Institute for Environment and Natural Renewable Resources (IBAMA). We thank M. Klenke (Manaus, Brazil) for providing us with valuable information about the wood industry.

Literature Cited

Acuña, C. de. 1891. Nuevo descubrimiento del gran río de las Amazonas. Colección de livros que tratan de America raros o curiosos. Vol. 2. Madrid. [Reprint of the original 1641 edition.]

Agassiz, L. & E. C. Agassiz. 1969. Voyage au Brésil, 1865–1866. Librairie Hachette & Cie., Paris. [Reprint.]

Alfaia, S. S. & N. P. Falcão. 1993. Estudo da dinâmica de nutrientes em solos de *várzea* da Ilha do Careiro no estado do Amazonas. Amazoniana 12(3/4): 485–493.

Bahri, S. 1993. Les systèmes agroforestiers de l'le de Careiro. Amazoniana 12(3/4): 551–563.

Bayley, P. B. 1986. Aquatic productivity in the central Amazon várzea in the context of fishery yield. Pages 351–359 *in* EMBRAPA, ed., Simpósio do Trópico Húmedo. Vol. 5. EMBRAPA/CPATU, Departamento de Difusão de Tecnologia, Brasília.

———— & M. Petrere Jr. 1989. Aquatic productivity in the central Amazon várzea in the context of fishery yield. Pages 385–398 *in* D. P. Dodge, ed., Proceedings of the International Large River Symposium. Canadian Special Publications in Fisheries and Aquatic Science 106.

Camargo, F. C. 1958. Report on the Amazon region. Pages 11–24 *in* Problems of humid tropical regions. Humid Tropics Research. UNESCO, Paris.

Cochrane, T. T. & P. A. Sanchez. 1982. Land resources, soils and their management in the Amazon region. A state of knowledge report. Pages 137–209 *in* S. B. Hecht, ed., Amazonia: Agriculture and land use research. Proceedings of the international conference sponsored by the Rockefeller Foundation. Centro Internacional de Agricultura Tropical, Cali.

Comissão de Desenvolvimento do Estado do Amazonas (CODEAMA). 1988. Cidade de Manaus. Pesquisa sobre orçamentos familiares (02.84–01.85). Comissão de Desenvolvimento do Estado do Amazonas, Manaus. Unpublished document.

Denevan, W. M. 1976. The aboriginal population of Amazonia. Pages 205–234 *in* W. M. Denevan, ed., The native population of the Americas. University of Wisconsin Press, Madison.

Goodland, R. J. A. 1980. Environmental ranking of Amazonian development projects in Brazil. Environmental Conservation 7(1): 9–26.

Goulding, M. 1980. The fishes and the forest. Exploitations in the Amazonian natural history. University of California Press, Berkeley.

Grenand, F. & P. Grenand. 1993. Histoire du peuplement de la várzea en Amazonas. Amazoniana 12(3/4): 509–526.

Gutjahr, E. 1995. Untersuchungen zur Optimierung der Ackernutzung in den Überschwemmungsgebieten des mittleren Amazonas. Studien zur Agrarökologie 21. Verlag Kovac, Hamburg.

Higushi, N., A. C. Hummel, J. V. de Freitas, J. R. Malinowski & B. J. Stokes. 1994. Exploração florestal nas várzeas do estado do Amazonas: Seleção de árvores, derrubada e transporte. Pages 168–193 *in* Eighth Updating Seminar on Harvesting and Wood Transportation, 8–13 May 1994, Curitiba, Brazil.

Hund, M. 1995. Möglichkeiten und Grenzen der landwirtschaftlichen Nutzung der Überschwemmungsgebiete (Várzea) am mittleren Amazonas unter besonderer Berücksichtigung der Viehhaltungs- und Dauerkultursysteme. Wissenschaftsverlag Vauck Kiel KG, Arbeiten zur Agrarwirtschaft in den Entwicklungsländern.

Instituto Brasileiro de Geografia e Estatística (IBGE). 1991. Anuário estatístico do Brasil. Instituto Brasileiro de Geografia e Estatística, Rio de Janeiro.

Irion, G. 1978. Soil infertility in the Amazonian rain forest. Die Naturwissenschaften 65: 515–519.

International Rice Research Institute (IRRI). 1985. Wetland soils, characterization, classification, and utilization. Proceedings of a workshop held 26 March to 5 April 1984. International Rice Research Institute, Manila, Philippines.

Junk, W. J. 1989. The use of Amazonian floodplains under an ecological perspective. Interciencia 14(6): 317–322.

————. 1993. Wetlands of tropical South America. Pages 679–739 *in* D. F. Whigham et al., eds., Wetlands of the world: Inventory, ecology, and management. Vol. 1. Kluwer Academic, Dordrecht.

———— & E. M. S. Honda. 1976. A pesca na Amazônia. Aspectos ecológicos e econômicos.- Anais do I. Encontro Nacional sobre Limnologia e Pesca Continental, Belo Horizonte-MG, 1975, Fundação João Pinheiro: 211-226.

————, P. B. Bayley & R. E. Sparks. 1989. The flood pulse concept in river-floodplain systems. Pages 110–127 *in* D. P. Dodge, ed., Proceedings of the International Large River Symposium. Canadian Special Publications in Fisheries and Aquatic Science 106.

Klinge, H., K. Furch, U. Irmler & W. J. Junk. 1981. Fundamental ecological parameters in Amazonia, in relation to the potential development in the region. Pages 19–36 *in* R. Lal & E. W. Russel, eds., Tropical agricultural hydrology. John Wiley, London.

Kohlhepp, G. 1984. Development planning and practices of economic exploitation in Amazonia. Recent trends in special organization of a tropical frontier region in Brazil. Pages 649–674 *in* H. Sioli, ed., The Amazon: Limnology and landscape ecology of a mighty tropical river and its basin. W. Junk, Dordrecht.

Landon, J. R., ed. 1991. Booker tropical soil manual: A handbook for soil survey and agricultural land evaluation in the tropics and subtropics. Longman, New York.

Le Cointe, P. 1922. L'Amazonie brésilienne. Ed. A. Challamel, Paris.

Lima, R. M. B. 1994. Descrição, composição e manejo dos cultivos mistos de quintal na várzea da Costa do Calderão, Iranduba, Amazonas. M.Sc. thesis, Universidade do Amazonas/Posgraduação INPA, Manaus.

Lima, R. R. 1956. A agricultura nas várzeas do estuário do Amazonas. Boletim Técnico 33. IAN, Belém.

Lourd, M. 1993. Os principais patógenos das plantas cultivadas na Ilha do Careiro. Amazoniana 12(3/4): 565–576.

Meggers, B. J. 1984. The indigenous peoples of Amazonia, their cultures, land use patterns and effects on the landscape and biota. Pages 629–647 *in* H. Sioli, ed., The Amazon: Limnology and landscape ecology of a mighty tropical river and its basin. W. Junk, Dordrecht.

Mertes, L. A. K. 1985. Floodplain development and sediment transport in the Solimões-Amazon River, Brazil. M.Sc. thesis, University of Washington, Seattle.

Ohly, J. J. 1986. Water-buffalo husbandry in the central Amazon region in view of recent developments. Animal Research and Development 24: 23–40.

———. 1987. Untersuchungen über die Eignung der natürlichen Pflanzenbestände auf den Überschwemmungsgebieten (várzea) am mittleren Amazonas, Brasilien, als Weide für den Wasserbüffel (*Bubalus bubalis*) während der terrestrischen Phase des Ökosystems. Göttinger Beiträge zur Land- und Forstwirtschaft in den Tropen und Subtropen, Heft 24.

Petrere, M., Jr. 1978a. Pesca e esforço, e captura por unidade de esforço. Acta Amazônica 8: 439–454.

———. 1978b. Pesca e esforço no estado do Amazonas. II. Locais, aparelhos de captura e estatística de desembarque. Acta Amazônica 8, Suppl. 2: 54 pp.

Petrick, C. 1978. The complimentary function of floodplains for agricultural utilization. Applied Sciences and Development, 12: 26–46.

Roosevelt, A. C. 1991. Moundbuilders of the Amazon: Geophysical archeology on Marajó Island. Academic Press, San Diego, CA.

Saint-Paul, U. 1994. Der neotropische Überschwemmungswald: Beziehung zwischen Fisch und Umwelt. Final report BMFT No. 0339366A. Federal Ministry for Science and Technology, Bonn.

Sioli, H. 1984. The Amazon and its main affluents: Hydrography, morphology of river courses, and river types. Pages 127–165 *in* H. Sioli, ed., The Amazon: Limnology and landscape ecology of a mighty tropical river and its basin. W. Junk, Dordrecht.

Sombroek, W. G. 1966. Amazon soils. Centre for Agricultural Publications and Documentation, Wageningen, The Netherlands.

Sternberg, H. O. R. 1956. A água e o homem na várzea do Careiro. Faculdade Nacional de Filosofia, Universidade do Brasil, Rio de Janeiro.

———. 1966. Die Viehzucht im Careiro-Cambixe-Gebiet. Ein Beitrag zur Kulturgeographie der Amazonasniederung. Heidelberger Geographische Arbeiten 15: 171–197.

Vieira, R. D. S. 1992. Várzeas amazônicas e a legislação ambiental brasileira. IBAMA, INPA, MPI-Plön, and FUA, Manaus.

Wallace, A. R. 1853. A narrative of travels on the Amazon and Rio Negro, with an account of the native tribes, and observations on the climate, geology, and natural history of the Amazon Valley. Ward, Lock & Co., London.

Walter, H. & S. W. Breckle. 1984. Ökologie der Erde. Band 2: Spezielle Ökologie der tropischen und subtropischen Zonen. Verlag G. Fischer, Stuttgart.

Weischet, W. 1977. Die ökologische Benachteiligung der Tropen. Verlag B.G. Teubner, Stuttgart.

Wilhelmy, W. 1969. Tropische Transhumance. Heidelberger Geografische Arbeiten 15: 198–207.

Zimmermann, J. 1987. Manaus importa alimentos e nas várzeas se produz fibras. Como explicar a contradição? Tübinger Geographische Studien 95: 207–219.

Changes in Soil Formation and Vegetation on Silt Bars and Backslopes of Levees Following Intensive Production of Rice and Jute

Miguel Pinedo-Vasquez

Human Influence on Várzea Landscape

In Amazonia, the establishment and development of plant communities on várzea flood-plains depends on the activities both of rivers and of people. River floods, swidden agriculture, timber extraction, and commercial fishing have long influenced the várzea landscape and the composition of its vegetation (Padoch & de Jong, 1990; Goulding, 1988). Over time, such human activities have created a várzea landscape where large human-disturbed areas predominate.

Agriculture is the principal human activity that leads to ecological changes in the Napo-Amazon várzea floodplain (Fig. 1). The inhabitants (ribereños) of this várzea region preferentially select the várzea over the terra firme areas to grow both subsistence and market crops. Although crops are planted along the entire gradient of várzea features, ribereños tend to plant crops most intensively on the following features: (1) natural levees (restinga alta and baja), (2) backslopes of levees (bajeales), and (3) silt bars or mud flats (barreales) (Padoch & de Jong, 1990). Along the Napo-Amazon várzea floodplain, backslopes of levees and mud flats were identified as areas used frequently and intensively for the commercial production of jute (*Urena lobata*) and rice (*Oryza sativa*), respectively.

Backslopes of levees and mud flats are preferred sites because of their characteristic soils and location with respect to other várzea landforms. While annual deposition of silt sediments make silt bars the best várzea environment for rice production, backslopes of levees are used mainly for planting jute, because of their proximity to backswamps, discharge channels, and oxbow lakes. Silt bars are newly formed features, and backslopes of levees are among the oldest parts of the várzea floodplain. Furthermore, the stability of these surfaces strongly controls the type of vegetation found on them and influences how human activity is expressed on these surfaces. For example, plant communities in backslopes of levees are dominated mainly by late successional tree species, whereas silt bars are dominated by early colonizer species such as grasses and herbs (Lamotte, 1990; Junk, 1984).

Plant succession on silt bars includes three defined stages. In newly formed mud

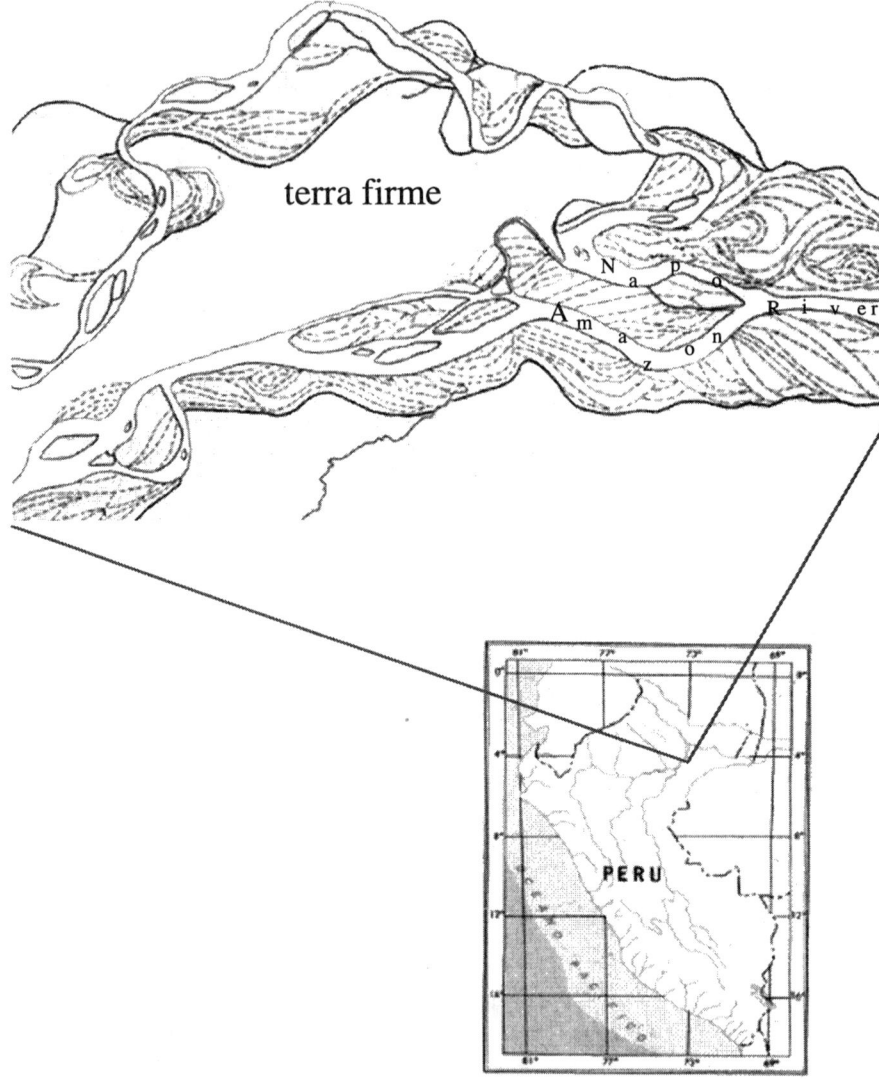

Figure 1. The Napo–Amazon várzea.

flats, plant communities in the first stage are dominated by floating grasses (*Paspalum fasciculatum*). After four to six years, in the second stage, giant grasses (*Gynerium sagittatum*) become dominant. And after eight to ten years, pioneer tree species (e.g., *Cecropia membranaceae*) begin to dominate the plant community in the third stage (Lamotte, 1990). Based on this knowledge of how plant succession occurs naturally on barreales, alterations to plant succession and community composition as a result of rice cultivation can be determined. Plant communities reestablishing in rice fields abandoned after different time periods were investigated to determine whether they followed the same successional patterns, and if not, what the new successional patterns were.

While vegetation on silt bars is characteristic of early stages of succession, vegetation on backslopes of levees has a more defined structural and floristic composition. Plant communities in this várzea feature are the result of longer time periods of succession, reflecting its greater stability in the várzea ecosystem (Kalliola et al., 1991). The impact of agriculture in these older and well-established plant communities appears to be similar to the impact on upland, terra firme communities. Changes in vegetation and soils after rice and jute cultivation are analyzed and discussed in this article. Data collected by measuring sedimentation and vegetation should explain the role played by humans in the establishment, development, and function of várzea ecosystems in the Napo-Amazon region.

Quantifying the Impact of Rice and Jute Cultivation on Sedimentation and Vegetation Pattern of Várzea Environments

The total area planted to agricultural crops and the historical time scale of when they were planted and abandoned was estimated for barreales and bajeales using question-naires and interviews. The questionnaires were applied to 270 ribereño households, 63 of which were selected for interviews. Local farmers were asked about the number of plots they farm, the total area they farm, and the number of times they had planted rice and jute in these fields. Using this information, abandoned fields where rice and jute had been continuously produced for six years and then abandoned for a variable number of years were identified and selected.

A total of 40 permanent study plots (20 in silt bars and 20 in backslopes of levees) were established in August 1991 in abandoned fields and in undisturbed areas. These plots were established in August because the Amazon River is normally at its lowest level in the Napo-Amazon várzea floodplain. Ten of the 40 permanent plots were established as control plots: 5 in undisturbed silt bars and 5 in undisturbed backslopes of levees. The other 30 plots (15 in silt bars and 15 in backslopes of levees) were established in fields that had been abandoned for either 2, 4, or 6 years following 6 years of continuous rice or jute cultivation. Each permanent plot established in the abandoned rice fields was 5 × 5 m in area; in the abandoned jute fields, 15 × 15 m.

Graduated poles and plastic boxes were installed at each corner (one in each of the four corners) of the 40 permanent plots to record the amounts and kinds of sediments that were deposited every year. Sediment deposition and erosion data were recorded in August of 1991, 1992, and 1993. The effects of past rice and jute production on sedimentation rates in barreales and bajeales were analyzed by comparing the kind and amount of sediments accumulating each year during the three years of observation. In each permanent plot, standard vegetation data were collected in the field in August of 1991, 1992, and 1993. All individual plants with stems ≥ 1.5 m in height were counted, tagged, and had their heights and diameters recorded once during each of the three years. Plant specimens were collected outside of each permanent plot of all species found in the plots.

The Impact of Rice and Jute Cultivation on Sedimentation Patterns

Rice and jute are planted in approximately equal quantities and with equal intensity in the Napo-Amazon várzea floodplain (Fig. 2). However, abandoned rice and jute farms accumulated different amounts and types of sediments during the 1991, 1992, and 1993 floods. Sediments of similar texture (i.e., silty and clayey) were deposited in both abandoned fields and undisturbed plots in backslopes of levees and silt bars by the three floods. In contrast, abandoned rice and jute fields accumulated different amounts of sediments during the same period.

The differences in the depth of sediments deposited in abandoned jute fields and in control plots were not found to be significant; however, the differences in the amount of sediments deposited in abandoned rice fields and control plots were significant (Fig. 3). In 1991, the mean depth of sediments deposited in the disturbed silt bars (8 cm) was less than one-half of that accumulated in the control plots (19 cm). Similarly, the mean depth of sediments deposited in abandoned rice fields by the 1992 flood was higher than in 1991. However, this pattern of accumulation in abandoned rice fields and undisturbed silt bars shifted following the 1993 high flood.

Following the 1993 high flood, the greatest volume of coarse sandy sediments accumulated in rice plots that had been abandoned for 2 years (21 cm) while the lowest depths were recorded in plots abandoned for 4 years (6 cm). The presence of dense shrub and tree pioneer vegetation in plots abandoned for 4 and 6 years correlated with the reduced depth of sediment accumulation after the 1993 high flood. While the pioneer vegetation remained alive during inundation, the herb-vine vegetation in the 2-yr-old abandoned plots was killed during the 1993 floods. Since these 2-yr-old abandoned plots did not have a dense cover of vegetation, they were exposed to strong river currents allowing the transportation and accumulation of heavy, large sized particles of bottom-load sand sediments.

Plant Community Development

While jute production did not produce significant changes in species composition in backslopes of levees, rice production had great impact on the species composition of silt bars (Fig. 4). None of the species found in the abandoned rice fields were found in the undisturbed barreales. In contrast, large numbers of species found in undisturbed bajeales were found in the abandoned jute fields. Change in species composition on barreales after rice farming, particularly a rapid sixfold increase in tree species, is notable. The average number of species with individuals ≥1.5 m in height found in the disturbed barreales according to their life forms were as follows: 4 shrub, 3 herb, 3 vine, and 1

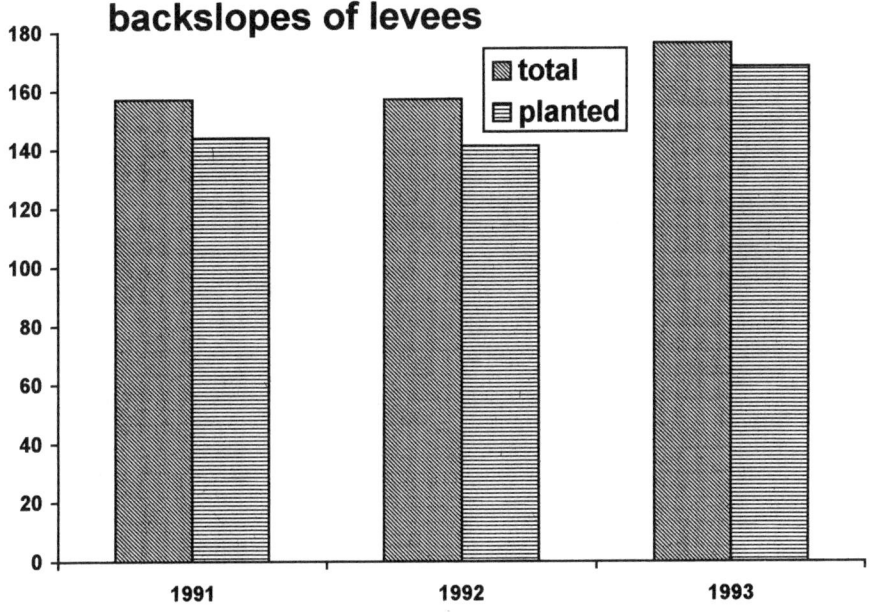

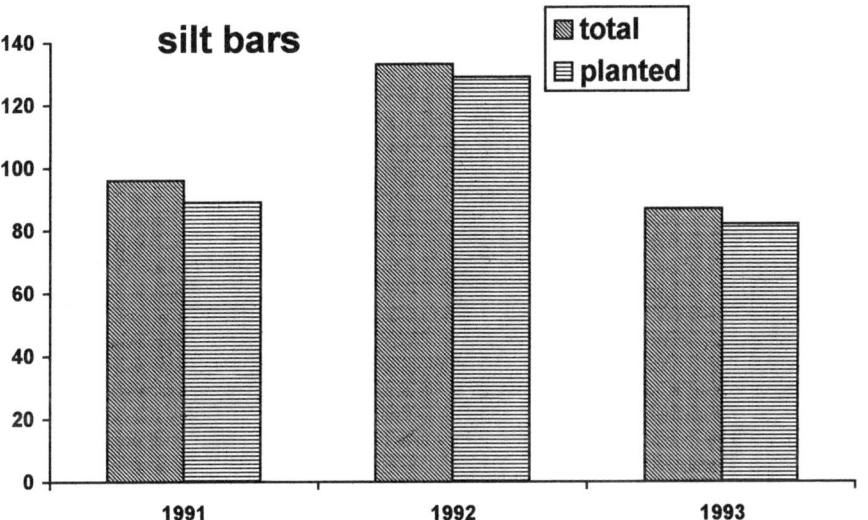

Figure 2. Total and planted area (ha) of backslopes of levees and silt bars.

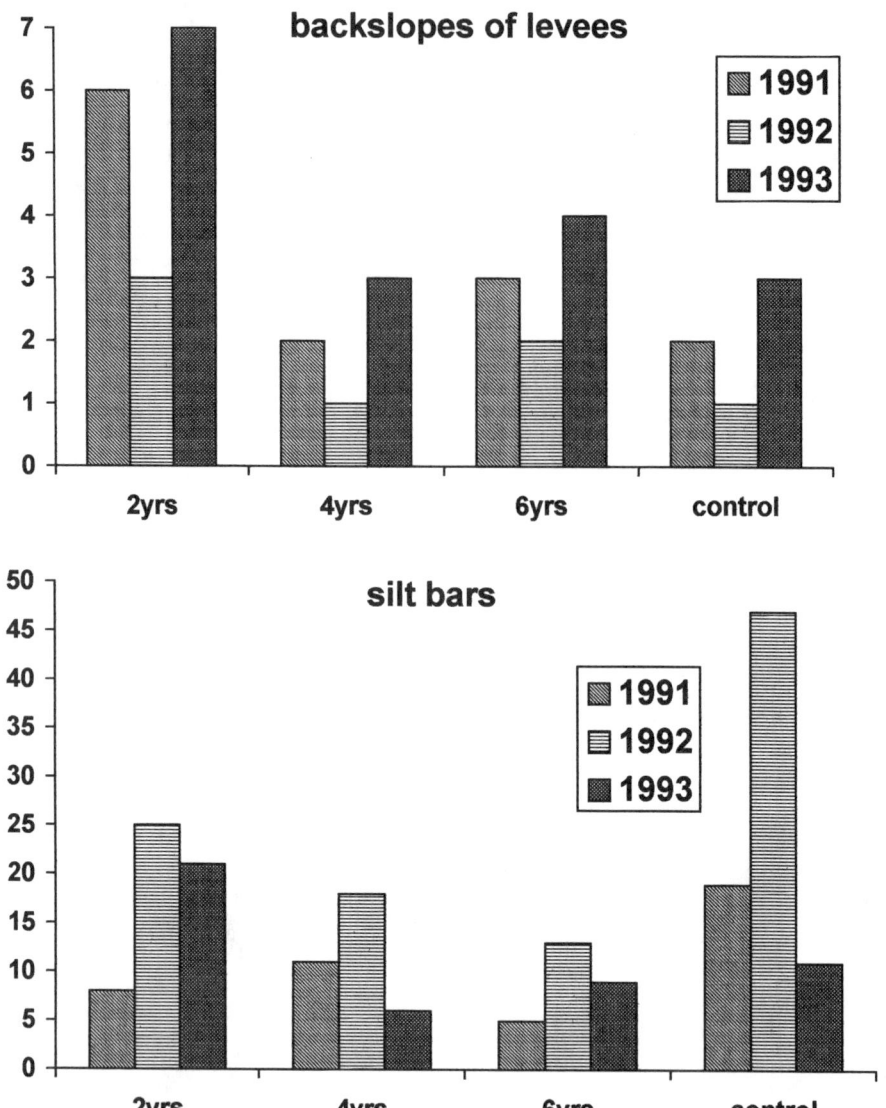

Figure 3. Mean depth (cm) of sediments deposited by the 1991, 1992, and 1993 floods on backslopes of levees and silt bars.

tree species in 2-yr-old abandoned plots; 5 shrub and 6 tree species in 4-yr-old abandoned plots; and 2 shrub and 5 tree species in 6-yr-old abandoned plots (Fig. 4). In contrast, very few species (4 floating grasses, 1 giant grass, and 1 vine species) were found in the silt bar control plots, and there was a noticeable lack of tree species.

Plant communities established in abandoned jute fields were initially dominated by grasses, herbs, and vines, which are quickly replaced by trees and shrubs after only 4

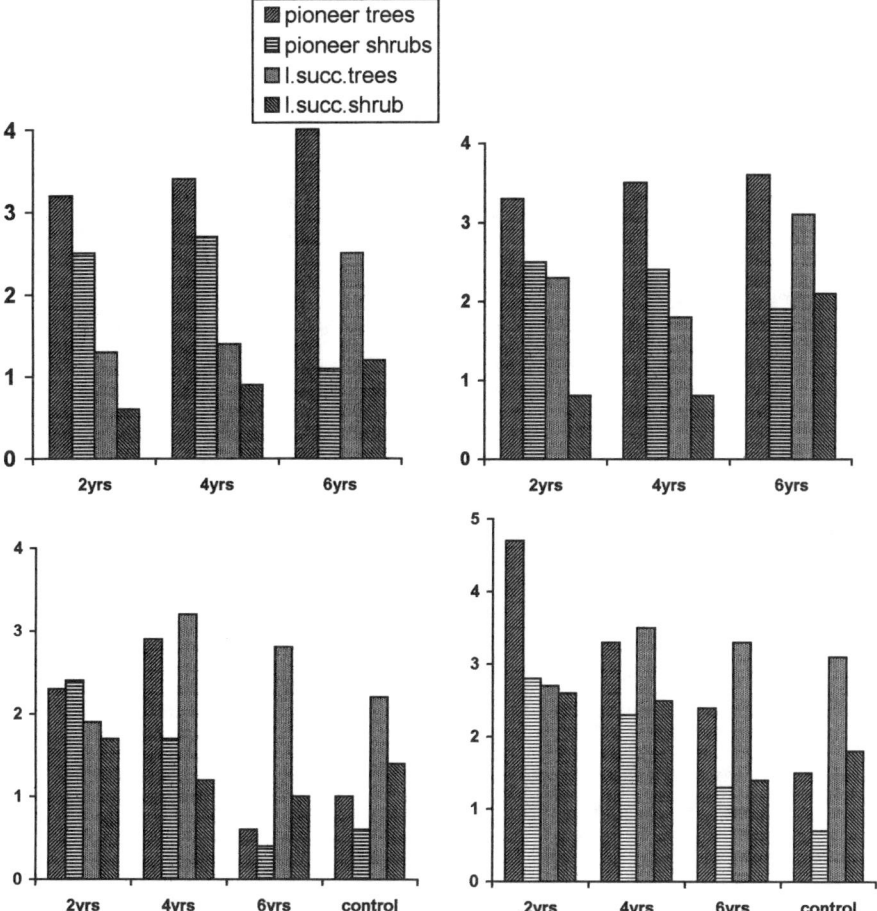

Figure 4. Mean annual height growth (m) and diameter increment (cm) per life form in silt bars (top) and backslopes of levees (bottom), for 1991, 1992, and 1993.

years of abandonment. While dense stands of herbs and vines were found in 2-yr-old abandoned rice fields, dense communities of grasses such as *Echinochloa polystachya* were established in 2-yr-old abandoned jute plots. The plant community established in a 2-yr-old abandoned jute field comprised 4 species of grasses, 4 of herbs, 3 of vines, 2 of shrubs, and 4 of trees. After 4 years of abandonment, the following plant species composition was recorded in jute fields: 2 vine, 4 shrub, and 7 tree species. A total of 4 shrub and 9 tree species were found in the jute fields abandoned 6 years earlier. Comparison of plant composition in 6-yr-old abandoned jute fields and in undisturbed backslopes of levee plots show them to have the same number of shrub and tree species. These results suggest that vegetation in disturbed bajeales, in contrast to disturbed barreales, tends to largely recover after 6 years of abandonment.

As the number of years of abandonment after rice and jute cultivation in barreales and bajeales increased, we noticed certain changes in stem density and shifts in species composition. While the average number of tree and shrub species ≥1.5 m in height tended to increase after 2, 4, and 6 years of abandonment, the number of clumps of grasses, stems of herbs, and vines decreased. Following 2 years of abandonment of jute fields, the establishment and mortality of *Echinochloa polystachya* dominated the community dynamics.

While populations of grasses declined after 2 years of abandonment, individuals of late successional tree and shrub species became dominant 4 and 6 years after abandonment of jute fields. Four years after abandonment, *Pseudobombax munguba* was the species with the highest percent of total individuals: 17% in 1991, 21% in 1992, and 22% in 1993. In both 6-yr-old abandoned jute fields and undisturbed plots, one of the three most numerous tree species was *Erythrina fusca*, which represented at least 14% of the total individuals found in 1991 and 1993 and 15% of those found in 1992.

The vegetation of the undisturbed silt bars was clearly dominated by *Paspalum fasciculatum*; this species represented 43% of the total individuals in 1991, 46% in 1992, and 51% in 1993. In contrast, after 2 years of abandonment, individuals of *Tessaria* spp. were the most important component of the community, representing 32% of total stems counted in 1991, 28% in 1992, and 19% in 1993. After 4 years of abandonment of rice fields, silt bar vegetation became dominated by individuals of *Cecropia membranaceae*, which constituted 27% of the total individuals in 1991, 30% in 1992, and 34% in 1993. Stems of this species also comprised the majority counted in the 6-yr-old abandoned rice fields (22% of the total in 1991, 23% in 1992, and 21% in 1993).

Vegetative communities on backslopes of levees that had been farmed recovered, sustaining species composition and diversity very similar to that of undisturbed plots (within 4–6 years); vegetative communities on silt bars, however, did not recover during the same time period. There were no similarities in species diversity between silt bars 2, 4, and 6 years after abandonment of farming and undisturbed silt bars.

Effects of Rice and Jute Cultivation on Sedimentation and Species Composition on Silt Bars and Backslopes of Levees

Intensive agricultural uses of the Napo-Amazon várzea have had significant effects on várzea formations and forests. These effects, however, are complex and variable depending on which areas of the várzea are used. While rice cultivation produced significant changes in the amount and kinds of sediments, very little such change occurred in areas where jute had been cultivated. Sand or silty-sand sediments were deposited in the abandoned rice fields, and silt or sandy-silt sediments were deposited in the undisturbed silt bar sites during the three years. Worbes et al. (1992) also found that when silt bars are covered by floating and giant grasses, they tend to accumulate further silt, and coarser sediments are deposited when these grasses are removed. Accumulation of sand or silty-sand sediments is one of the main reasons why local farmers are forced to abandon rice fields on silt bars.

Vegetation and soils of silt bars tend to deteriorate under rice cultivation faster than do backslopes of levees under jute cultivation. Areas used for rice cultivation became less productive when abandoned because these sites evolve into more open "beachlike" areas due to the accumulation of coarse sands after floods. Several factors identified during this study explain why silt bars tend to become beaches following rice cultivation. The following four factors were found to be the most important: (1) the location of silt bars close to the main river channel, (2) their shape and size, (3) the age and stage of succession of their vegetation, and (4) the intensity with which people convert barreales into rice fields.

Mud flats are mainly narrow, elongated, newly deposited silt or silty-sand sediments near the river channel. These várzea features occupy a niche between playas (sand bars) and caños (discharge channels). Due to their location, mud flats occupy some of the lowest gradients of the várzea floodplain and are exposed to the direct impact of river currents. Variations in current speed determine the kind and amount of sediments deposited or eroded by floods during the year. In the years when current speed is very strong, sandy or sandy-silt sediments tend to be deposited; in years when the current is not strong, silt sediments are deposited (Dumont et al., 1990). These patterns, however, tend to change when the floating grass vegetation is removed.

During floods in areas of silt bars where there is no vegetation, erosion and/or sedimentation rates tend to increase. Sandy sediments were very rarely found to be deposited on silt bars that were covered by dense grass vegetation. *Paspalum fasciculatum* is the most dominant and abundant floating grass species that protects barreales from high current speed and subsequent deposition of sandy sediments. Although herb, vine, shrub, and tree vegetation becomes established in abandoned rice fields, its function in controlling sedimentation was found to be very limited. Because these species do not produce a fibrous, dense mass of roots and stems comparable to grasses, they are not as effective a buffer in controlling river current speeds and sedimentation rates. Irion (1984) noted that the presence of dense communities of floating grasses facilitates the formation of silt bars along the Amazon floodplain. The probability that silt bars will then remain part of the várzea landscape and survive floods is greater when the established plant community is dominated by floating grasses.

Although grass species stabilize a site, their presence reduces the ability of other plant species to colonize silt bars, and therefore grasses remain longer. Grasses have a significant effect in decreasing the colonization rate of the site by other types of vegetation such as herbs, vines, and shrubs. Worbes et al. (1992) reported that very few plant species can compete with grasses in silt bars even though seeds of a large number of species are present on these surfaces. It was recorded during this study that vegetation other than grasses becomes established in barreales only when grasses are removed either by humans or by the river currents.

In contrast to rice cultivation on silt bars, the conversion of backslopes of levees to jute fields does not lead to an increase in sedimentation or to a change in sediment texture following floods. On backslopes of levees, similar clay-silt sediments were deposited in the plots of abandoned jute fields and in the control plots during a three-year period. There was also no significant difference in the average depth of clayey-silt sediments deposited in abandoned jute fields and control plots following the 1991, 1992, and 1993 floods.

Patterns of Plant Recovery of Barreales and Bajeales after Abandonment from Rice and Jute Cultivation

On silt bars, plant succession on abandoned farmlands does not lead in the short term to the formation of the same kind of vegetative community that dominated undisturbed barreales. Up to 6 years after abandonment, these areas resembled a typical riverine forest where fast-growing tree and shrub species form dense stands. Their plant species composition differed from that observed in barreales that had not been used for agriculture. Such change in successional patterns is part of the reason why there are no similarities in species composition between the original vegetation established on silt bars and that which was established following farming.

Lack of similarity in species composition between the vegetation of abandoned rice fields and the vegetation of undisturbed silt bars is a result of two main factors: (1) Weeding and other cultural practices before and during rice farming reduced dramatically the productive material and microhabitats for grasses such as *Gynerium sagittatum*; and (2) when rice fields were abandoned, new microhabitats were created in which the grasses were not as competitive with other woody and nonwoody early colonizer species. The end result of the factors mentioned above is: (1) the elimination of the phase when giant grasses dominate, and (2) the establishment and development of vegetation following a sequence of herbs/vines, then pioneer shrubs, and finally pioneer trees. This contrasts with the typical pattern of floating grasses that give way to giant grasses and then, finally, pioneer trees.

While plant succession after rice production resulted in the formation of different vegetation communities on barreales, succession on bajeales led to the recovery of both the species composition and structure of a mature plant community typical of undisturbed backslopes of levees. Patterns of plant succession and recovery occurring on abandoned bajeales are similar to those reported for abandoned fields in terra firme formations in Amazonia (Saldarriaga, 1986). In the first years after abandonment, swidden-fallow fields on terra firme sites were reported to be dominated by herbs, vines, grasses, and fast-growing shrubs and trees. Gradually, after many years of abandonment (36–38 years), the original primary tree and shrub species come to dominate the plant community (Uhl, 1987).

Patterns of establishment and development of plant communities following rice and jute production differ. While vegetation typical of undisturbed backslopes tends to reestablish on disturbed sites, vegetation on silt bars did not recover to predisturbance levels following farming activities. The following summarizes the principal stages of plant succession, recorded during this study, that led to vegetation recovery on disturbed backslopes of levees: Stage 1 is marked by grasses and herbs after 2 years of abandonment; stage 2 involves pioneer trees and shrubs after 4 years of abandonment; and stage 3, late successional tree species after 6 years of abandonment. The dynamics of plant establishment on silt bars after rice farming are characterized by the following stages: Stage 1 begins with herbs and vines after 2 years of abandonment; in stage 2, pioneer shrub and tree species occur after 4 years of abandonment; and stage 3 is characterized by the growth of pioneer tree species after 6 years of abandonment.

Literature Cited

Dumont, J. F., S. Lamotte & F. Kahn. 1990. Wetland and upland forest ecosystems in Peruvian Amazonia: Plant species diversity in the light of some geological and botanical evidence. Forest Ecology Management 33/34: 125–139.

Goulding, M. 1988. Ecology and management of migratory food fish of the Amazon basin. Pages 71–85 *in* F. Almeda & C. M. Pringle, eds., Tropical rainforests: Diversity, and conservation. California Academy of Sciences, San Francisco.

Irion, G. 1984. Sedimentation and sediments of Amazonian rivers and evolution of the Amazonian landscape since Pliocene times. Pages 201–214 *in* H. Sioli, ed., The Amazon: Limnology and landscape ecology of a mighty tropical river and its basin. W. Junk, Dordrecht.

Junk, W. J. 1984. Ecology of the várzea, floodplain of Amazonian whitewater rivers. Pages 215–244 *in* H. Sioli, ed., The Amazon: Limnology and landscape ecology of a mighty tropical river and its basin. W. Junk, Dordrecht.

Kalliola, R., M. Puhakka, J. Salo, H. Tuomisto & K. Ruokolainen. 1991. The dynamics, distribution and classification of swamp vegetation in Peruvian Amazonia. Annales Botanici Fennici 28: 225–239.

Lamotte, S. 1990. Fluvial dynamics and succession in the lower Ucayali River basin, Peruvian Amazonia. Forest Ecology and Management 33/34: 141–156.

Padoch, C. & W. de Jong. 1990. The impact of forest products trade on an Amazonian place and population. Pages 151–158 *in* G. T. Prance & M. J. Balick, eds., New directions in the study of plants and people. Advances in Economic Botany 8. New York Botanical Garden, Bronx.

Saldarriaga, J. G. 1986. Forest succession in the upper Rio Negro of Colombia and Venezuela. Ph.D. thesis, University of Tennessee.

Uhl, C. 1987. Factors controlling succession following slash-and-burn agriculture in Amazonia. Journal of Ecology 75: 377–407.

Worbes, M., H. Klinge, J. Revilla & C. Martius. 1992. On the dynamics, floristic subdivision and geographical distribution of várzea forests in Central Amazonia. Journal of Vegetation Science 3: 553–564.

Spatial Heterogeneity and Temporal Variability of Some Amazonian Floodplain Soils

Daniel J. Zarin

Introduction

When compared with large regions of species-rich and nutrient-poor terra firme forest in the Amazon basin, the floodplain environments of Amazonia are often considered to be ecologically homogeneous and agriculturally fertile. While this generalization may be both accurate and useful at a very coarse scale of analysis, it is an insufficient basis for analysis of ecological constraints, existing land uses, and potential management alternatives for any of the floodplain types that exist within the confines of the Amazon basin.

In this paper, I review some of the ways in which Amazonian floodplain soils vary across both space and time. Spatial and temporal variability in soil quality in Amazonian floodplains may affect their productive potential. At present, these soils support a myriad of land uses, including the continuous or near-continuous cultivation and harvesting of annual crops and perennial fruits, the extraction of nonfruit native forest products (wood, heart-of-palm, palm fiber, etc.), and ranching. Physical and hydrological, rather than chemical, characteristics can be the most important constraints on the agricultural usefulness of many floodplain environments, even where the soils are nutrient poor. For instance, it is widely recognized that mechanized agriculture is inappropriate in the floodplain environment, because of the sensitivity of soil physical structure. For similar reasons, cattle ranching is also an ecologically problematic land use for many floodplain soils. Yet ranching intensity is increasing in some floodplain areas (Bahri, 1993; Guillaumet et al., 1993), often with concurrent increases in erosion and decreases in biomass productivity and biodiversity (Junk, 1989; Goulding, 1993).

A number of attempts have been made to classify the floodplain types by flooding frequency, water turbidity and dissolved organic content, vegetative cover, and successional stage (e.g., Prance, 1979; Junk, 1989; Kubitzki, 1989; Worbes et al., 1992). Much of the data reported in this paper comes from analyses of the soils of what Prance (1979) has classified as seasonal and tidal várzea (Table I). Those soils have been mapped by RADAMBRASIL (1:1,000,000) primarily as dystrophic-to-eutrophic hydromorphic gley soils and eutrophic-dystrophic alluvial soils, depending on soil texture and nutrient status (Projeto RADAM, 1974). Those Brazilian classifications would fall in the Entisol soil order of the U.S. Soil Taxonomy, generally within either the Aquent or Fluvent suborders (Soil Survey Staff, 1988). Less frequently inundated soils contig-

Table I. Key to the principal types of Amazonian forests subject to inundation

I. Periodically inundated forest	
A. Flooded by regular annual cycles of rivers	
i. White[a] water	I. Seasonal várzea
ii. Black[b] and clear[c] water	2. Seasonal igapó
B. Flooded by tidal movements	
i. Salt water	3. Mangrove
ii. Fresh water backup	4. Tidal várzea
C. Flooded by irregular rainfall (flash floods)	5. Floodplain forest
2. Permanently inundated forest	
i. White water	6. Permanent swamp forest
ii. Black and clear water	7. Permanent igapó

Source: Prance, 1979: 29.

[a] Whitewater rivers are muddy brown to whitish in color, owing to the heavy load of Andean sediment. These rivers are rich in nutrient cations and are neutral to basic in pH.

[b] Blackwater rivers drain the Amazon lowlands and are low in pH and nutrient poor, though high in organic acids, the residue of incomplete decomposition of organic matter.

[c] Clearwater rivers originate in the pre-Cambrian cratons and lack both the nutrient-rich sediment load of the whitewater rivers and the organic acid load of the blackwater rivers.

uous with the tidal and annual floodplains may be variously classified as Mollisols, Alfisols, Ultisols, and Inceptisols, with sediment provenance and flooding frequency being the principal determinants of the taxonomic order. Where a marked dry season occurs, pockets of Vertisols may also develop within the regularly inundated várzea (Roosevelt, 1980).

Spatial Trends

The provenance of suspended and dissolved loads within the basin's rivers is a primary determinant of several fundamental soil characteristics within the active floodplains (Hoag, 1987; Konhauser et al., 1994). Table II demonstrates that, for the Peruvian Amazon, sediments deposited by rivers draining the Andean regions that contain sedimentary limestone formations (Table II, groups I and II) tend to be moderately acid to neutral in pH and therefore very low in aluminum saturation of the cation exchange complex. Exchangeable calcium and magnesium in those floodplain soils are moderate to high, as is "available" phosphorus. In contrast, sediment deposited by the rivers that are restricted to within basin lowland drainage or that drain the exclusively acid igneous and volcanic formations of the northern Ecuadorian Andes (group III) are characterized by low pH values and correspondingly high levels of aluminum saturation. Exchangeable calcium and magnesium, and "available" phosphorus, are present in moderate to low concentrations in those acidic soils. Exchangeable potassium and "available" micronutrient levels are apparently less affected by sediment provenance. Analogous

Table II. Some chemical characteristics of soils on floodplains of the Peruvian Amazon

Group[c]	River	pH[d]	Al sat.[e] (%)	Exchangeable[a] (cmol/kg)			Available[b] (ppm)			
				Ca	Mg	K	Mn	Cu	Zn	P
I	Aguaytia	7.7	0	27.2	3.1	0.47	175	4.6	3.0	15
	Cumbaza	6.4	0	6.7	1.1	0.20	87	1.5	0.8	5
	Cushabatay	7.8	0	46.2	2.6	0.47	169	6.4	3.8	13
	Mayo	7.5	0	39.7	9.7	2.02	0	6.2	4.5	145
II	Cashiboya	5.6	0	29.7	7.0	0.69	130	4.9	3.2	29
	Napo	5.6	2	5.6	2.0	0.27	19	5.2	1.9	20
	Nucuray	5.1	3	12.4	2.0	0.23	73	4.6	4.1	19
	Paranapura	5.8	1	5.0	1.0	0.21	32	1.9	1.9	27
	Pastaza	6.4	0	3.7	6.1	0.25	32	12.0	2.3	17
	Samiria	5.6	1	10.2	2.8	0.24	69	1.9	2.5	25
	Tapiche (lower)	5.1	5	13.7	3.5	0.48	38	1.9	3.6	27
	Tapiche (upper)	5.2	12	9.5	2.0	0.26	30	2.2	1.4	13
	Utoquinea	5.9	0	15.7	3.2	0.35	164	3.9	4.2	30
III	Blanco	4.7	49	4.9	1.9	0.26	13	1.0	1.4	8
	Mazan	4.2	78	0.8	0.5	0.23	8	2.3	1.1	6
	Nanay	4.2	69	5.6	2.0	0.27	6	0.7	1.6	20
	Putumayo	4.9	46	2.9	0.9	0.29	17	2.6	1.8	13
	Tamshiyacu	4.6	68	1.5	0.3	0.30	30	2.7	2.3	6
	Tigre	5.1	9	3.6	1.3	0.08	44	2.3	3.5	10
	Yavari Mirim	4.0	78	1.8	0.5	0.36	35	2.3	3.3	6
Intergroup differences[f]		***	***	**	0	n.s.	n.s.	n.s.	n.s.	0

Note: Data for 0–20 cm depth (Hoag, 1987).

[a] Exchangeable cations determined by extraction with $1N$ NH_4OAc at pH 7.

[b] Available nutrients determined by extraction with Melich III solution.

[c] Rivers were grouped as follows, on the basis of characteristics of their primary watersheds (Hoag, 1987: 27–28): I, Northern Peruvian Andes: predominantly Paleozoic–Mesozoic sandstones, slates, and limestones containing acid igneous and metamorphic batholiths; II, Southern Ecuadorian Andes: predominantly Paleozoic–Mesozoic metamorphosed sedimentary rocks, basalt, and igneous rocks overlain in some areas by volcanic ash of recent origin but flanked to the east by Tertiary calcareous sandstones and limestones; II, Western edge of the Peruvian basin: tributaries that dissect or are in close proximity to calcareous sandstone and limestone outcrops; III, Within-basin drainage: polycycled, unconsolidated Quaternary deposits with no inputs of calcareous sediments or solutes; III, Northern Ecuadorian Andes: acid igneous and volcanic lithology.

[d] pH in water.

[e] Aluminum saturation by cation extraction with $1N$ KCl.

[f] Kruskal-Wallis test (Sokal & Rohlf, 1995): *, $P < 0.05$; **, $P < 0.01$; ***, $P < 0.001$; n.s., not significant at $P < 0.05$.

groupings may also be assigned to the floodplains of the basin with the territorial boundaries of Brazil (Stallard & Edmond, 1983).

Broad-scale spatial patterns are apparent in the mineralogy and major-element chemistry of Amazonian floodplain soils as one travels downriver in the middle and lower Amazon (Martinelli et al., 1993). Trace element distribution apparently does not follow the same pattern (Fernandes et al., 1994). Along the main channel, Martinelli et al. (1993) report downstream increases for quartz and kaolinite, and decreases for total and exchangeable calcium and magnesium (Fig. 1). Calculations based on X-ray diffraction data demonstrate that the trends illustrated in Figure 1 (B and C) for total and exchangeable calcium and magnesium are paralleled by decreases in the concentration of their principal host minerals, namely, smectite-vermiculite and Ca-plagioclase (Martinelli et al., 1993). Those trends are likely to result from three ongoing, interactive processes.

First, although the Andean region comprises only 12% of the area of the basin, it contributes 80% of the sediment load (Gibbs, 1967). The major rivers draining the Andes are therefore the most turbid in the basin. Those rivers that drain the carbonates and minor evaporites of the Peruvian Andes and the reduced shales and minor carbonates of the Bolivian Andes are also the most nutrient rich, because of the favorable chemical composition of the sediment's source material (Hoag, 1987; Stallard & Edmond, 1983). As the water from the western rivers flows into the main channel and travels downstream, the sediment load is diluted by mixing with sediment- and nutrient-poor rivers of the lowland basin and the cratons. As a result, there is less suspended sediment available to be deposited on the floodplains downstream, and that which is available is, on average, depleted in nutrients.

Second, biogeochemical transformation of alluvial sediment probably occurs in transport and certainly occurs while sediment is being stored on the floodplains (Johnsson & Meade, 1990; Martinelli et al., 1993). High-activity minerals, such as feldspars, smectite, and vermiculite, are broken down into their low-activity constituents, such as kaolinite and quartz, with a corresponding loss of mineral nutrients and a decrease in nutrient-adsorption potential (cation-exchange capacity). Because the floodplains are very active and up to 30% of the water in the mainstem passes through the floodplains (Richey et al., 1989), downstream deposits necessarily inherit the altered signature of soil stored, weathered, and subsequently eroded and transported from upstream floodplains.

Figure 1 (opposite). Downriver trends in soil characteristics along the main channel of the Amazon River, Brazil. **A.** Low-activity mineral concentrations: $R^2 = 0.40$ for quartz; $R^2 = 0.39$ for kaolinite. **B.** Total concentration of major nutrient cations: $R^2 = 0.43$ for CaO; $R^2 = 0.45$ for MgO; $R^2 < 0.01$ for KO. **C.** Exchangeable nutrient cation concentrations (determined by extraction with 0.05 N HCl): $R^2 = 0.61$ for Ca; $R^2 = 0.60$ for Mg; $R^2 = 0.20$ for K. (Data from Martinelli et al., 1993. Distances are downriver from Vargem Grande [ca. 16 km upriver from the confluence of the Içá and Solimões Rivers]. Only those data reported by Martinelli et al. for samples collected during Jul–Aug 1985 are used in this graph, in order to avoid confounding spatial and temporal homogeneity. Each data point represents 1 composite sample. Sample depth = 0–10 cm.)

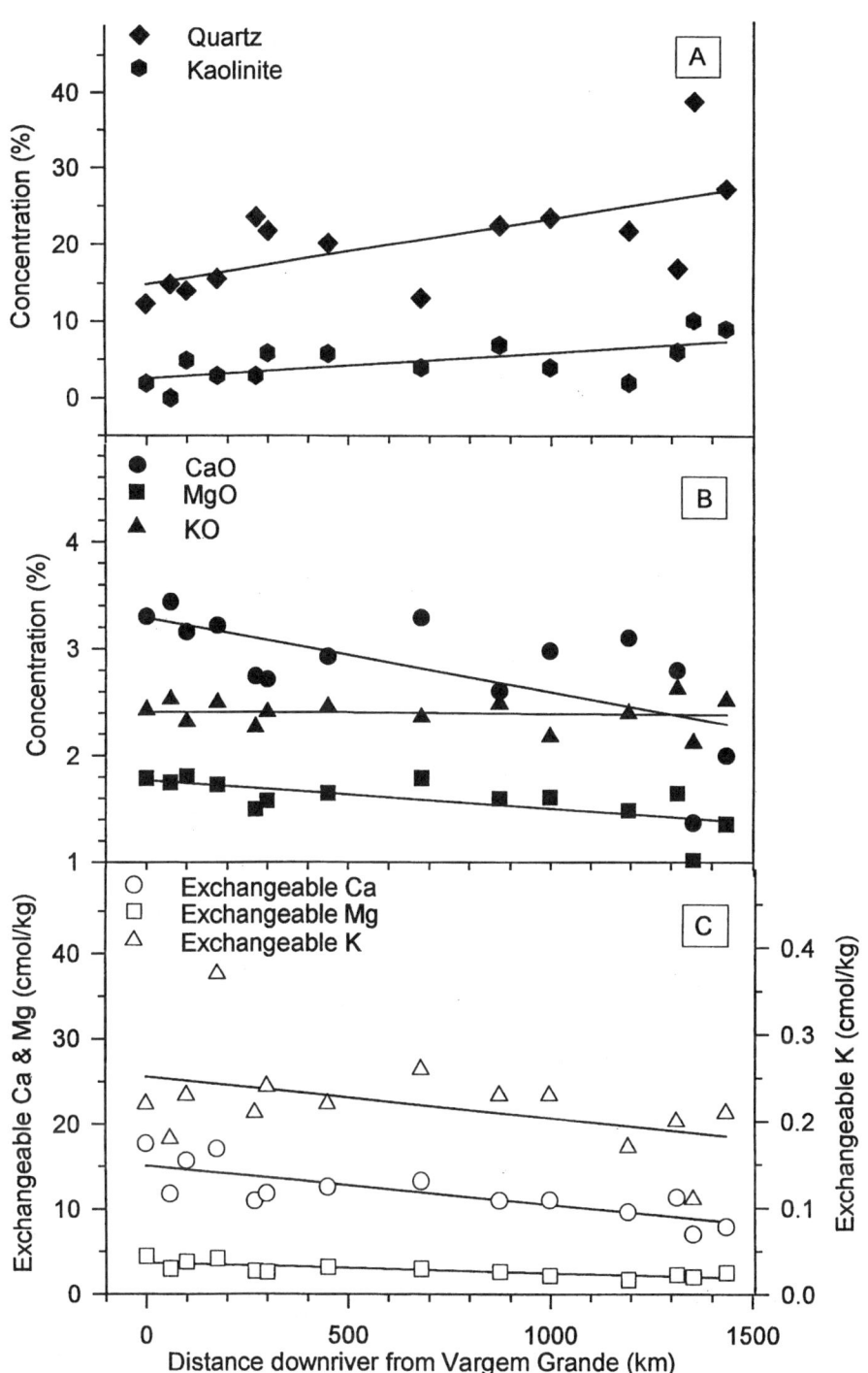

Third, biological uptake of suspended and deposited nutrients continually diminishes the available nutrient pools as they travel downstream (Martinelli et al., 1993; Sholkovitz & Price, 1979). Although some portion of those pools is returned to the river and will travel downstream, the return is not 100%, because some of the nutrients accumulated in organic matter will be buried in the várzea (Winkler-Prins, 1993), and some will be sequestered in long-term storage in refractory organic matter, which eventually may be transported and buried offshore (Damuth & Kumar, 1975). In many parts of the floodplain, human populations pull nutrients out of the downstream flow in the form of marketed agricultural and forest products, thereby further impoverishing the downstream depositional potential. At other points along the river, unknown levels of human inputs in the form of sewage and municipal and industrial waste may also alter nutrient status. Those anthropogenic impacts on floodplain biogeochemistry may be important, at least locally.

Temporal Variability

Shifts in várzea soil chemistry within a nine-month period are apparent in data reported by Victoria et al. (1989) for both main channel and tributary collection sites. Although the shifts are not statistically significant when all collection sites are considered as one sample (Victoria et al., 1989), demonstrable shifts are apparent for some variables at selected individual collection sites over time (Fig. 2). Such shifts appear greatest at the Japurá River site for the four variables illustrated in Figure 2. However, lack of within-site replication at each collection period makes it impossible to determine whether any of the differences, over time, are significant or not. For that reason, speculation on what might control the observed shifts is probably unwarranted. Nonetheless, it should be noted that oxygen isotope data reported by Richey et al. (1989) indicate that the relative magnitude of different water sources to the floodplain areas varies within the course of a year. Such variation may also influence floodplain soil chemistry.

Temporal variability is also apparent in soils of the high and low várzea environments of Ilha do Careiro (Fig. 3), at the confluence of the Negro and Solimões Rivers (Alfaia & Falção, 1993). Increases in exchangeable potassium and calcium between May and July of 1987 may have resulted from nutrient input from rising water that apparently inundated the low várzea during May and June. The very small sample size, once again, limits our ability to determine whether the temporal variability shown in the figure represents actual significant differences over time in the population the sample purports to represent.

A number of factors may influence intra-annual variations in soil fertility characteristics on the várzea. Those factors include flooding frequency and duration, which may be controlled by seasonality of precipitation and/or tidal cycles; dissolved and suspended loads, which may be altered by the changes in the provenance of water to any given location on the várzea (Richey et al., 1989); and periodic pulses in *in situ* nutrient cycling such as fruiting, leaf drop, and organic matter decomposition. The relative importance of these various factors and the constraints that each may impose upon the utilization of várzea resources remain unknown.

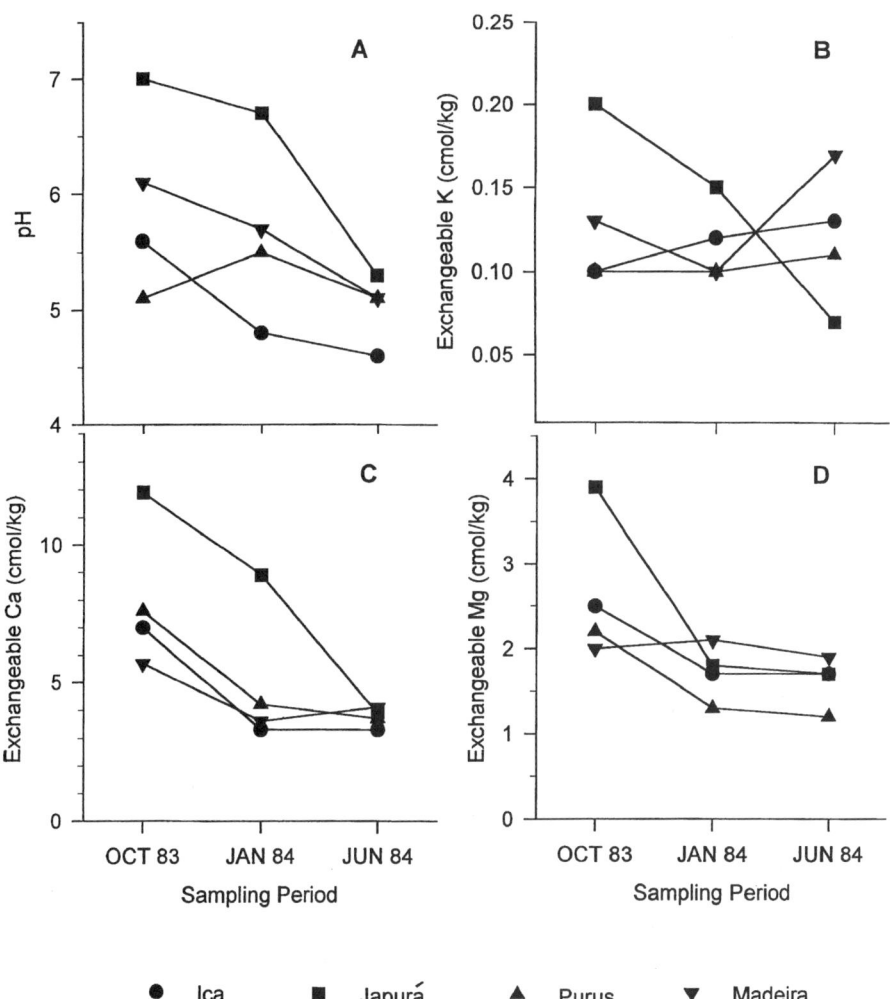

Figure 2. Temporal variability in some characteristics of várzea soils of four Amazonian tributaries. The Içá and Japurá Rivers originate in the northern Andes. The Purus River originates in the sub-Andean trough and the Madeira originates in the Bolivian Andes. Sampling periods correspond to the dry season (Oct–Nov), early rising water (Jan–Feb), and early falling water (Jun–Jul). (Data from Victoria et al., 1989. Figures for pH are of pH in water. Exchangeable cations determined by extraction with 0.05 N HCl. Each point represents 1 composite sample. Sample depth = 0–60 cm.)

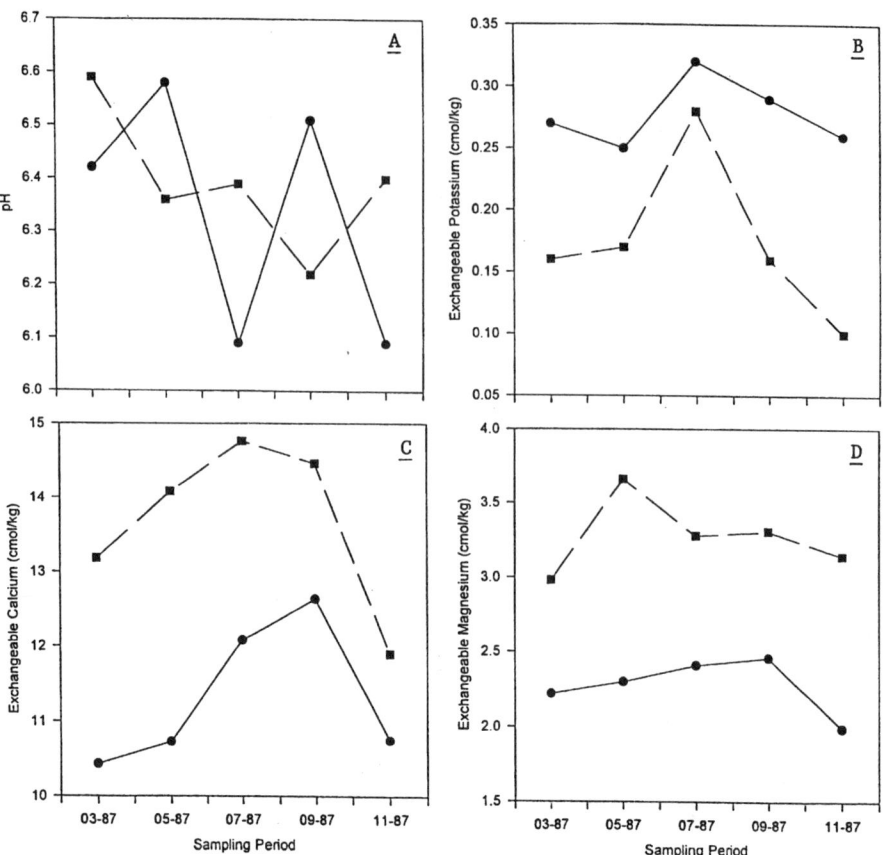

Figure 3. Temporal variability in some characteristics of high and low várzea soils, Ilha do Careiro, Brazil. Circles denote data from high várzea sites; squares denote data from low várzea sites. (Data from Alfaia & Falção, 1993. Figures for pH are of pH in water. Exchangeable cations determined by extraction with 0.5 N HCl. Each data point for high várzea represents the mean of 3 composite samples; each data point for low várzea represents the mean of 2 composite samples. Sample depth = 0–40 cm. Alfaia and Falção [1993] report statistically significant differences [Neuman-Keuls test] over the sampling period for pH in the high várzea and exchangeable K in the low várzea, but the small sample size does not justify emphasis of those conclusions.)

Literature Cited

Alfaia, S. S. & N. P. Falção. 1993. Estudo da dinamica de nutrientes em solos de várzea da Ilha do Careiro no estado do Amazonas. Amazoniana 12: 485–493.

Bahri, S. 1993. Les systèmes agroforestiers de l'ile de Careiro. Amazoniana 12: 551–563.

Damuth, J. E. & N. Kumar. 1975. Amazon Cone: Morphology, sediments, age and growth pattern. Geological Society of America Bulletin 86: 863–878.

Fernandes, E. A. N., E. S. B. Ferraz & H. Oliveira. 1994. Trace elements distribution in the Amazon floodplain soils. Journal of Radioanalytical and Nuclear Chemistry, Articles 179: 251–258.

Gibbs, R. J. 1967. The geochemistry of the Amazon River system: Part I. The factors that control the salinity and the composition and concentration of the suspended solids. Geological Society of America Bulletin 78: 1203–1232.

Goulding, M. 1993. Flooded forests of the Amazon. Scientific American 268: 114–120.

Guillaumet, J. L., M. Lourd, S. Bahri & A. A. dos Santos. 1993. Os sistemas agrícolas na Ilha do Careiro. Amazoniana 12: 527–550.

Hoag, R. E. 1987. Characterization of soils on floodplains of tributaries flowing into the Amazon River in Peru. Ph.D. dissertation, North Carolina State University, Raleigh.

Johnsson, M. J. & R. H. Meade. 1990. Chemical weathering of fluvial sediments during alluvial storage: The Macuapanim Island point bar, Solimões River, Brazil. Journal of Sedimentary Petrology 60: 827–842.

Junk, W. J. 1989. The use of Amazonian floodplains under an ecological perspective. Interciencia 14: 317–322.

Konhauser, K. O., W. S. Fyfe & B. I. Kronberg. 1994. Multi-element chemistry of some Amazonian waters and soils. Chemical Geology 111: 155–175.

Kubitzki, K. 1989. The ecogeographical differentiation of Amazonian inundation forests. Plant Systematics and Evolution 162: 285–304.

Martinelli, L. A., R. L. Victoria, J. L. I. Dematte, J. E. Richey & A. H. Devol. 1993. Chemical and mineralogical composition of Amazon River floodplain sediments, Brazil. Applied Geochemistry 8: 391–402.

Prance, G. T. 1979. Notes on the vegetation of Amazonia III. The terminology of Amazonian forest types subject to inundation. Brittonia 31: 26–38.

Projeto RADAM. 1974. Mapa Exploratorio do Solos. Folha 22.

Richey, J. E., L. A. K. Mertes, T. Dunne, R. L. Victoria, B. R. Forsberg, A. C. N. S. Tancredi & E. Oliveira. 1989. Sources and routing of the Amazon River flood wave. Global Biogeochemical Cycles 3: 191–204.

Roosevelt, A. C. 1980. Parmana: Prehistoric maize and manioc subsistence along the Amazon and Orinoco. Academic Press, New York.

Sholkovitz, E. R. & N. B. Price. 1979. The major-element chemistry of suspended matter in the Amazon estuary. Geochimica et Cosmochimica Acta 44: 163–171.

Soil Survey Staff. 1988. Keys to soil taxonomy. SMSS Technical Monograph no. 6. Cornell University, Ithaca, NY.

Sokal, R. R. & F. J. Rohlf. 1995. Biometry: The principles and practice of statistics in biological research. W. H. Freeman, New York.

Stallard, R. F. & J. M. Edmond. 1983. Geochemistry of the Amazon 2. The influence of geology and weathering environment on the dissolved load. Journal of Geophysical Research 88: 9671–9688.

Victoria, R. L., L. A. Martinelli, J. E. Richey, A. H. Devol, B. R. Forsberg & M. N. G. Ribeiro. 1989. Spatial and temporal variations in soil chemistry on the Amazon floodplain. GeoJournal 19: 45–52.

Winkler-Prins, A. 1993. A case study of floodplain soils of the Amazon estuary. Association of American Geographers Annual Meeting, 8 April 1993, Atlanta, Georgia.

Worbes, M., H. Klinge, J. D. Revilla & C. Martius. 1992. On the dynamics, floristic subdivision and geographical distribution of várzea forests in central Amazonia. Journal of Vegetation Science 3: 553–564.

Section 5:
Land Resource
Management

Introduction

Nigel J. H. Smith

The Amazon floodplain contains a wealth of fish, shellfish, wildlife, and plant resources, many of which are unique to the area. Riverine inhabitants have long gathered from this environment for a wide array of products, including fish and game, fruits and nuts, building materials, fuelwood, and medicinal plants. Floodplain forests have helped supply the burgeoning urban populations with relatively cheap, high-quality protein by providing important seasonal habitats for the region's fisheries. But the economically and ecologically important floodplain forests are increasingly threatened by expanding agricultural activities, particularly ranching. The cattle and water buffalo stampede up the Amazon is being driven largely by an expanding market for beef in the growing towns and cities of the region. Landscape transformations on the floodplain, such as the raising of livestock along the margins of lakes, may also be adversely affecting some fish species.

These dramatic ecological and cultural changes may undercut the region's fisheries and other natural resources if left unchecked. After more than two decades of development focused on the uplands, policymakers and investors are looking increasingly to the Amazon floodplain as the "new venue" for development activities. The broad, alluvial soils of the Amazon represent the last great agricultural frontier in Latin America. But if this potential is not used wisely, it could be largely destroyed. While the enormous potential of the Amazon floodplain as a source of food, fish, and other natural products has long been recognized, no systematic effort has been made to understand its diverse ecological and cultural attributes as a foundation for rational development.

Any plan to conserve biodiversity and to manage natural resources needs to consider the interplay of land use systems and their internal dynamics. A balanced approach to conservation and development would recognize that cattle and water buffalo are here to stay on the várzea, with improved management, and would include other options for generating income by small, medium, and large landholders.

Whether the goal is to protect forests, to raise native animals for food production, or to increase agricultural yields, an essential prelude to any intervention is an intimate understanding of how people are using resources in the environment. The research efforts described in the four papers in this section will help build a knowledge base for more effective efforts to improve the quality of life for people who depend on the Amazon floodplain, directly or indirectly, for a living.

Floodplain Forests and Agricultural Intensification

The sinuous forests of the Amazon floodplain are a rich resource for agricultural improvement. Seasonally flooded forests contain populations of several economically important plants, such as açaí, burití, and tapereba, all of which produce nutritious fruits. Açaí also produces a much desired heart-of-palm and is widely cultivated in the Amazon, as is tapereba; the yellow fruits of the latter, known as "yellow mombim" in English, are used to make refreshing drinks and ice cream. Future attempts to raise yields and other qualities of these trees will require extensive tapping of genetic resources. The loss of wild populations of these fruit trees on the Amazon floodplain not only undercuts the income-generating capacity and nutrition of local residents but also forecloses options for improvement programs. If the local importance of floodplain forests is better understood by planners, then it will be easier to promote the conservation of such habitats.

Future agricultural development on the Amazon floodplain, as in other regions, also hinges on the periodic introduction of new crops. Many species of floodplain forest trees linger on the threshold of domestication. Home gardens contain a wealth of forest species at various stages of proto-domestication, but they are largely ignored in agricultural development. Currently, numerous species are only collected, rather than cultivated. If floodplain forests disappear, options for some future crops that are well adapted to periodic flooding will be lost.

Agrobiodiversity and Tree Farming

One of the common misconceptions about Amazonia is that it is largely virgin. Far from being in a pristine state, significant tracts of both upland and floodplain forests have been either cleared or altered in some way, as Roosevelt discusses in this section. People are often cast as the enemy of biodiversity. But in fact, cultures can sometimes enrich the biodiversity of the environments they are living in: They may plant trees or otherwise deliberately encourage them around habitation sites, and some disturbance of mature forest can actually increase species diversity by creating ephemeral plant communities.

Some sections of the Amazon floodplain forest have been enriched over time with economically important plants, especially açaí and burití palms. Tapereba was introduced to the Amazon region long ago but is now naturalized in floodplain forests. One way to reduce pressure on the remaining várzea would be to promote agroforestry in its various forms. In their paper here, Padoch and Pinedo-Vasquez look at the cultural forests that are most noticeable in the Amazon's estuarine area. They examine how people have enriched forests and thus increased agrobiodiversity, and they've found that this has important implications for development policy.

One of the oft-overlooked aspects of agricultural development for floodplains is tree farming. Although policymakers easily grasp the notion that alluvial soils would be appropriate for intensive cereal or vegetable gardening, they still consider trees too vulnerable to waterlogging. Yet, many wild and domesticated plants are adapted to the floodplain of the Amazon, and they offer a promising avenue for agricultural intensification, as Roosevelt discusses in this section. Tree crops could offer an attractive alternative to cattle ranching and would increase the biodiversity of cleared areas.

The idea of tree farming on the Amazon floodplain is not new. As Roosevelt points out, the Marajoara culture at the mouth of the Amazon apparently domesticated or semi-domesticated several fruit trees that have since reverted to the wild. On Careiro Island near Manaus, polycultural orchards are losing ground to the onslaught of cattle ranching. Some of the papers in this section hold that a greater appreciation for the role of tree crop farming would help agroforestry gain a stronger footing along the entire Amazon floodplain.

People as Agents of Landscape Transformation

The hand of human beings in altering the landscape is increasingly clear. Cultures both ancient and modern continue to manipulate the environment to improve access to resources and to increase agricultural productivity. Some of these interventions have implications for future development strategies.

People have altered watercourses in places on the Amazon floodplain for a variety of purposes—to improve access to timber and fruit trees, to facilitate the sale of crops, to generate power. Raffles explains how small-scale farmers fashion canals in the Amazon estuary. People have been reshaping their landscapes for thousands of years. In their paper here, Anderson, Marques, and Nogueira discuss a site in São José where some estuarine water courses were modified in the colonial period to generate tidal power for sugar cane mills. An innovative project at the Museu Goeldi is investigating the feasibility of restoring this ancient, long-abandoned practice of harnessing tidal energy to provide a source of electricity for people in remote areas.

Lessons from Environmental Archaeology

At first glance, it might appear that the study of ancient cultures has little to offer modern developers. But, as Roosevelt shows us, environmental archaeology provides information relevant to the land's potential ability to support human populations. This discipline teaches us about the societies that exploited Amazonian resources in the past, yielding data on the relationship between human culture and various types of land use.

One poignant lesson from the past occupation of the Amazon is that it is indeed possible to accommodate large and healthy populations while avoiding massive ecological destruction. Even at their most intensive, native management methods were compatible with reasonable biodiversity conservation goals: All of the economic species documented as being intensively used in prehistoric times are still around today. The pre-Conquest Amazon floodplain was far more densely settled than today's floodplain, but much more forest was left standing, primarily because the indigenous population of that period did not raise livestock. The path to sustainable development incorporates lessons from the past and a blend of indigenous and scientific knowledge. Neither Indians nor rural peasants are conservationists in the sense of directing their activities to preserve "nature" in a pristine state, but they are knowledgeable resource managers, and their input is needed in development plans.

Policy Considerations

Two overriding messages need to be brought to the attention of policymakers grappling with conservation and development issues along the Amazon floodplain. First, habitats are highly heterogeneous, and development strategies need to be fine-tuned to the nuances in elevation and drainage conditions. The Amazon floodplain is a complex mosaic of different soils, vegetation communities, and flooding regimes. An agricultural-development blueprint cannot be uniformly applied across the entire floodplain; it must be tailored to local ecological and socioeconomic conditions.

Second, local peoples—independent households, organized communities, and medium- and large-scale operators involved in intensive cereal production and ranching—should be consulted about natural resource management, and they need to be actively involved in the design and implementation of development and conservation measures.

Given the heterogeneity of floodplain habitats, it seems logical that conservation should encompass a broad range of levels of disturbance. It should be possible to maintain some habitats where human interventions are kept to a minimum or excluded entirely (provided local residents agree and there is political support in urban areas). Other conservation areas would be subject to varying degrees of disturbance. How such areas can be managed properly is a major research agenda. The private sector will be critical here: Much of the remaining floodplain forest in the middle Amazon is in the hands of ranchers. It is therefore essential to embrace everyone who has a stake in the várzea's future, not just small-scale farmers.

The Usefulness of Traditional Technology for Rural Development: The Case of Tide Energy near the Mouth of the Amazon

Scott Douglas Anderson, Fernando Luiz Tavares Marques, and Manoel Fernandes Martins Nogueira

A Problem: Decentralized, Low-Cost Generation of Electricity

Every day, literally hundreds residents on the floodplains near the mouth of the Amazon travel by boat to cities and towns in the region to charge automotive batteries. They use the batteries in their homes primarily to power televisions and occasionally to supply lighting for short periods in the evening. Even with this restricted use, it is typical for a battery to need recharging on a weekly basis. This takes 24 hours and so requires a second trip to town to pick up the recharged battery and, for those who can afford it, a second battery at home in reserve (see Fig. 1). Because of limited storage capacity, this electricity cannot be used for other domestic needs, such as refrigeration or pumping water, and its use to produce income, in agriculture or small-scale industrial or commercial activities, is out of the question. The demand for electric power is hardly being met.

Like many rural residents elsewhere in the region, they do not receive electricity from central generation plants because it is considered uneconomical to distribute power to them merely for household consumption and whatever incipient income-producing activities they might have. (The cost of installing posts, lines, and transformers in the region is on the order of US$10,000 per kilometer.) The problem of distribution to these rural residents is further aggravated by the fact that they live dispersed on floodplains covered by forests and cut by rivers, which transmission lines would have to penetrate and bridge (Nogueira et al., 1993).

A decentralized source of power, diesel, does exist in the region and is commonly used by floodplain residents to power small boats. But it is rarely used by them to supply their household electrical demand, and then for only a few hours a night by some better-off families. (The decentralized supply of electricity using diesel to scattered urban centers in Amazonia is heavily subsidized by the federal government.) Nor are diesel

Figure 1. Top. A floodplain resident takes a battery to be recharged in town. The battery is used chiefly to power a black-and-white television. The small car battery shown, the most commonly used, needs to be recharged every 5–10 days. Even though many people have diesel engines in boats, they rarely use this source of power to generate electricity because of its high cost of operation. **Bottom.** One of the many shops along the waterfront of Belém that specialize in recharging batteries for home use. This shop generally handles 20–30 batteries a day and occasionally up to 100 on a Friday. A 24-hour recharge costs US$2–4, depending on the size of the battery. Because most users recharge a battery only after it is heavily drawn down—that is, when the image on the television screen shrinks—such batteries usually last less than a year.

engines frequently used for income-producing purposes on the floodplain. Although available, diesel technology is not widely employed because people consider its operating costs—principally fuel, but also maintenance—to be too high in relation to its benefits.

The problem, then, is to find a source of energy to generate electricity that is both decentralized and low-cost and that is also within the technical and financial reach of floodplain residents near the mouth of the Amazon.

We believe that such a source of energy may be found in the waters near the mouth of the Amazon itself: the energy of the tide. This solution traces its origins to the end of 1987, when the first author, following up on indications from local residents, encountered sites with the remains of tide-powered sugar cane mills in the county of Igarapé-Miri near the southern mouth of the Amazon. Returning over the following two years, we began to study mill sites in this area because of our long-term interest in their remains as archaeological artifacts of great value in understanding the economic and social history of the region. However, we also had a more immediate goal: to determine whether the technology for using the tide to power mills in the past could aid in solving energy needs in the present.

A Site with the Remains of a Tide-Powered Mill

A place where tide-energy technology was applied is the São José site (Fig. 2) on the Furo Anapuzinho. The most important natural features there are a channel (*furo*) subject to the tide from both ends and a smaller tributary stream (*igarapé*) that fills and empties through its mouth with the rise and fall of the tide in the channel. As shown in the figure, there are the remains of works found on the banks of both the stream and the channel. If such a site can be confirmed as one with the remains of a tide-energy system, it can yield much information that will be useful in evaluating its potential and in understanding traditional technology.

To confirm a site as having had a tide-energy system, one critical feature is the vertical placement of the remains of works in relation to the tide. It is necessary that they be at a level within the range of the tide where they could have influenced the flow of the tidewater through them. In the case of the São José site, the floors of the works are set almost exactly at the midpoint between the low and high tide, which substantiated the possibility that they were part of a tide-energy system.

A second critical feature is the distribution of the remains within the context of the streams at a site. This must be compatible with the basic functions performed, which were, minimally, to retain water in a reservoir at high tide and, when the tide is lower, to release it and produce mechanical energy. In addition to interpreting the remains themselves, we were fortunate to obtain functional descriptions of the works from a witness to the operation of the mill in the late 1920s and, independently, from an heir to the site, who had seen the mill in a much more intact state and received additional information about it from his father, its last operator. Their descriptions, which were in close agreement, were compatible with those basic functions.

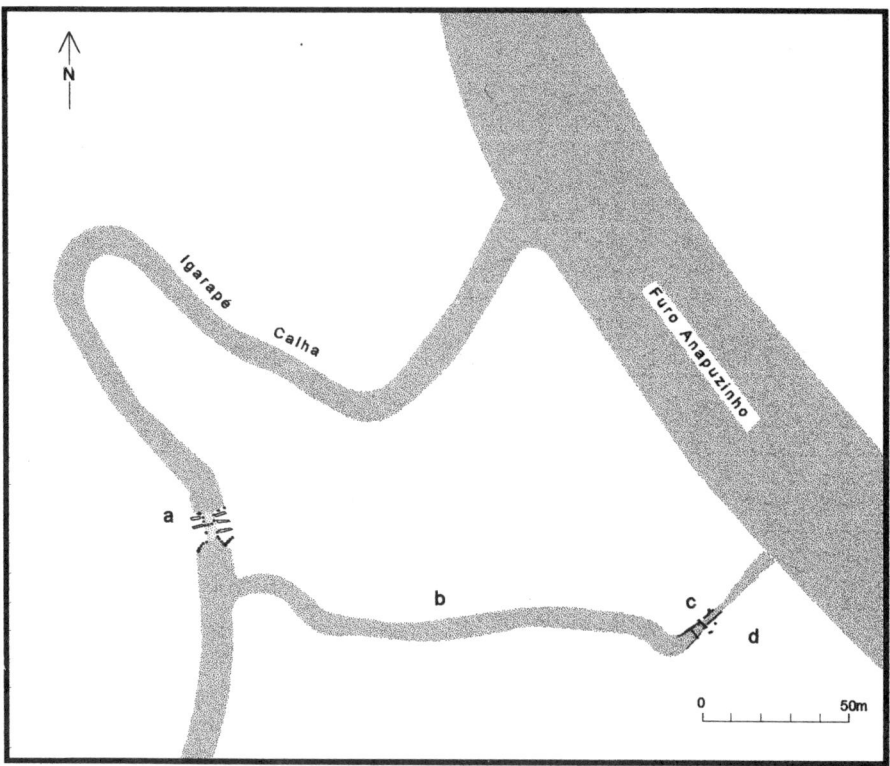

Figure 2. A plan of the São José site where the remains of a tide-powered mill are found. The remains include **(a)** a wooden, earth-filled structure with a brick and wood floor that crosses and narrows a tide-filled stream; **(b)** a distinct, regular depression in the ground, 140 m long, leading from the stream to **(c)** a narrow, straight depression with wood-lined walls and floor that opens onto a tide channel. The remains of a wooden frame and panel are located at the beginning of this depression and, near its midpoint, two massive wooden pillars are found alongside. To the right of the depression are **(d)** a large iron roller and other metal pieces of good-sized, sugar cane-grinding equipment.

According to our local informants, the functions of the works at the São José site were as follows: The wooden, earth-filled structure that crosses and narrows the tide stream was a dam that retained water in the stream at high tide; the distinct, regular depression in the ground was a canal; the canal carried water from the stream above the dam to the narrow, straight, wood-lined depression, a sluice, that opened onto the tide channel; the remains of the wooden frame and panel at the head of the sluice were a floodgate that could be opened to release water and turn a water wheel supported on wooden pillars.

We believe that sites in this region can be confirmed, with confidence, as having had a tide-energy system if these two critical features, vertical placement and distribution of remains, are satisfied. And just from this confirmation a site would be indicated as having a stream with storage capacity and range of the tide great enough to warrant reexploitation. In addition, a closer examination of a confirmed site may reveal even more.

For example, the wood used in the works could be identified from samples taken from the remains, even after being exposed to sun, rain, and tide for perhaps 100 years. Pillars and heavy beams were made from massaranduba (*Manilkara* sp.), and lighter posts and beams as well as boards were made from *acapu* (*Vouacapoua* sp.) The use of these materials in the past supports current local opinion as to what are some of the most durable woods available in the region for construction in wet conditions.

We also found that the São José site has natural advantages as a place to build on the poorly consolidated floodplain soils of the region. This was revealed while probing the subsoil at the site when we observed that the dam and the sluice were built on small upcroppings of very hard clay. These upcroppings also explained the specific layout of the remains, since, once the decision was made to site the dam and sluice on them, the route of the canal, which passed through soft alluvial soils, was also determined, physically linking the functions to be performed at the two superior building sites.

Most interesting and important is the energy potential at the site. To estimate that potential, we measured the volume of water that entered the stream where the dam crossed it and the range of the tide in the channel near the mouth of the stream. To our surprise, we determined later that the range of the tide recorded at the site was almost 30% higher than the range of the same tide recorded earlier the same day at the port of Belém, closer to the sea. The reason for this may be due to local narrowing and shallowing of the tide channel—a funnel effect on the tide, so to speak. Regardless, this measure, and others, indicated that the São José area also has the advantage of a relatively higher tide-energy potential in relation to the benchmark of the nearest major port with tide records and tables.

These natural features and the remains clearly reveal that the choice and use of the São José site took advantage of specific local knowledge of natural resources that would still be very useful in building any tide-energy system. This shows the type of information that can be obtained by carefully examining a site that is confirmed as having had a tide-energy system, even if the remains of that system are poorly preserved.

The Technology of a Tide-Powered Mill

The São José site's well-preserved remains offered the opportunity to obtain a greater understanding of the technology of the tide-energy system there. We sought to reconstruct, on paper and with models, the original form of the energy system at the mill and, from that, to determine in detail how that system operated. This was based on what could be substantiated from the remains themselves and also on what was compatible both with the functions that had to be performed and with information supplied by our informants. A result of this is the artist's reconstitution of the original form of the remains of the energy system at the São José site, shown in Figure 3.

On the basis of this and other, more detailed reconstructions (Anderson & Marques, 1990), we believe that the energy system of the mill operated in the following way: Water of the rising tide pushed open the two lateral gates of the dam and filled the

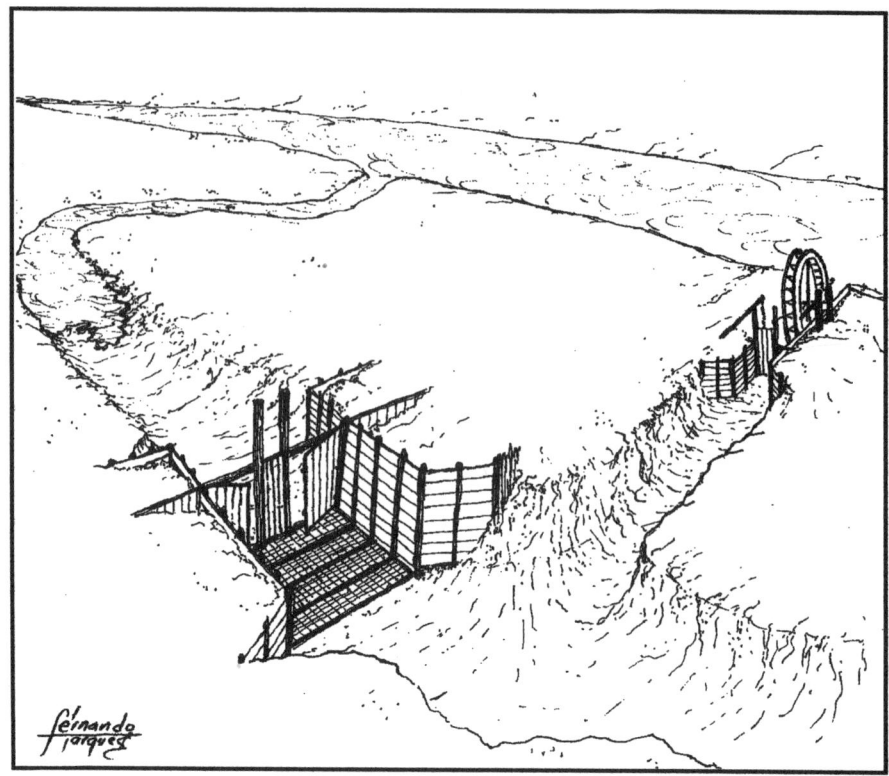

Figure 3. An artist's reconstruction (not to scale) of the remains of the tide mill at the São José site shown in Figure 2. The tide is shown as though it were rising in the channel in the background and flowing into the stream at the left. In the foreground is a wood and earth dam, which crosses the tide stream, with two lateral gates that swing open in an upstream direction and a central gate that opens vertically. Upstream from the dam, to the right, a canal leads to a sluice, with a vertically opening floodgate at its head and an undershot water wheel supported on pillars near its midpoint. The height of the gates in the dam was 1.75 m, the length of the sluice 23 m, and the diameter of the water wheel approximately 8.3 m.

stream. When the tide began to fall, the current reversed and closed the gates automatically, retaining the water in the stream and canal above the dam. The canal conducted the water held in the stream to the head of the sluice, closed by the floodgate. When the tide fell below the floor of the millrace in the sluice, the floodgate would be opened manually to release water into the millrace and turn an undershot, wooden water wheel. The water wheel could turn until the tide rose above the floor of the millrace and stopped the wheel, and the cycle would then begin again.

The reconstitution of the dam alerted us to solutions to two problems related to using a tide stream in the region as a reservoir. The first problem is leakage of water through and under the dam; this was resolved by a wooden wall through the dam in line with the gates and an impermeable wood and brick floor between the walls of the dam upstream from the gates. The second is the problem of accumulation of sediments in the stream used as a reservoir; we believe this was resolved by periodically opening the central gate in the dam at low tide to release the water in the reservoir rapidly and flush out sediments accumulated on the stream bed.

The reconstitution of the sluice showed the specific solutions applied in the mill to control the release of water and produce mechanical energy. The floodgate at the head of the sluice was raised by a long wooden lever that held the gate open approximately 15 cm above the millrace. The water wheel was 8.2–8.4 m in diameter and turned in a millrace 93 cm wide. As noted above, the level of the floor of the millrace, and therefore the bottom of the water wheel, was set very close to the midpoint between the high and low tides. This meant the wheel was powered with a head of water equal to one-half the range of the tide and that it could turn free of the tide during one-half of a tide cycle, that is, for slightly over six hours.

With these factors in hand, it was possible to calculate measures of performance of the tide-energy system at the site. For a 3.3-meter tide, they yielded a flow of 350 liters per second through the floodgate, which, after losses, would have produced about four horsepower on the water wheel. Against the resistance of grinding sugar cane, the wheel would have turned about six revolutions per minute, which corresponds to the rate found in steam-powered mills in the region today.

The reconstitution of the works and their operation at the São José site demonstrates the proven traditional technology used in the mill to meet the functional requirements of a tide-energy system. This includes retaining the water of the high tide with a dam in a natural stream, transporting it in an artificial canal, controlling its release with a floodgate, and converting its potential energy into mechanical energy with an undershot water wheel in a sluice. Except for metal hardware, this was all accomplished using locally available materials, labor, and skills.

The Reality of Tide Power in the Past

These solutions to the functional problems of a tide-energy system were not limited to the mill at the São José site, however. In fact, the detailed knowledge obtained at this one site made it possible to interpret more poorly preserved remains at other sites in the county of Igarapé-Miri. In this way it has been possible to confirm a total of 15 sites there as having had tide-energy systems (Anderson & Marques, 1992).

In addition, we later encountered sites of tide mills nearer Belém with remains of cut stone, rather than wood, as the principal building material. A view of the remains at such a site is shown in Figure 4. Specifically, our systematic search for sites with the remains of tide-energy systems in the county of Barcarena has yielded 14 sites. Historical references indicate that these remains may date back to the late 1700s (Marques, 1993: 89). In addition, five other sites with remains of this type have been brought to our attention in a nonsystematic way, bringing the total number of such sites to 19.

The location of confirmed sites with remains of tide-energy systems is shown in Figure 5. The sites extend over a distance of roughly 150 km, in a zone of fresh water with mean spring tides having a range of at least 2.5 m. At most of these sites the remains show the dam-canal-sluice solution found at the São José site, but at two sites the dam

Figure 4. A view of the remains of a sluice in a tide mill at the Fazendinha site on Cutijuba Island near Belém. It was built of cut sandstone blocks interfilled irregularly with rows of bricks and partially plastered with a lime-and-sand stucco. Over each end of the millrace are massive vaults, which give lateral support to the walls. Vertical grooves found in the wall on each side of the upstream vault could have secured a floodgate. The stairs go from the millrace up to ground level. By way of comparison with the sluice in Figure 3, this sluice is 24.5 m long and held a water wheel at least 7 m in diameter.

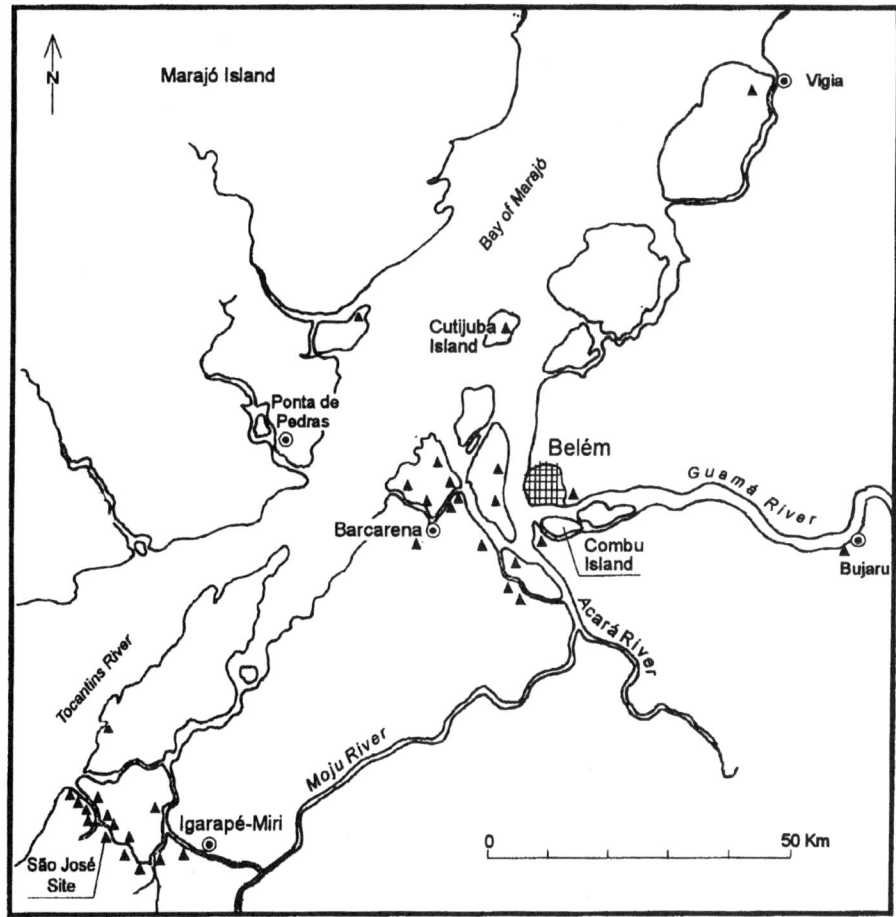

Figure 5. Solid triangles indicate sites where the remains of 34 tide mills have been located near the southern mouth of the Amazon. Of this total, the 15 near Igarapé-Miri were built of wood and earth as shown in Figure 3. The remains of 19 closer to Belém were constructed of cut stone and brick, of the type shown in Figure 4. The concentration of sites around Igarapé-Miri and Barcarena reflects, in part, the fact that these counties have been more thoroughly searched for the remains of mills. Historical references indicate that the remains of tide mills may also be found along the Moju, Acará, and Guamá Rivers.

and sluice are adjoined and cross a tide stream in a single structure. All of the cut–stone remains are found on hard substrates, but that is not always the case for the wood and earth remains.

At most sites the dams are very poorly preserved, no doubt because they were a barrier to navigation and also offered, especially in the case of the stone dams, desirable and easily accessible building materials. Nevertheless, at several sites near Igarapé-Miri, the placement of posts at the dam site is almost identical to that of São José, indicating similar solutions to problems of leakage through the dam and of removal of sediments from the reservoir. Sluices are often better preserved, and the lengths and widths of their millraces, similar to those of São José, suggest that the size of the water wheels employed in them was also of the same order. This general pattern of the works can be explained by the fact that these energy systems were built to utilize a common resource, the tide, with a common magnitude, 2–3 m, for a common purpose, grinding sugar cane.

The evidence of these 34 sites shows that the use of the tide to produce energy was widespread at the southern mouth of the Amazon over a long period of time. Moreover, it appears from remains that the technical solutions employed to utilize this source of energy at these sites were similar to the São José site, about which we have considerable information. Thus, on the basis of this one, well-preserved site, we feel confident that we know many important and specific features of the proven technology that was employed in tide-energy systems throughout the region.

A Project to Develop Tide-Generated Electricity

The successful use of tide energy near the mouth of the Amazon in the past substantiates the possibility of using it there again. Moreover, knowing in detail the technology that was used to exploit this resource, we believe that traditional technology can and should be the starting point for developing any system to use tide energy in the region.

Presently we are implementing a project to develop technology to produce tide-generated electricity near the mouth of the Amazon River. The very low operating and maintenance costs of a tide system may make it an attractive alternative to diesel or battery-supplied electricity, despite higher initial investment costs. Also, those initial costs could be divided among several users. For example, measurements of the energy potential taken at the São José site indicated that, after losses, on the order of six kilowatt-hours of electricity could be generated there, per tide (Anderson et al., 1993: 273). This would be enough to meet the domestic needs of five rural households, if efficient lights, televisions, and refrigerators were used. Since, at this writing, we are only at the beginning stages of this project, we have no results to present as yet, but can show how we have proceeded so far, using traditional technology as a basis for developing new technology.

In terms of the basic functions to be performed by a tide-generated electric system, we opted, as was done in the past, to use a tide stream as a reservoir and to build a dam across it to retain the water of the high tide. In choosing a site for the dam, we were not concerned that it be exceptional in terms of tide range or substrate, as is the São José site, since we want the new system to be viable at common sites, which were also used in the past, so as to benefit a potentially greater number of residents of the flood-plain. (Of course, for residents, the identification and use of superior sites would be of great interest.) On the other hand, we were concerned that the volume of the stream be at least that at the São José site, and by measuring the total flow of water in tide streams near Belém we confirmed three with the requisite volume, one of which was selected as the site of the project.

As for the dam itself, we opted for a structure with wooden walls and earth fill, as was used at the São José site, but supported on piles driven into the soft substrate, which is a technique used commonly in the region today for building docks and bridges on floodplain soils. A view of this laboratory dam in the final stages of construction is shown in Figure 6. Except for beams of massaranduba crossing the dam, we did not use the types of wood used in dams in the past, since they are expensive and this dam is expected to have a useful life as a laboratory limited to just a few years (although those woods might be of interest to someone building a dam to use for an indefinite period). Also, we do not plan to use wood and brick to waterproof the floor above the dam, since we want to test waterproofing with much cheaper, heavy plastic sheeting set between two layers of boards.

As in the past, the dam will have the capacity to permit the tide to pass through it and fill the stream and then to close automatically, as well as the capacity to permit the

Figure 6. A laboratory dam to test alternatives for using the tide to generate electricity, under construction in April 1995 on Combu Island, near Belém. The choice of material used in its construction, pilings and walls of wood with earth fill, was based on the remains of dams found near Igarapé-Miri, such as the one shown in Figure 3, which successfully retained water of the high tide in streams in the past. The openings in the dam between the pillars can be closed with planks as needed and will permit the testing of different kinds and sizes of floodgates and turbines.

rapid emptying of the reservoir to flush out sediments. To do this we could have used upstream-swinging gates and a vertically opening gate similar to those at the São José site, but we want to test filling and flushing gates set horizontally in frames in the dam, as these are less expensive. The gates will pivot on sealed ball bearings, which, like plastic sheeting, is a modern option not available to builders of traditional mills.

To transport the energy of the tide from the dam to the user, we plan to test two options. The first, already installed, employs aluminum cables to transmit energy as electricity, which will first have to be generated at the dam site. The second will use, as in the past, a canal to transport water to the user and will generate electricity there. The earth used for fill in the dam was removed so as to leave a long trench that can be enlarged to make the canal. Besides the advantage of requiring only local labor and materials, it appears now that the canal option may also cost less to install and maintain.

To perform the functions of releasing water and producing mechanical energy, we have not opted for traditional solutions. Based on our calculations of the performance of the traditional system, the use of a water wheel would be less efficient than a modern turbine. In addition, the knowledge of techniques for building and balancing such wheels has been lost in the region, and the quantity and quality of material would be quite expensive. Overall, modern turbines are so efficient and relatively inexpensive that it is impractical to use a sluice with a floodgate and water wheel as was done in the past.

Regarding less-material aspects of this project, some important gaps in our knowledge of traditional technology have become evident. For instance, although we are confident that the vertically opening gate in the dam was used to flush sediments out of the reservoir, we do not know how often it was necessary to do this: weekly? monthly? more frequently at certain seasons? Nor do we know what collateral effects the storing of water in a once free-flowing stream might have on nearby vegetation or on human health. Users in the past must have had answers to these questions, but since they left no written or oral record that we have found, these are all questions that will have to be studied.

In elaborating this project, we have not simply attempted to replicate successful past technology in the present, nor have we used it superficially, merely as a cosmetic for a contemporary solution. Rather, to develop a new technology we divided our problem into fairly independent functional parts in order to be able to compare alternatives. Although we tended to favor components taken from traditional technology, the criteria being that they offer solutions proved in the region and accessible to local residents, we were not committed to them. Detailed knowledge of both traditional and contemporary technology gave us an informed basis for choosing among components, using the additional criteria of cost and efficiency. This allowed us not only to select but also to reject components of traditional technology and, when there was not a clear choice, to make provision for testing alternatives and innovations. The case of tide energy shows how it is possible and advantageous to combine components available from both traditional and contemporary technology to develop a new technology.

The Potential of Tide Power along the Nearby Atlantic Coast

It should be mentioned that the possibility also exists for using the tide as a source of energy along the Atlantic coast above and below the mouth of the Amazon River. This is substantiated, in part, by the range of the tide reported for major ports there, given in Figure 7. Although some ranges are less than that in Belém, this is not necessarily a limiting factor. With a large volume of water (such as is available from the ocean), modern turbines can generate energy efficiently with a head as low as one meter. (A more serious constraint might be that of holding a sufficient volume of water at high tide, which is a question of local topography.) Moreover, just as the case of the São José site, at certain advantageous places along the coast, the range of the tide may be considerably higher than at the nearest major port of reference—as, for instance, at the northern mouth of the Amazon below Macapá and near São Luiz.

This possibility is also substantiated, as it is near the mouth of the Amazon, by the

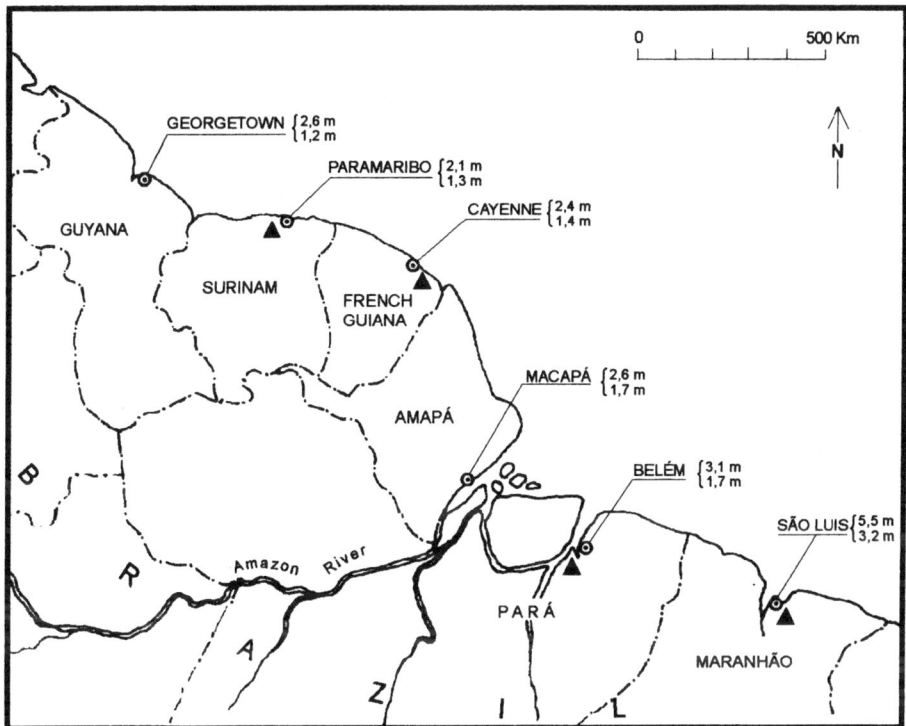

Figure 7. The range of the tide at major ports along the Atlantic coast above and below the mouths of the Amazon. The upper figure given after each port is the mean range of the spring tides there, and the lower figure is the mean range of the neap tides, calculated from navigation charts. Solid triangles indicate places where tide energy was employed in the past.

use of tide energy in the past. Referring to Suriname, a visitor in 1798 reported the following:

> To build the sugar factories, whose mills are powered by water, they do not seek small rivers or streams, but rather the banks of rivers reached by the tide, and at those places they open a large canal, thirty to thirty-six feet in width, and with a depth corresponding to the low tide; this canal is filled by the water of the rising [tide], and there they hold it by means of gates until the tide has fallen a span or a span and a half, and as soon as it is this way the factory begins to grind until the waters of the rising [tide] have a greater force than those of the canal. [Barata, 1846: 187]

The first author has also received confirmation that a tide-powered sugar cane mill was located on the Approuague River in French Guiana (Y. Le Roux, pers. comm., 1991). Finally, in 1992 at the Tamancão site near São Luiz, where reference is made locally to a water-powered mill, the first two authors inspected cut-stone remains subject to the influence of the tide and distributed in a way perfectly compatible with the functions of a tide-energy system.

This leads us to conclude that tide energy was used in the past at *all* of these places along the Atlantic coast, and could be used again, if there were a demand for decentralized, low-cost electricity and the technology to generate it. The study of the remains at those sites might well establish a basis for developing new technology to do just that.

On the Usefulness of Traditional Technology

Great changes have occurred in Amazonia over the last 30 years as new social, economic, and technical models were introduced and maintained in an attempt to develop the region. Traditional ways, when not discouraged, were at best disregarded and, intentionally or not, they were often undermined, overwhelmed, and destroyed. The marginalization of many traditional means of livelihood has led to serious social and economic disruption, in particular on the floodplain. Other papers in this volume report some of the environmental and human problems—predatory fishing, overexploitation of timber, and conflict over land and fishing rights—now faced there as a result of such marginalization.

New technologies should be developed that would offer better livelihood alternatives to residents of the floodplain and to meet other of their needs. One way to do this is to base new technologies on those already used in the region. Such traditional technologies may offer advantages of exploiting available natural resources, using them in proven ways, and requiring only those skills and materials that are accessible to local residents. As exemplified by the case of tide energy, traditional technology can call attention to natural resources that are not being used, indicate places where resources are found in abundance, show equipment and operations necessary to exploit them, and demonstrate specific solutions to problems of their use.

However, if a traditional technology is no longer flourishing, as is often the case, its usefulness for developing a new technology may be limited. This would be the case, of course, if the technology uses a natural resource for which there is no longer a social demand. However, if a technology has declined or fallen into disuse because it uses a resource inefficiently or incompletely, parts of it may be replaced or it may be extended by other technology. The tide-energy project demonstrates both of these possibilities. In fact, the case of tide energy suggests that those components of traditional technology that are now more useful act on a resource while closer to its natural state; whereas those components that are now less useful, and need to be complemented or enhanced, act on a resource in a more finished or final state, which is determined by social demand and can change considerably.

In any case, the more detail that is known about a traditional technology, the more useful that technology will be as a base for a new technology. The drawback is that such detail is easily lost over time and with disuse. The tide-energy project was fortunate in being able to apply archaeological techniques to recent, well-preserved remains of a functionally straightforward system, to recover the components of a past technology. Still, important information concerning the timing of certain activities and the ecological impact of the technology was not recovered, even with the aid of a witness. Imagine, then, the difficulty of reconstituting much more complex livelihood systems—for example, from much older and often perishable remains. For this reason it is significant that researchers are now turning to the study of traditional technologies presently or recently employed on the floodplain (as has been done among indigenous peoples on the uplands) in order to register, preserve, and possibly reuse them. Other papers in this volume illustrate these research activities, concerning, for instance, traditional forestry, fish, and game management practices.

Although in some ways exceptional, this case of tide energy is not an exception. Rather it is an example that illustrates the opportunities, limitations, and requisites for bringing knowledge of a traditional technology to bear on a current problem of rural development. Because of this potential, traditional technologies no longer can be dismissed as dead ends but, rather, should be regarded, and used, as points of departure.

Acknowledgments

The authors wish to thank the Ford Foundation for support, in 1988, that made it possible to study the remains of mills in Igarapé-Miri; the Conselho Nacional de Desenvolvimento de Ciência e Tecnologia – CNPq for a 1989 grant to study remains in Barcarena; and the Ford foundation, again, for a 1994 grant to build a laboratory dam to test alternatives to use the tide to generate electricity.

Literature Cited

Anderson, S. D. & F. L. T. Marques. 1990. Engenhos movidos a maré: Atividades de levantamento no campo no município de Igarapé-Miri, Pará, de novembro de 1988 a novembro de 1989. Final report on archaeological fieldwork submitted to the Ford Foundation, Rio de Janeiro.

————— & —————. 1992. Engenhos movidos a maré no estuário do Amazonas: Vestígios encontrados no município de Igarapé-Miri, Pará. Boletim do Museu Paraense Emílio Goeldi, série Antropologia 8(2): 295–301.

—————, **M. F. M. Nogueira & F. L. T. Marques.** 1993. Tide-generated energy at the Amazon estuary: The use of traditional technology to support modern development. Renewable Energy 3(2/3): 271–278.

Barata, F. J. R. 1846. Diário de uma viagem que fez à colonia hollandeza de Surinam. Revista Trimestral de História e Geographia 8: 157–193.

Marques, F. L. T. 1993. Engenhos de maré em Barcarena, Pará: Arqueologia de seus sistemas motrizes. M.Sc. thesis, Pontífica Universidade Católica do Rio Grande do Sul, Rio Grande do Sul.

Nogueira, M. F. M., C. U. da S. Lima & R. R. P. Ribeiro. 1993. The use of small hydroelectric power plants in the Amazon. Renewable Energy 3(4): 907–911.

Farming above the Flood in the Várzea of Amapá: Some Preliminary Results of the Projeto Várzea

Christine Padoch and Miguel Pinedo-Vasquez

Recent research into the pre-Conquest history of the várzea, especially its estuarine zone, has shown that large, socially complex populations flourished for many years in the Amazon floodplains (Roosevelt, 1991, and this volume). Areas that now are used for extensive cattle ranching and that support few people and almost no farming were once home to stable communities that practiced intensive agriculture.

The complex societies that prospered on the várzeas of Marajó and neighboring islands had disappeared by the time of the Conquest; no historical accounts describe them. Archaeological evidence reveals, however, that the várzea residents of Marajó and bearers of Marajoara culture depended primarily on the intensive cultivation of seed crops and the capture of aquatic fauna, with the use of roots, fruits, and game meat playing subsidiary roles in their diets (Roosevelt, 1991).

Exactly how farming was done by the Marajoara may forever remain unclear, primarily because of the difficulties of finding data in a zone as dynamic as the várzea. The large mounds that have been excavated on Marajó Island, and are the most distinctive environmental features of Marajó culture, were largely used as house sites (Roosevelt, 1991). Roosevelt argues, however, that in ancient times the Marajoara might well have dealt with the problems of periodic flooding of their lands by planting crops on raised fields that since have been covered by deposits of silt and sand. The Marajoara probably also controlled flooding and drainage by constructing canals and ditches.

Reducing flood danger to crops by raising the elevation of planting surfaces is characteristic of pre-Conquest farming in many flood-prone areas of South America. These include the altiplano of the Andes (Erickson, 1992), the Llanos de Mojos of Bolivia (Denevan, 1966; Erickson, 1980), and the coastal plains of the Guianas (Boomert cited in Roosevelt, 1991). It is also a characteristic of farming methods found in the várzea of present-day Amapá.

This chapter reports on research being conducted by the Projeto Várzea on technologies developed and used by the present residents of the estuarine várzea located not far from historic Marajoara. Although these farmers are far removed in time and culture from the mound builders of Marajó, their technologies may, indeed, point to ways in which the ancient várzea farmers might have coped with the difficulties of farming the floodplain.

Research Sites and Their Settlement Histories

The floodplain areas of Amapá, Brazil's northernmost state, have, until recently, been among the least researched of várzea regions. This project that we began in 1993 has joined several others (see Raffles, this volume) in attempting to fill this gap.

The Projeto Várzea is an interdisciplinary applied-research effort centered in two várzea areas near Amapá's capital city of Macapá. The Pedreira-Ipixuna várzea lies to the northeast of the city, and the Carvão-Ajudante-Mazagão Velho site, to the southwest (Fig. 1). These two sites are distinguished from other areas of the tidal floodplain in several ways that clearly affect agricultural potential.

Among important environmental factors is the strong seasonality of rainfall. The sites are subject to a dry period (10.2 mm/month) from June to December and a wet period (280 mm/month) from December to May. Both sites experience a complex flood regime that combines daily tidal floods (*mareas*) and seasonal floods (*lançantes*). Mareas vary in height with lunar variation, and lançantes differ with variable rainfall

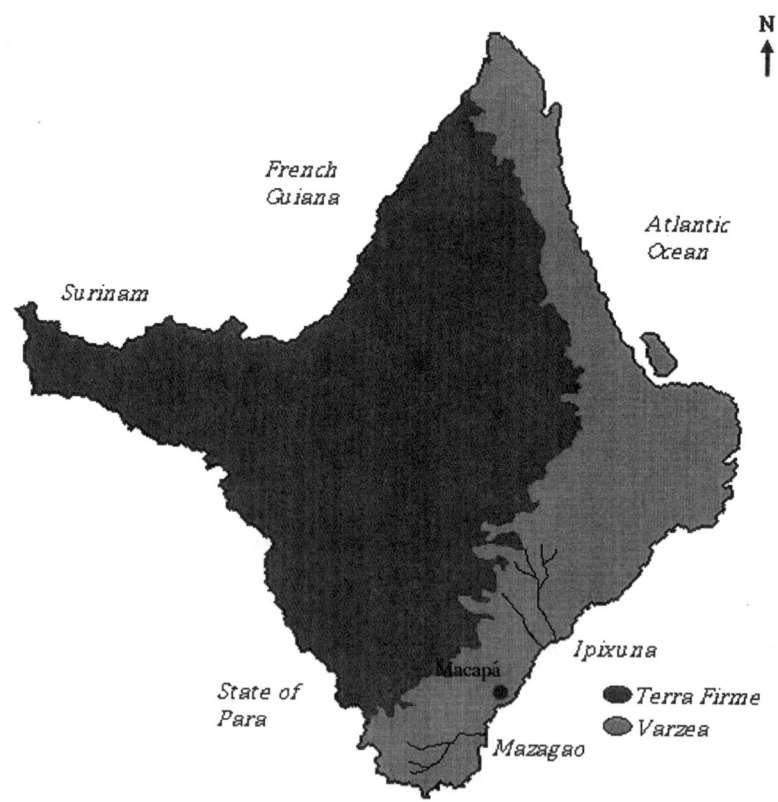

Figure 1. Study sites.

upriver. These environmental factors join with economic and political variables to create conditions that call for considerable expertise if agriculture is to be rewarding.

Currently, large scale landowners are not looking much beyond the extraction of várzea timbers or palm hearts, or, alternatively, cattle raising, as ways of making a profit in the floodplains. Historically, however, these várzea floodplains have experienced a series of different types and intensities of land and resource use by different communities. Today's small farmers (*caboclos*) continue to practice several kinds of locally adapted agriculture.

Little is known of the pre-Conquest inhabitants and the uses they made of várzea lands and other resources at either the Carvão-Ajudante-Mazagão Velho or the Pedreira-Ipixuna sites. Eighteenth-century government reports mention that indigenous folk living in or near the várzea between the present sites of Macapá and Mazagão Velho grew manioc and extracted timber (Muniz, 1916). Both várzea floodplains are now populated by caboclos of mixed African, Amerindian, and European descent.

Of the two várzea sites, the history of the Carvão-Ajudante-Mazagão Velho area is far better known. The site of the present village of Mazagão Velho was a location of an early Portuguese colonization project. In 1770 construction began of a village that was to house several dozen Portuguese families brought by the Crown from a failed colony at Mazagão in Morocco. By 1772, 89 of these families, accompanied by 76 African slaves (Muniz, 1916: 414), had taken up residence. The houses for the settlers were provided by the colonial government and were built by skilled and unskilled labor brought from other parts of Amazonia; many of those workers were Amazonian natives.

The town of Nova Vila de Mazagão was established for several reasons. Perhaps its most important function was to help secure the region—which had at various times been contested by France, Spain, and Holland—for the Portuguese. But the settlement was also to provide an economic return to the colonial government that had financed its establishment. The colonists at Mazagão, like those of the early Amapá várzea towns of Macapá and Vila Vistosa were to produce rice to help feed themselves as well as population centers in Pará and Maranhão.

In the early years, Mazagão did indeed produce rice in quantities that allowed for substantial export (Bastos, 1947: 48); the details of how rice was produced in colonial times are not available. But early agricultural successes were short-lived, as was the town's municipal health. An epidemic that may have been plague ravaged the population in 1781. After that year, agricultural production ceased to grow and by the end of the century, Mazagão had declined considerably in economic activity and production. It also ceased to play an important political role. The town has since suffered several reversals of fortune, but it now has a population of approximately 424 inhabitants.

The settlements of Carvão and Ajudante have a character different from Mazagão Velho, which, despite its small size, has the appearance of a substantial town. The two rather dispersed settlements were established well after Mazagão Velho. Residents of Carvão who know a good deal of its history believe that it was once the site of a nineteenth-century *quilombo*, a settlement of slaves who had escaped bondage and established an independent community.

The Pedreira-Ipixuna region is known to have been populated by caboclos at the beginning of this century, when commercial logging started. Little historical material is available on its subsequent use and settlement. It is today a small but widely dispersed community.

Extraction of a variety of nontimber forest products as well as commercial fishing for several large species of catfish and shrimp have been important sources of cash for inhabitants of both várzea floodplains since 1950. Harvest of açaí (*Euterpe oleracea*) fruits and heart of palm from both managed and unmanaged stands became a major commercial activity since 1960. Currently, açaí products (fruits and heart of palm), timber, and shrimp are important cash products marketed from the várzea of both study areas. However, agriculture remains the central economic activity for almost every várzea household; currently corn is the main cash crop in Carvão-Ajudante-Mazagão Velho, and bananas in Pedreira-Ipixuna.

Research on many aspects of agricultural production is being done by Projeto Várzea researchers working with local farmers. One focus of this work is on methods and techniques used by caboclos for producing crops in the risky, flood-prone environments that the Amapá várzea presents.

Farming the Floodplain

Throughout the world and throughout history, floodplains have been favored for agricultural production because of their fertile soils, proximity to river transport and aquatic resources, and because floodwaters periodically cleanse farm fields of weeds and insect pests. But the beneficial floods also bring the danger of loss of crops, loss of life and property, and sometimes loss of the very fields because of erosion.

Farmers of the Amazon várzea have dealt with these problems in a variety of physical, social, and political ways (Chibnik, 1995). Probability of losing all agricultural production is largely mitigated by wise choice of flood-resistant crops, careful zonation of production according to small but crucial variations in elevation, precise timing of agricultural operations, and a series of other locally developed technologies.

The importance of any of these strategies varies widely along the Amazon and its tributaries. The estuary, where tides threaten twice daily but where seasonal variations in flood height are not large, presents both different opportunities and different problems to farmers than those presented by upper Amazon sites, where seasonal changes of 10 m or more in water level are not uncommon (Padoch & Pinedo-Vasquez, 1991; Denevan, 1984).

Farming the várzea sites of Amapá demands considerable skill. Low-lying areas along the Amazon and its many tributaries are inundated twice a day by high tides averaging 1.2 m. Many areas, however, are inundated less predictably. Interactions between seasonal river level changes and tides are complex. High river levels from December to April, together with high tides, raise waters to about 2.3 m above low water levels, inundating all várzea lands. The combination of exceptionally high tides and rain-swollen rivers can occasionally push waters even higher.

Apart from the height of the river, the time any area is flooded strongly affects agricultural success. Several crops selected and planted by caboclos in várzea areas withstand a short flood even if it comes frequently, but cannot survive a long period under water. The combination of high mareas with high lançantes in the main river and tributaries can lead to persistence of high waters over a long time, producing substantial crop loss. The speed of river currents also changes frequently throughout the year. Again, crops that may survive a simple rise in water level may be lost to a strong erosive current that can uproot both trees and herbaceous species.

Varying river levels are associated in the research sites with varying quality of river water. Amazon River waters carry considerable silt, whereas most of the tributaries are black waters and carry little. When Amazon waters rise they inundate the lower reaches of the tributaries and deposit silt on their várzeas. When the Amazon is low, the acidic, clearer waters of the Mazagão, Mutuaca, Ajudante, and Pedreira Rivers bathe the mouths of the tributaries.

Rising above the Flood

Small agriculturists of the estuary must cope with a complex and often risky flood situation. Like many other floodplain farmers, Amapá's várzea dwellers are careful in their selection of planting sites, crops and varieties, and seasons. Even small differences in elevation, variety, or date can be crucial in deciding where to make fields and what to plant within any field. Apart from searching out favorable sites, estuarine smallholders also manipulate and alter sites; they *make* favorable sites by changing the level of the planting surface. Several appropriate technologies for elevating land surfaces have been developed.

An important ecological function of the vegetation established in várzea environments is the capture of sediments and protection of these deposits from erosion by tidal waters. Through these natural processes, the level of lands in the estuary rises gradually until it reaches above the level of tidal floods. In both our research sites, farmers help speed up natural processes of sediment retention and rise in land levels in palm forests.

Although most species of riverine vegetation tend to accumulate sediments and build up soils, some species do so much more rapidly than others. The caboclos of Pedreira-Ipixuna and the Mazagão Velho area consider two common palms: buriti (*Mauritia flexuosa*) and açaí (*Euterpe oleracea*) as the most important plants for capturing sediments and building up land. Both of these palms are managed by várzea populations throughout the Amazon for a multitude of economic uses including their fruits, heart of palm, building materials, and handicrafts (see Hiraoka, this volume; Anderson, 1988; Kahn & Mejia, 1990; Padoch, 1988). They are also both managed by Amapá's várzea dwellers for their capacity to accumulate sediments and raise planting surfaces above tidal flood levels.

Management of buriti is conducted in largely natural forests, while açaí is managed in both natural stands and in planted areas such as house gardens (*quintais*). Although both palm species tend to raise soils over flood levels, each species does so in different ways. Açaí tends to form clumps and sprouts, while buriti produces modified hydromorphic roots. Açaí sprouts and buriti roots both retain suspended sediments and protect accumulated sediments from erosion by tidal waters.

Caboclos use several strategies and techniques to manage açaí and buriti and speed up the processes that raise the level of their lands above the flood. In managing buriti stands, caboclos first limit access to selected buriti areas. They set up signs along all the possible routes into a particular buriti stand, signaling their claim on the patch and restricting entrance by others. By limiting access, caboclos hope to prevent disturbance of sediments that have settled around buriti trunks. During the high-water season, when fruit collectors visit buriti forests, they are allowed to enter the claimed buriti stands in canoes only during high tides, and then only keeping to the courses of certain well-defined streams.

Natural processes in the buriti palm forest are accelerated by stimulating the production of hydromorphic roots. As roots are covered by sediments, new ones are formed. Caboclos therefore continually cover the roots by carefully overlaying existing ones with sediment-covered buriti leaves gathered in the forest. With their roots covered, buriti palms will produce ever more roots which are then again covered. This succession of formation of new roots is repeated until the level of the soils reaches an elevation above the range of tidal waters. When the land is no longer subject to tidal inundation, several leguminous tree species tend to became established and compete with the buriti. Most caboclos know that when these trees become dominant in forest areas where buriti once thrived, the land is high enough and they can then fell the forest to make fields.

The time required to build up soils over the level of tidal floods depends on where the selected buriti forest is located in relation to the main channel of the Amazon River. While in areas near the main channel this process takes 5–7 years, in distant sites it may take up to 11 years. This difference reflects the distinct volumes of sediments suspended in floodwaters. With distance from the main channel, the suspended load is reduced.

While the formation of new roots is stimulated by management of buriti palms, açaí stands are managed even more actively. About twice per year clumps of the palm are thinned and pruned, and several stems and many leaves cut from each clump. This slash is then arranged to cover the ground around and between palm clumps. Thinning encourages new sprouts; depositing the slash on the ground helps build up the elevation of the area. The leaves and poles are soon covered by sediments left by the high tides.

The length of time that is required to raise soil levels over that of the daily floods in açaí groves depends on the number of clumps per area and the frequency with which stems and leaves are removed and laid down. Areas with a density of approximately four clumps of seven adult individuals each (>2.5 m in height) in an area of 10 m^2 and where pruning of stems and leaves was done once or twice a year can, according to local farmers, be raised above flood levels in 3–5 years. More frequent cutting or bringing in materials from outside the immediate area can result in faster build-up. The raising of soil levels in managed açaí forests depends more on the accumulation of such organic materials than on the build-up of sediments transported by tidal waters as is the case in buriti forests.

Apart from using açaí plant material to build soils, caboclos make and maintain small drainage channels around their selected açaí stands. Although some channels are formed naturally by the movement of the tidal waters, most of them are modified. Várzea farmers alter them by cleaning or removing plant materials from their margins and in some cases by removing sediments. These channels are used to drain tidal water more quickly and to reduce moisture in the newly formed soils.

When the managed areas reach a level above the tidal waters, the açaí stands are

cut to make fields or to extend and diversify house gardens. Most caboclos who manipulate stands of açaí palms to raise soil levels around their houses do so to produce a
larger variety of fruits, medicinals, timbers, and other valuable species including ones
that are not flood resistant. For instance, most caboclos families have managed to produce papaya (*Carica papaya*) and avocado (*Persea americana*), two nonresistant fruit species,
in their house gardens.

Manipulating the natural processes of sediment accumulation and rise in soil levels,
Amapá caboclos not only transform inundated areas into lands that rise above the flood,
they also help make especially fertile soils. Particularly in managed açaí stands, the
mixing of plant materials with river-borne sediments enriches the soils that are then
available for the planting of various more demanding crops.

Intricately Zoned Slash-and-Burn

Although caboclos indeed know how to manipulate natural soil-formation processes
and thus change land levels, most várzea agricultural fields are not uniform and some
parts are usually subject to flooding. Techniques used in making the fields where they
raise the majority of their crops appear be "standard" shifting cultivation techniques.
We have found, however, many often subtle but apparently important variations in the
system that allow the farmers to survive and succeed in the várzea environment. Perhaps
the most important in the Amapá várzea is the extraordinary care given to determining
zones of the field appropriate for planting particular crops and varieties according to
their resistance to differing periods of inundation. Planting techniques used by caboclos
are designed to take advantage of very small differences in topography and exposure to
tidal floods.

Rather than ploughing, hoeing, or otherwise leveling their fields, farmers carefully
observe and select sites and segregate crops and varieties that are adapted to the different
conditions in their flood-prone fields. When clearing new farms, farmers leave felled
stems of trees (usually açaí) lying in the field. They then return several times to the field
and monitor marks on the felled trunks to see how far the flood reached. Notice is also
taken of zones where floodwaters remain for extended periods or where rainwater may
accumulate. The areas near drainage channels and in small depressions tend to be those
where water remains longest and deepest. Decisions on what to plant and where thus
follow extensive and repeated monitoring.

Care is also taken to preserve and make use of any slight topographic differences
that might exist. For instance, small mounds of organic material, largely leaf litter formed
naturally around tree stumps, are protected from burning by clearing all around them
or by laying fresh buriti petioles all around the mound to block the fire.

Following the identification and marking of distinct areas within the field, caboclos
select the type of crop to be planted in each area. Crops are selected on the basis of
their tolerance to short-term inundation and to longer-term standing water. Our preliminary research in 36 fields planted in 1994, found an average of 13 flood-tolerant
and 7 nontolerant species in each field. Of the 13 flood-tolerant species, 4 withstood
both short-term tidal floods and standing waters and the other 9 were resistant only to
tidal floods. Many of the most flood-sensitive crops, such as tomatoes, peppers, and

watermelons, are preferentially planted around tree stumps and palm clumps, taking advantage of both the slight elevation and the organic matter–enriched soils.

Farming on Platforms and Rafts

Although flood-sensitive crops are planted on the highest land in swiddens, Amapá farmers also produce many vegetables, medicinals, condiments, and flowers on raised platforms. Two of the most common kinds of platforms made at our research sites are those that stand on solid supports and those that float on rafts. The firmly anchored platforms are made in varying sizes and can be attached to the house or located farther away on poles. An average platform in the two study areas is 12.5 m² (5 × 2.5 m) in size and stands on 6–8 poles 1.45 m high above mean water level.

The second kind of platform used by caboclos to plant and protect crops are floating platforms or rafts. These platforms are built on three or four logs averaging 60 cm in diameter and 5 m in length. These logs tend to be of species with very low-density wood such as açacu (*Hura crepitans*). Most caboclos use vines and tree bark to tie the logs together and attach them to 6–8 poles to make a raft. On this raft a platform is made using burití palm trunks or sawnwood.

A mixture of air-dried fresh sediments and dried organic matter is used as a planting medium on both kinds of platforms. The most common material used as organic matter at the two sites is composted açaí seeds. On an average-sized platform, caboclos tend to use about 120 kg of dried soil sediments and 150 kg of dry composted açaí seeds.

Caboclos commonly employ three distinct techniques to compost açaí seeds, depending on the season and abundance of seeds. The most frequently used technique requires storing the seeds in a hole in the ground or in an open container, keeping them wet, and covering them with banana, açaí, or burití leaves. Caboclos tend from time to time to mix the seeds, rotating those on the bottom to the top and removing roots from any that are germinating. Açaí seeds processed using this technique are ready for use as a planting medium after 90–120 days. This technique is commonly employed during the fruit season, when seeds are available in great abundance.

The second processing technique consists of collecting açaí seeds in a sack and keeping the sack under the water for an average period of 20 days. Although the seeds do not decompose while they remain under the water, they do lose their capacity to germinate and can be used on planting platforms. This rapid technique is mostly used during the dry season, when fruit and seed production is reduced but the need for the planting material is high.

In the third technique, farmers accumulate seeds in drainage ditches near the kitchen where waste water and other refuse is thrown out. Açaí seeds processed by this technique are ready for use in 10–12 days. Although this technique is faster than the other two, most caboclos do not favor it because they believe that the açaí seeds are apt to be contaminated by fungi that may damage cash crops, such as onions, and that these fungi are later extremely difficult to control.

Processed açaí seeds are the principal source of nitrogen in the planting medium used on platforms. Açaí seeds also increase the availability of calcium, phosphorous, and potassium for plant uptake. These four nutrients are highly available when sediments are mixed with processed açaí seeds (N = 1.6 mm/gr; Ca = 23 mm/gr; P = 3.7 mm/gr; K = 14.6 mm/gr). Because of the abundance of these nutrients, species that demand high concentration of NPK are first planted on the platforms; these include tomatoes, bell peppers, and onions. Gradually they are replaced by crops that are less demanding of nutrients.

Caboclos in the two sites plant and produce crops in the same soils for three to four years. After this period they clean the platforms and replace the planting medium with a fresh mixture of soil and açaí seeds.

Each kind of platform is used by caboclos principally to plant and protect crops from floods. Many farmers also insist that platforms help protect sensitive plants from pests and diseases. Households in the two sites produce most of their herbaceous medicinal, food, and ornamental plants on platforms. These production techniques are also commonly used by caboclos living in flood-prone shanty towns and suburban villages near the cities of Macapá and Santana. For instance, the bulk of the medicinal and ornamental plants produced by residents of the Macapá suburb of Fazendinha are produced on platforms. The residents of this várzea village not only produce small quantities of medicinal and ornamental herbs for their own use, but many are also successfully producing vegetables such as onions in considerable quantities for the Macapá and Santana markets.

Locally Developed Technologies: Past and Future

The archaeological work of Roosevelt and others is providing us with increasing evidence that pre-Conquest settlements in the várzea were large, complex, and dependent on agriculture. The change and dynamism of the floodplain that is its source of fertility and abundance has also, unfortunately, caused the loss of much of the evidence of how it was farmed in the past. The agriculturists of Marajoara surely did not farm the same way today's caboclos farm in Pedreira-Ipixuna or the Carvão-Ajudante-Mazagão Velho site. In broad outline, however, some of the practices used by contemporary farmers may suggest how early várzea dwellers coped with várzea environments. Raising land levels by managing and accelerating natural processes of silt accumulation, modifying natural drainage channels, selecting and monitoring zones of differing flood susceptibility, farming on extensive platforms—all of these may have been components in Marajoara's suite of production technologies. Examining areas of *terra preta* marking abandoned settlements at Igarapé Grande near the Pedreira-Ipixuna research sites, we have found evidence of past agricultural practices. Digging through várzea soils we have encountered profiles where highly organic, leaf-filled strata alternate with layers of river sediments. These might suggest that past inhabitants of Amapá's várzeas used soil management techniques similar to those described above for managing açaí stands.

The technologies described above are limited to small-scale farming enterprises. However, they are adaptable, and have been adapted successfully, to intensive commercial production of high-value crops such as onions or ornamentals, and are being used for other purposes.

Platforms, despite their limited size, are especially important tools in certain phases of more extensive agricultural undertakings. For instance, Ipixuna banana producers, plagued by the devastating Moko disease, have used platforms to temporarily keep planting material in an environment that is relatively easy to maintain in disease-free conditions. Once the banana suckers have sprouted leaves, they are transplanted into fields.

Adaptation of traditional technologies to changing conditions is a hallmark of caboclo resource management. The Projeto Várzea sees as its main task to aid in this process by identifying, analyzing, experimenting with, and helping upgrade the technologies used by Amapá várzea farmers. The practices described above give only a hint of the richness of the caboclo repertoire.

Literature Cited

Anderson, A. 1988. Use and management of native forests dominated by açaí palm (*Euterpe oleracea* Mart.) in the Amazon estuary. Pages 144–154 *in* M. Balick, ed., The palm — Tree of life: Biology, utilization, and conservation. Advances in Economic Botany 6. New York Botanical Garden, Bronx.

Bastos, A. de Almeida. 1947. Uma expressão ao Amapá. Impresa Nacional, Rio de Janeiro.

Chibnik, M. S. 1995. Risky rivers: The economics and politics of floodplain farming in Amazonia. University of Arizona Press, Tucson.

Denevan, W. M. 1966. An aboriginal cultural geography of the Llanos de Mojos de Bolivia. Ibero-Americana 48. University of California Press, Berkeley.

———. 1984. Ecological heterogeneity and horizontal zonation of agriculture in the Amazon floodplain. Pages 311–336 *in* M. Schmink & C. H. Wood, eds., Frontier expansion in Amazonia. University of Florida Press, Gainesville.

Erickson, C. 1980. Sistemas agricolas prehispanicos en los llanos de Mojos. America Indígena 40(4): 731–755.

———. 1992. Landscape management in the Andean highlands: Raised field agriculture and its environmental impact. Population and Environment 13(4): 285–300.

Kahn, F. & K. Mejia. 1990. Palm communities in wetland forest ecosystems of Peruvian Amazon. Forest Ecology and Management 33: 169–179.

Muniz, J. da Palma. 1916. Limites municipaes do Estado do Pará: Mazagao. Annais da Bibliotheca e Archivo Publico do Pará, vol. 9. Belém.

Padoch, C. 1988. Aguaje (*Mauritia flexuosa*) in the economy of Iquitos, Peru. Pages 214–224 *in* M. Balick, ed., The palm — Tree of life: Biology, utilization, and conservation. Advances in Economic Botany 6. New York Botanical Garden, Bronx.

——— & M. Pinedo-Vasquez. 1991. Floodtime on the Ucayali. Natural History (May): 48–57.

Roosevelt, A. C. 1991. Moundbuilders of the Amazon: Geophysical archaeology on Marajó Island, Brazil. Academic Press, San Diego.

Exploring the Anthropogenic Amazon: Estuarine Landscape Transformations in Amapá, Brazil

Hugh Raffles

Introduction

This paper describes the early stages of an ongoing research project in the Amazon estuary, reviewing data from interviews carried out with local residents between June and August of 1994. I briefly present four examples of landscape management undertaken by local farmers who have cut streams (*igarapés*) from tributaries of the main river channel. It should be emphasized that although these interventions may have initially occurred on what appears to be a fairly small and localized scale, erosion of the river banks and the widespread use of these techniques have given them a significance such that today's landscape is, by all accounts, quite different from that of even 30 years ago. Furthermore, although research up to now has focused on those manipulations specifically associated with the community I call Guariba, it is likely that comparable interventions are widespread in the estuarine floodplain (*várzea*).

My initial research shows that local people have radically transformed the landscape of this part of the Amazon estuary over the past 30 years. Major waterways previously thought to be "natural" are, in fact, anthropogenic, opened by farmers to improve access to agricultural fields and forest products. These interventions are an important issue in the local politics of the estuarine floodplain. Both *ribeirinhos* (smallholders) and *fazendeiros* (large landowners) in the area have found manipulation of the river system to be a significant means for securing access to and control over resources.

This research supports the many recent studies of human–environment interactions in Amazonia that have emphasized the transformative impact of "traditional" Amazonians on what has conventionally been viewed as an overwhelmingly inhospitable environment (Parsons & Denevan, 1967; Denevan, 1970, 1992; Posey, 1985; Denevan & Padoch, 1987; Irvine, 1989; Balée, 1989, 1993, 1994; Hecht & Posey, 1989, 1990; Roosevelt, 1989, 1991, 1993; cf. Meggers, 1954, 1971, 1973, 1985). Studies of upland forest management have noted manipulations of fundamental ecosystem attributes, including soil formation processes (Hecht & Posey, 1989, 1990) and the species composition and density of plant communities (Posey, 1985; Irvine, 1989; Balée, 1989, 1993, 1994). Data showing that supposedly "natural" environments are physically constructed have challenged assumptions about the "pristine" tropical forest and the passivity of local inhabitants. Indeed, the assertion of "a biocultural origin that would not have existed without past human interference" for large areas of these forests (Balée,

1993: 231) forms an important element in the scholarly response to long-standing and still influential environmental determinist explanatory paradigms of Amazonian social relations. However, as I contend below, the accumulation of this type of data, and the change in academic discourse which it signals, has broader implications than those that have so far been articulated by Amazonianist scholars.

The Study Area

Guariba is a riverine community of approximately 25 families. It is located on the tributary I call Rio Guariba, close to its confluence with the Rio Amazonas, a journey of about four hours by small motor-launch northeast along the northern channel of the Amazon estuary from the city of Macapá, the state capital of Amapá. The estuarine zone of the Amazonian floodplain is characterized by a highly dynamic fluvial regime (Sponsel, 1992; Moran, 1993: 96–97). Residents with houses and fields along the banks of the major tributaries in the study area are accustomed to moving back from the river margin every few years as the rivers expand or shift their bed. Yet despite the intriguing social questions raised by these unstable conditions, and the considerable recent interest in várzea ecology and resource management practices, the estuarine zone of the floodplain remains understudied (Hiraoka, 1992). Little research has focused on either current or historical resource use in this geographic area (for notable exceptions, see Wagley, 1953; Lima, 1956; Meggers & Evans, 1957; Parker, 1981; A. Anderson et al., 1985; A. Anderson & Jardim, 1989; S. Anderson, 1991; Roosevelt, 1991; Motta-Maués, 1993; Padoch & Pinedo-Vásquez, this volume).

Prior to the arrival of Europeans in the sixteenth century, complex mound-building chiefdom cultures supported by intensive maize and manioc cultivation, hunting, and fishing are thought to have flourished on the estuarine floodplain (Roosevelt, 1989, 1991, 1993). As chronicler of Orellana's pioneering voyage along the Amazon in 1540–1541, Gaspar de Carvajal marveled at the large numbers of people living along the banks of the lower reaches of the river (Medina, 1934). Yet the Conquest and subsequent demographic disaster transformed a largely self-contained agricultural economy into one in which extraction and production were organized above all for the market. As several commentators have noted, Amazonian ribeirinho society — with resource management practices characterized by the flexible interplay of swidden agriculture and extraction for household consumption and exchange — emerged from the accumulated rubble of colonial adventure and the Cabanagem revolt only with the increasing autarky that followed the collapse of the rubber "boom" in the early years of this century (Ross, 1978; Weinstein, 1985; cf. Sweet, 1992; Whitehead, 1993a, 1993b).

This history should be linked to the persistence of the extractive economy in the region. With few exceptions, from the expulsion of the Jesuits in 1759 until the arrival of the military government in the 1960s and 1970s, large-scale economic penetrations of the Amazon were limited to short-cycle extractive episodes driven by the demands of distant metropolitan markets for tropical exotica (Bunker, 1985; Hemming, 1987; Padoch & de Jong, 1990; Schmink & Wood, 1992). The well-documented rubber period was therefore only one, if the most significant, instance of a pattern whose legacy is still apparent in attempts to invigorate an international trade in nontimber forest products (Raffles, 1993).

An outline of economic history in the study area corresponds with much of what is known of the estuarine várzea. As with other floodplain communities relatively close to the urban markets of Belém and Macapá, modern economic and environmental histories in Guariba have been closely tied to the rise and fall in demand for extractive products. In the early twentieth century, there was strong demand for timber, particularly *muritinga* (*Maquira coriaceae*), *cedro* (*Cedrela odorata*), and *ucuúba* (*Virola* spp.) (see Almeida Lustosa, 1976). By the 1940s, while timber extraction continued, there was also a significant market in oilseed extraction of *muru muru* (*Astrocaryum murumuru*), *andiroba* (*Carapa guianensis*), and *pracaxi* (*Pentaclethra macroloba*), animal skins, and, briefly, rubber (*Hevea brasiliensis*) under the multiple stimuli of the U.S.-sponsored "*batalha da borracha*" (Bunker, 1985; Hecht & Cockburn, 1989). Much of the trade in the area was—and still is—organized under structures derived from the credit-and-supply *aviamento* system, which originated during the rubber period and tied the area to capital in Belém and Macapá, intermediate links in a chain that led across the Atlantic to Europe (Bunker, 1985; Dean, 1987; Schmink & Wood, 1992).

In the 1950s, bananas were introduced in Guariba and there was an increasing commercialization of fishing, particularly of catfish and shrimp. By the 1960s, as in much of the Amazon estuary, local people were becoming increasingly involved in extraction of both fruit and palmito from the much-loved *açaí* palm (*Euterpe oleracea*), establishing the basis of today's livelihood patterns in which ribeirinhos rely heavily on açaí, fish, and shrimp for household consumption and as sources of income.

This latter period, from 1960 to the present day, is of particular interest for reconstructing a history of landscape change in Guariba. With minor forest products, timber, and bananas growing in importance in local and regional markets, the first fazendeiro settled in the area, bringing four families recruited as labor from his home district in the island municipality of Afuá, Pará.

Landscape Transformations I: Interventions in the 1960s

The landscape that local people remember from the 1960s is startlingly different from that visible today. Communication between the two major tributaries in the area—nowadays possible by at least two direct routes—required a time-consuming and risky passage via the main channel of the Amazon. According to local informants, the complex network of streams and broad waterways that typify the area today did not exist in the 1950s and early 1960s. It is from this period, when permanent modern settlement took place and people began to focus on the extraction of forest products, that the first interventions can be documented. Informants have described two episodes in detail.

IGARAPÉ DA VALA   In the early 1960s, the local fazendeiro organized members of the four families living in Guariba into work-teams to dig the narrow channel that has today grown into Igarapé da Vala. These workers had the unenviable task of removing the dense vegetation of the aroid *aninga* (*Montrichardia arborescens*) and *pirí* or *tabua* grass (*Cyperus giganteus*) to open a channel approximately 2 m wide—that is, wide enough for a canoe to pass twice-daily at high tide. They uprooted the aninga, which reproduces vegetatively, and cut the pirí close to the ground. The subsequent erosion that produced the broad channel visible today has occurred without further human intervention. As in all cases so far documented, the intent of this management has been to alter the fluvial regime at the microlevel and initiate a process of accelerated stream increment (Fig. 1).

I estimate that Igarapé da Vala today extends for somewhat more than 2 km and is close to 50 m across at its widest point. It enables collectors of forest products to pass from the Rio Guariba into an area in which they have consistently harvested and managed timber and açaí over the past 30 years (Fig. 2).

IGARAPÉ PRACAXI   Igarapé Pracaxi is a smaller channel today than Igarapé do Vala. It is one of several in the area that ribeirinhos constructed independently of the local *patrão* (patron/landowner) during a period in which they were obligated—by debt, intimidation, and personal ties—to sell their produce exclusively either to him or to members of his household.

Working over a period of several months, a ribeirinho family constructed this igarapé from a narrow, seasonal channel (*rego*). These days it is permanently navigable and extends approximately 1000 m. As with Igarapé da Vala, the purpose was to improve access to forest products. In this case, people were keen to extract large timber

Figure 1. Igarapé de Vala.

Figure 2. Igarapé de Vala.

species for sale at the sawmill the local patrão operated at the mouth of the river during this period.

Landscape Transformations II: Recent Interventions

The techniques involved in landscape management have changed over time, in response to changes in local social relations. A critical moment in the environmental history of this area occurred when fazendeiros introduced large numbers of water buffalo in the 1970s, apparently with fiscal support under the POLAMAZONIA program. Heightened tension locally can be correlated with their appearance. These large animals are difficult to control. They swim across rivers and are liable to enter fields and destroy crops. Ribeirinhos also complain of their negative impacts on water quality and fish harvests. The exacerbation of lateral erosion due to the persistent grazing by buffalo on aninga is highly probable.

IGARAPÉ GUARIBA   Despite the generally negative effects of buffalo on ribeirinho livelihood and the unremitting antagonism expressed toward them by local people, ribeirinhos have been able to generate some compensatory effects from the animals through creative management. Specifically, in opening channels since the 1970s ribeirinhos have utilized buffalo belonging to fazendeiros to reduce the demands on human labor. The most important igarapé connecting the two major tributaries in the area—linking communities on these rivers with a recently completed unpaved road to Macapá—was opened by a team of ribeirinhos who first cut a 2 m wide channel through *campo alagado* (*pirizal*) and then repeatedly drove buffalo through the opening. Informants have suggested that the majority of igarapés formed in recent years have been opened in this way.

IGARAPÉ ARIRANHA   In 1974, a team of 30 ribeirinhos, organized under their own management, spent around 30 working days clearing a 5 m wide channel approximately 1 km through a dense covering of aninga and pirí, to help a family get easier access to their banana field. As with Igarapé Pracaxi, this route had previously been passable only in a small canoe, with great difficulty, and only in the rainy season. The rest of the year farmers would travel a certain distance on foot—backs laden with bananas—and continue a little further by small canoe, before finally switching to a larger vessel. In 1984, unable to take full advantage of the market for bananas because their crops were spoiling in other, more distant fields, farmers extended the same channel by another 500 m more, creating access to several more plots.

Political Economy and Resource Conflict

The different forms of labor mobilization apparent during the history of landscape management in Guariba can be linked to the local and regional political economy. The changing structure of landholding in the area has been accompanied by changes in the relations of production that have involved different forms of labor organization. At times, channel-building has been organized by fazendeiros under the quasi-coercive structures of the patrão system. More recently, ribeirinhos have organized labor under more cooperative arrangements, including forms of *mutirão* (reciprocal labor). Notwithstanding these differences, both large and small farmers have built channels as a way of taking advantage of market opportunities in particular agricultural and extractive commodities. In turn, if we were able to aggregate data across the area supplying Macapá, we would likely find that urban markets in forest and agricultural products have been influenced by the shape of the landscape, which increases the flow of goods and could, in certain cases, accelerate the depletion of extractive resources. Moreover, in the local context, changes in the landscape have complicated landholding structure by disturbing property boundaries and affecting the value of individual holdings.

The close relationship of local economic history to the shape of the landscape makes it apparent that stream manipulations are an important factor in local resource conflict. The resolution of ongoing conflict in Guariba over açaí, for example, is likely to depend on the capacity of people in the area to exercise control over the shape of the landscape. Açaí has a central symbolic and nutritional place in the ribeirinho diet and importance as a subsistence and cash crop (Anderson & Jardim, 1989). However, for more highly capitalized landowners, value lies in the frequently destructive harvesting of the palm heart for export. Local conflicts have centered on the increasing scarcity of the palm. Igarapé Pracaxi, described above, is an example of a disputed channel that, if extended, could provide access to remaining "wild" stands of *Euterpe oleracea*. Despite pressure from local landowners, ribeirinhos have prevented the landscape from being changed in this way, and, at present, the stream ends in an impassable thicket of aninga (Fig. 3).

Figure 3. Igarapé Pracaxi.

Ecological Considerations

The fluvial landscape in this area is subject to three types of "non-natural" disturbance: human stream construction through manual labor, stream-opening in which both humans and buffalo are involved in the initial clearing process, and erosion induced by the physical impact of buffalo and their repeated predation of aninga and pirí.

In the undisturbed system, erosion is limited by dense stands of aninga and pirí which can be present either monospecifically or in association. When these are subject to continuous disturbance, degraded areas are colonized initially by an unidentified *cipó* (vine) and then by a dense covering of the thorny shrub *aturiá* (*Drepamocarpus lunatus*). This second plant is aggressive and highly competitive, and although it can have an important stabilizing influence on stream margins, it makes land unsuitable for either pasture or cultivation. An alternative pathway after the loss of aninga is a landscape denuded of all vegetation except close-cropped grasses, and subject to rapid erosion. It is assumed that this latter system develops in locations with higher densities of buffalo in which aturiá cannot become established.

Significant loss of vegetation and soil from the system appears to occur only after repeated disturbance. The level of "maintenance" required to keep newly cut streams open, therefore, becomes an important question to be addressed as this research develops. Such maintenance may be particularly labor demanding in the early "establishment phase" of stream-building, when repeated interventions may be required to prevent recolonization by aninga and pirí, as well as aturiá. It is clear from visits to Guariba that the opening of channels increases the range over which buffalo are able to graze and trample. Buffalo may reduce or even remove the need for continued human management. On the other hand, as we have seen, buffalo play a considerable role in exacerbating and accelerating the erosion of the landscape.

The ecology of stream management in Guariba is thus complicated by two interrelated processes: the widespread, destructive activities of the buffalo herds and the powerful erosive forces of the rivers themselves. In this context, it is worth noting that human interventions are concerned above all with the *location* of streams as modes of access to resources, rather than with the *rate* at which streams open and the landscape changes. However, farmers do exert control over the rate and scale of stream-opening and development by their selection of location in relation to tidal flows of varying strengths. One reason for the relatively rapid growth of Igarapé Guariba, for instance, appears to be its position as a link between two rivers.

Discussion: Situating an 'Anthropogenic Amazon'

As I described in the introduction to this paper, the last ten years have seen an accumulation of studies exploring the varied ways in which local people in the Amazon basin have transformed the regional landscape. This work has broad implications. It challenges us to reconsider how we distinguish between "extraction" and "production"—terms that occupy a central position in discussions of forest management and resource use. Similarly, it raises questions about the meanings of definitive terms such as "occupation" and "ownership," and so offers an opportunity to reframe some of the debate over property rights in the region. However, to make such connections effectively and to link them to specific agendas, we need to explore the relationships between local resource management practices and regional (and global) relations of power. To do this, it helps to consider how changing "ways of seeing" the natural world are involved in the production and circulation of power.

Although the fluvial interventions described above have occurred on a relatively small scale, their impacts locally have been of disproportionate significance, transforming the physical landscape and fluvial map over a considerable area. Informants in other parts of the Amazon estuary, including the islands of Marajó and Caviana, Pará, and at Jarí, Amapá, as well as upstream on the middle Amazon, have described similar techniques involving the management of streams. We can also find examples of parallel interventions, such as the "artificial canal" at Igarapé-Miri, traversed by Henry Walter Bates and Manoel Buarque, among many others, that connects the Rio Mojú with the Tocantins river-system (Bates, 1910 [1863]: 61; Buarque, 1919: 4). Such manipulations, carried out independently by local people or, as at Igarapé-Miri, constructed in more formal engineering projects supported by state funding, do not appear to be unusual. One may therefore (conservatively) speculate—and here the immediate concern is only with fluvial manipulation and not any possible changes in species composition or density of vegetation—that, along with a considerable proportion of the upland forest (Balée, 1989; Denevan, 1992), significant areas of the estuarine floodplain may be classifiable as cultural landscapes.

It is now more widely accepted that historical Amazonian populations exercised a transformative impact on their environment. Estimates of the extent of anthropogenic terra firme forest in the region vary from Balée's (1989) 11.8% to Denevan's rather more provocative claim that "there are no virgin tropical forests [in the Americas] today, nor were there in 1492" (1992: 375). Yet despite the increasing familiarity of the argument, there has been an absence of sustained consideration of its implications beyond an assertion of human transformative capacity in relation to earlier formulations of local passivity in the face of overwhelming natural forces. Antecedents of this latter view can be located in such classic expedition narratives as Alfred Russell Wallace's *Travels on the Amazon and Rio Negro* (1912 [1853]) and H. W. Bates's *The Naturalist on the River Amazons* (1910 [1863]), both of which are of tremendous importance in their locking of the European gaze on Amazonian nature, on its abundance, and its potential for profit.

Although the two naturalists traveled together and described shared experiences for at least part of their narratives, there are important differences between their accounts. To judge from his *Travels*, Wallace's experience in the Amazon—culminating in the tragic fire which destroyed his collections and left him marooned in the middle of the Atlantic Ocean—was an almost entirely unhappy one. We find him at odds with disconcerting insects, unnerving sickness, misfiring rifles, and people he often characterizes as lazy or downright dishonest natives. To top it all off, his beloved brother Herbert dies of yellow fever in Belém, despite Bates's best efforts to save him (Wallace, 1905: vol. 1, 282). Wallace's account of local people is decidedly ambivalent (cf. Camerini, 1996). Despite his logistical dependence on their support and a fascination with the economic botanical aspects of local resource management (see Wallace, 1853), Amazonians hardly figure in the *Travels* until the final 30-page material-culture compendium, "On the Aborigines of the Amazon." When not emerging from the forest bearing specimens for his increasingly burdensome collection, non-Europeans are largely subsumed by an enterprise dedicated to the systematization of the natural world.

Amazonians are far more evident in Bates's evocative account, yet they are consistently characterized by their "indolence," a trait routinely related to the luxuriance of the tropical climate and vegetation. In a characteristic and recurrent motif, Bates elides nature and human nature:

> The lower classes are as indolent and sensual here as in other parts of the province [Pará], a moral condition not to be wondered at in a country where perpetual summer reigns, and where the necessaries of life are so easily obtained. [1910 (1863): 83–84]

A judgment of this type slides seamlessly into an interpretation of indigenous resource management practices as casual and wasteful—in pointed, if unspoken, contrast to the disciplined productivity of the English agrarian landscape:

> Around the shed were planted a number of banana and other fruit trees; amongst them were the never-failing capsicum-pepper bushes, brilliant as holly-trees at Christmas time with their fiery-red fruit, and lemon trees; the one supplying the pungent, the other the acid for sauce to the perpetual meal of fish. There is never in such places any appearance of careful cultivation, no garden or orchard; the useful trees are surrounded by weeds and bushes, and close behind rises the everlasting forest. [Bates, 1910 (1863): 76]

Bates and Wallace share a view of the forest that draws on the intimate nineteenth-century relationship between race and natural science (see, e.g., Livingstone, 1994). For both men, the Amazon is a site of almost unlimited productive potential. Foreshadowing the blithely optimistic rhetoric deployed by the developmentalist Brazilian state of the 1970s, the perceived inability of the natives to rise out of their slothful condition is—with a depressing insouciance—contrasted with the possibilities that limitless fertility offers to farmers of a different stock. The dream here is the realization of a vision of landscape imported wholesale across the Atlantic. Wallace, for instance, in an especially bullish moment, writes:

> When I consider the excessively small amount of labour required in this country, to convert the virgin forest into green meadows and fertile plantations, I almost long to come over with half-a-dozen friends, disposed to work, and enjoy the country; and show the inhabitants how soon an earthly paradise might be created, which they had never even conceived capable of existing. . . . In the whole Amazon, no such thing as neatness or cultivation has ever been tried. Walks and avenues, and gardens have never been made; but I can imagine how much beauty and variety might be called into existence from the gloomy monotony of the forest. [1912 (1853): 230, 232]

And Bates, on a similar theme:

> The incorrigible nonchalance and laziness of the people alone prevent them from surrounding themselves with all the luxuries of a tropical country. They might plant orchards around their houses, grow Indian corn, and rear cattle and hogs, as intelligent settlers from Europe would certainly do, instead of indolently relying solely on the produce of their small plantations, and living on a meagre diet of fish and farinha. [1910 (1863): 143]

The underlying narrative here is one in which the architecture of the forest dwarfs those living within it both physically and morally. Such a view forms part of a genealogy for an environmental determinist academic discourse on Amazonia that was to dominate the anthropological literature from Julian Steward in the 1940s through the diffusionary theories of Betty Meggers in the 1960s and 1970s, all the way down to the still-rumbling functionalist debates around the cultural response of indigenous Amazonians to a postulated protein scarcity (Steward, 1948; Meggers, 1971; Gross, 1975; Sponsel, 1992). Meggers expressed this position with axiomatic elegance: "The level to which a culture can develop is dependent upon the agricultural potentiality of the environment it occupies" [1954: 815]. The sanguine nineteenth-century assessment of the region's productive potential has been radically inverted in this work, and perceived limitations are attributed to environmental rather than racial determinants. However, in the vision of a local society fatally constrained by its proximity to tropical nature we find a striking and fundamental continuity with the older literature.

Not only physically and morally intimidated by the power of Nature, local people are further reduced to a state of cultural impoverishment by their inability to respond other than through reactive adaptation. In the tradition of pioneering European naturalists like Wallace and Bates, modern academics trained in the Western tradition and entering the forest with a vision molded by discourses of native peripherality saw precisely what they expected (Said, 1978). In turn, through an objectivist representational mode expressed in the language of the natural sciences, they tended to remove the human subject and reproduce the primacy of biological systems over social systems (Nugent, 1994). Indeed, the consistency of this narrative throughout a voluminous literature establishes Amazonian studies as perhaps unique in the extent to which social science academic production has been colonized by natural scientific modes of explanation.

The scattered appearance over the past few years of an alternative narrative in the Amazonian literature, which I am indicating by the shorthand of an "anthropogenic Amazon," provides the context for an evaluation of the data on landscape change presented in the first half of this paper. In this rereading of Amazonian realities, the Nature–Culture relationship is recast with an emphasis on local human agency to produce what may be termed a progressive (or optimistic) narrative (Cronon, 1992). Research of this type has stressed the ability of local people to transform their environment in ways that, while previously hidden, are now visible. To generalize, where previous natural scientists saw only degraded forest, scholars now discover sophisticated horticultural production systems; where earlier archaeologists interpreted extensive earthworks, intricately patterned pottery shards, and thick deposits of anthropogenic *terra prêta do índio* as the ephemera of a transient, immigrant culture, current researchers find evidence of complex, long-term human occupation.

In this context, it is useful to identify the links between the denaturalization of landscape that is being undertaken in Amazonian studies and the work of contemporary art historians and cultural geographers. In both cases, researchers are questioning the transparency of landscape. However, where Amazonianist scholars have been concerned with demonstrating the biocultural or artifactual identity of landscape in the region, geographers and cultural critics working on these issues have been primarily interested in understanding the ways in which the overarching *concept* of landscape is itself a cultural product. Drawing especially on the work of literary and cultural critics such as Raymond Williams (1973, 1977) and John Berger (1972), humanist cultural geographers, for instance, have

> begun to problematize the term "landscape" as a reference to relations between society and the environment . . . and they have argued that it refers not only to the relationships between different objects caught in the fieldworker's gaze, but that it also implies a specific way of looking. [Rose, 1993: 87]

For a cultural geographer such as Denis Cosgrove (1989: 121), landscape is, in fact, a "way of seeing." Following Berger (1972), he understands the idea of landscape as a "visual ideology," a move which, by pointing to the class character of landscape representation—that is, to its masking of the contestation and exploitation inhering in rural social relations—finds easy resonance with Williams's (1973) influential exploration of agrarian themes in English literature.

Scholars working in this area have pointed to several more or less coherent discourses on landscape. For instance, a history of the "landscape idea" in Europe has been traced through painting, poetry, prose, garden design, architecture, and civil engineering from the Renaissance to the present day (Clark, 1984; Cosgrove, 1984; Cosgrove & Daniels, 1988; Cosgrove & Petts, 1990; Bender, 1993; Mitchell, 1994; Schama, 1995). Through analysis of a variety of representational media, a narrative of the increasing importance of landscape as a stable, ahistorical, and quintessentially natural stage on which human labor is itself naturalized has been detailed and tied to the invention of linear perspective and scientific cartography. Landscape is here understood as a systematization of an aesthetic of Nature: "a way of composing and harmonizing the external world into a 'scene,' a visual unity . . . [denoting] a new relationship between humans and their environment" (Cosgrove, 1989: 121). Historian Simon Schama's recent foray into landscape studies indicates that European landscape aesthetics may be considerably more variegated than geographers seem prepared to allow (Schama, 1995). Furthermore, in drawing attention to the rich history of Chinese landscape aesthetics, W. J. T. Mitchell (1994) has effectively pointed out that the emphasis on the decisive moment of the Renaissance in the geographical literature can lead to the critical effacement of non-European traditions. However, the important point survives: Through its organization as a way of representing social reality, landscape is constituted as a field of knowledge and, as such, is involved in the production and reproduction of relations of power.

A partial corrective is in order here. A social constructivist contention that what was once nature is now culture, and, by implication, what was once forest is now "landscape," is unnecessarily binary (Demeritt, 1994). A legitimate emphasis on cultural construction runs the risk of marginalizing the active role of nature in the dynamic interaction of people with their environment—something that it is perhaps easier to overlook from the position of the critic than from the midst of a practice grounded in ethnography (see Katz, 1992). Stressing the importance of culture and ideas does not have to erase the empirical forest. Indeed, in the shift from a natural to an anthropogenic landscape it is not the forest that has changed. Instead, as I have tried to suggest with examples from Bates and Wallace, it is our way of seeing—and talking about—Amazonia which has changed (for current destabilized views of nature in ecological science, see Oliver & Larson, 1990; Zimmerer, 1994).

The term "anthropogenic Amazon," as I am using it, therefore contains two interlinked elements: (1) an understanding of the region as significantly artifactual in a biophysical sense, that is, it asks that the literature on human transformation in Amazonia be considered as a coherent, if emergent, body of work with a collective significance; and (2) status as an alternative representation of the region which requires a reassessment of the place of social relations in the interaction of local people and the environment. The recent literature describing the impact of local populations on the Amazonian landscape has so far focused on the first of these issues. Yet, although dependent on this literature, our understanding of the anthropogenic Amazon as a representation and an element of discourse is of equal importance. As with "landscape," the concepts "nature" and "Amazonia" are both terms that carry a discursive weight. Such apparently transparent categories can also be historicized and disaggregated to reveal the ways in which they stand as (both academic and popular) representations of social practice, and as such constrain and influence that practice.

Conclusion

In this short paper I have described interventions carried out by local people to improve access to forest products and agricultural fields by cutting channels in the estuarine várzea of Amapá. By drawing attention to the historical, political-economic, ecological, and representational dimensions of these practices, I am suggesting that valuable insights can be gained by using landscape as an analytical optic in natural resource–oriented research. The preliminary data on landscape transformation presented here support those claims of an anthropogenic origin to large areas of the Amazon basin recently advanced by a number of researchers. As a whole, this work represents a notable shift in academic discourse within Amazonian studies. I have argued that Amazonianists concerned with demonstrating the creativity and resilience of local populations should engage more directly with other scholars addressing similar issues of the relationships between nature, culture, landscape, and power.

The increasingly substantial data-set supporting arguments for an anthropogenic Amazon is of considerable theoretical interest both within and beyond Amazonian studies. Such assertions of human landscape transformations are more than mere curiosities: They connect with the ways in which ideas of Nature are constructed and, concurrently, with their complex filtering through the academy to the wider popular and policymaking communities. Viewing the landscape as a form of cultural expression and a locus from which to read larger social relations provides an opportunity for Amazonianists to explore the impact of ideas of Amazonian Nature on the ways in which both outsiders and Amazonians themselves perceive the region. Engaging in work of this type means that we take the question of representation seriously, as practiced by ourselves, our peers, and those (with) whom we study. In doing so, we open up new and fertile ground for scholarly research and intervention (see, in this context, the recent work done by William H. Fisher [1996] and Candace Slater [1996]).

"Society," art historian T. J. Clark (1984: 6) has observed, "is a battlefield of representations." In proposing that discussion of the anthropogenic Amazon be situated in terms of state and international discourses of "nature" and, by implication, "environment" as they relate to Amazonia, I have tried to show that the question of representation is closely bound up with key political concerns.

There is a final twist to this argument: Reading the biophysical landscape of the region as a cultural expression of its inhabitants challenges researchers and policymakers to rethink local people's transformative resource management practices. These practices can be understood not only—in the manner of the first part of this paper—as interventions in ecological processes and local relations of power, but, equally, as the source of alternative representations to be deployed in debates on the future of the region. If we can understand the Amazonian landscape as a cultural entity, we can recognize it as an expression of local agency. While itself a product of academic discourse, the anthropogenic Amazon is also a form of indigenous self-representation, and as such it creates an opportunity for local people to influence regional political debate.

Acknowledgments

This research was carried out with the generous assistance of the Joint Council of Latin American and Caribbean Studies of the Social Science Research Council and the American Council of Learned Societies with funds provided by the Ford Foundation; the National Science Foundation Cultural Anthropology Program (award #SBR-9529972); the Program in Agrarian Studies at Yale University; the Yale Center for

International and Area Studies; and the Yale Tropical Resources Institute. Thanks to everyone in Guariba for warmth, generosity, and all that açaí! Special thanks to Christine Padoch, Miguel Pinedo-Vasquez, Dan Zarin, and everyone involved with Projeto Várzea, without whom I would never have made it to Amapá. My gratitude to Sharon Simpson, Nancy Peluso, Waldir Pereira, Fernando Rabelo, Jaime Rabelo and Marcirene Machado and family, Michael Reynolds, Trish Shandley, Dan Vogt, and Kristiina Vogt. For insightful comments and suggestions on an earlier draft of this paper, thanks to Cathy Corman, Amity Doolittle, Kate Dudley, Heinzpeter Znoj, and two anonymous reviewers, none of whom, of course, bears any responsibility for the final product.

Literature Cited

Almeida Lustosa, A. de. 1976. No estuário Amazônico: 'À margem da visita pastoral.' Conselho Estadual de Cultura do Pará, Belém.

Anderson, A. B. & M. A. G. Jardim. 1989. Costs and benefits of floodplain forest management by rural inhabitants of the Amazon estuary: A case study of açaí palm production. Pages 114–129 *in* J. Browder, ed., Fragile lands of Latin America: The search for sustainable uses. Westview Press, Boulder.

———, **A. Gély, J. Strudwick, G. L. Sobel & M. G. C. Pinto.** 1985. Um sistema agroflorestal na várzea do estuário amazônico [Ilha das Onças, Município de Barcarena, Estado do Pará]. Acta Amazonica, 15(1–2) suppl.: 195–224.

Anderson, S. D. 1991. Engenhos na várzea: Uma analise do declinio de uma sistema de produção tradicional na Amazonia. *In* P. Lenna & A. Engracia de Oliveira, eds., Amazonia: A fronteira agricola 20 anos depois. Museu Paraense Emílio Goeldi/ORSTOM, Belém.

Balée, W. 1989. The culture of Amazonian forests. Pages 1–21 *in* D. A. Posey & W. Balée, eds., Resource management in Amazonia: Indigenous and folk strategies. Advances in Economic Botany 7. New York Botanical Garden, Bronx.

———. 1993. Indigenous transformation of Amazonian forests: An example from Maranhão, Brazil. L'Homme 33(2–4): 231–254.

———. 1994. Footprints of the forest: Ka'apor ethnobotany—The historical ecology of plant utilization by an Amazonian people. Columbia University Press, New York.

Bates, H. W. 1910 [1863]. The naturalist on the River Amazons. J. M. Dent, London.

Bender, B., ed. 1993. Landscape: Politics and perspectives. Berg Press, Oxford.

Berger, J. 1972. Ways of seeing. BBC Books, London.

Buarque, M. 1919. Tocantins e Araguaya. Imprensa Oficial do Estado do Pará, Belém.

Bunker, S. G. 1985. Underdeveloping the Amazon: Extraction, unequal exchange, and the failure of the modern state. University of Chicago Press, Chicago.

Camerini, J. 1996. Wallace in the field. Osiris, n.s. 11: 44–65.

Clark, T. J. 1984. The painting of modern life: Paris in the art of Manet and his followers. Princeton University Press, Princeton, NJ.

Cosgrove, D. E. 1984. Social formation and symbolic landscape. Croom Helm, London.

———. 1989. Geography is everywhere: Culture and symbolism in human landscapes. Pages 118–135 *in* D. Gregory & R. Walford, eds., Horizons in human geography. Macmillan, London.

——— **& S. Daniels, eds.** 1988. The iconography of landscape: Essays on the symbolic representation, design and use of past environments. Cambridge University Press, Cambridge.

——— **& G. Petts, eds.** 1990. Water, engineering and landscape: Water control and landscape transformation in the modern period. Belhaven Press, London.

Cronon, W. 1992. A place for stories: Nature, history, and narrative. Journal of American History 78(4): 1347–1376.

Dean, W. 1987. Brazil and the struggle for rubber: A study in environmental history. Cambridge University Press, Cambridge.

Demeritt, D. 1994. The nature of metaphors in cultural geography and environmental history. Progress in Human Geography 18(2): 163–185.

Denevan, W. M. 1970. Aboriginal drained-field cultivation in the Americas. Science 169: 647–654

———. 1992. The pristine myth: The landscape of the Americas in 1492. Annals of the Association of American Geographers 82(3): 369–385.

——— & C. Padoch, eds. 1987. Swidden-fallow agroforestry in the Peruvian Amazon. Advances in Economic Botany 5. New York Botanical Garden, Bronx.

Fisher, W. H. 1996. Native Amazonians and the making of the Amazon wilderness: From discourse of riches and sloth to underdevelopment. Pages 166–203 in E. M. DuPuis & P. Vandergeest, eds., Creating the countryside: The politics of rural and environmental discourse. Temple University Press, Philadelphia.

Gross, D. R. 1975. Protein capture and cultural development in the Amazon basin. American Anthropologist 77: 526–549.

Hecht, S. B. & A. Cockburn. 1990. The fate of the forest: Developers, destroyers and defenders of the Amazon. Penguin, London.

——— & D. A. Posey. 1989. Preliminary results on Kayapó soil management techniques. Pages 174–188 in D. A. Posey & W. Balée, eds., Resource management in Amazonia: Indigenous and folk strategies. Advances in Economic Botany 7. New York Botanical Garden, Bronx.

——— & ———. 1990. Indigenous soil management in the Latin American tropics: Some implications for the Amazon basin. Pages 73–86 in D. A. Posey & W. L. Overal, eds., Ethnobiology: Implications and applications. Proceedings of the First International Congress of Ethnobiology, Belém, 1988. Volume 2. Museu Paraense Emílio Goeldi, Belém.

Hemming, J. 1987. Amazon frontier: The defeat of the Brazilian Indians. Macmillan, London.

Hiraoka, M. 1992. Caboclo and ribeirinho resource management in Amazonia: A review. Pages 134–157 in K. H. Redford & C. Padoch, eds., Conservation of neotropical forests: Working from traditional resource use. Columbia University Press, New York.

Irvine, D. 1989. Succession management and resource distribution in an Amazonian rain forest. Pages 223–237 in D. A. Posey & W. Balée, eds., Resource management in Amazonia: Indigenous and folk strategies. Advances in Economic Botany 7. New York Botanical Garden, Bronx.

Katz, C. 1992. All the world is staged: Intellectuals and the projects of ethnography. Environment and Planning D: Society and Space 10(5): 495–510.

Lima, R. R. 1956. A agricultura nas várzeas do estuário do Amazonas. Boletim Tecnico do Instituto Agronomico do Norte (Belém) 33: 1–164.

Livingstone, D. N. 1994. Climate's moral economy: Science, race and place in post-Darwinian British and American geography. Pages 5–34 in A. Godlewska & N. Smith, eds., Geography and empire. Blackwell, Oxford.

Medina, J. T. 1934. The discovery of the Amazon according to the account of Friar Gaspar de Carvajal and other documents. American Geographical Society, New York.

Meggers, B. J. 1954. Environmental limitations on the development of culture. American Anthropologist 56: 801–824.

———. 1971. Amazonia: Man and culture in a counterfeit paradise. Aldine-Atherton, Chicago.

———. 1973. Some problems of cultural adaptation in Amazonia with emphasis on the pre-European period. Pages 311–320 in B. J. Meggers et al., eds., Tropical forest ecosystems in Africa and South America. Smithsonian Institution, Washington, DC.

———. 1985. Aboriginal adaptation to Amazonia. Pages 307–327 in G. T. Prance & T. E. Lovejoy, eds., Key environments: Amazonia. Pergamon Press, Oxford.

——— & C. Evans Jr. 1957. Archaeological investigations at the mouth of the Amazon. Bureau of American Ethnology, Bulletin 167. Smithsonian Institution, Washington, DC.

Mitchell, W. J. T. 1994. Imperial landscape. Pages 5–34 in W. J. T. Mitchell, ed., Landscape and power. University of Chicago Press, Chicago.

Moran, E. F. 1993. Through Amazonian eyes: The human ecology of Amazonian populations. University of Iowa Press, Iowa City.

Motta-Maués, M. A. 1993. "Trabalhadeiras" e "camarados": Relações de gênero, simbolismo e ritualização numa comunidade amazônica. Universidade Federal do Pará, Belém.

Nugent, S. 1994. Invisible Amazonia and the aftermath of Conquest: A coda to the quincentenary celebrations. Journal of Historical Sociology 7(2): 224–241.

Oliver, C. D. & B. C. Larson. 1990. Forest stand dynamics. McGraw-Hill, New York.

Padoch, C. & W. de Jong. 1990. Santa Rosa: The impact of the forest products trade on an Amazonian place and population. Pages 151–158 *in* G. T. Prance & M. J. Balick, eds., New directions in the study of plants and people: Research contributions from the Institute of Economic Botany. Advances in Economic Botany 8. New York Botanical Garden, Bronx.

Parker, E. P. 1981. Cultural ecology and change: A caboclo várzea community in the Brazilian Amazon. Ph.D. dissertation, University of Colorado.

Parsons, J. J. & W. M. Denevan. 1967. Pre-Columbian ridged fields. Scientific American 217(1): 93–100.

Posey, D. A. 1985. Indigenous management of tropical forest ecosystems: The case of the Kayapó indians of the Brazilian Amazon. Agroforestry Systems 3: 139–158.

Raffles, H. 1993. Indigenous species in the Amazonian economy: Ecological and socioeconomic perspectives on market-oriented models of rain forest development. CEDAR Research Papers 8. University of London, London.

Roosevelt, A. C. 1989. Resource management in Amazonia before the Conquest: Beyond ethnographic projection. Pages 30–61 *in* D. A. Posey & W. Balée, eds., Resource management in Amazonia: Indigenous and folk strategies. Advances in Economic Botany 7. New York Botanical Garden, Bronx.

———. 1991. Moundbuilders of the Amazon: Geophysical archaeology on Marajó Island. Academic Press, New York.

———. 1993. The rise and fall of the Amazon chiefdoms. L'Homme 33(2–4): 255–283.

Rose, G. 1993. Looking at landscape: The uneasy pleasures of power. Pages 86–112 *in* Feminism and geography: The limits of geographical knowledge. Polity Press, Cambridge.

Ross, E. 1978. The evolution of the Amazonian peasantry. Journal of Latin American Studies 10: 193–218.

Said, E. W. 1978. Orientalism. Vintage, New York.

Schama, S. 1995. Landscape and memory. HarperCollins, London.

Schmink M. & C. H. Wood. 1992. Contested frontiers in Amazonia. Columbia University Press, New York.

Slater, C. 1996. Amazonia as Edenic narrative. Pages 114–131 *in* W. Cronon, ed., Uncommon ground: Rethinking the human place in nature. W. W. Norton, New York.

Sponsel, L. E. 1992. The environmental history of Amazonia: Natural and human disturbances, and the ecological transition. Pages 233–251 *in* H. K. Steen & R. P. Tucker, eds., Changing tropical forests: Historical perspectives on today's challenges in Central and South America. Forest History Society, Durham, NC.

Steward, J. H. 1948. Culture areas of the tropical forests. Pages 883–899 *in* J. H. Steward, ed., Handbook of South American Indians, volume 3. The tropical forest tribes. Bureau of American Ethnology, Bulletin 143. Smithsonian Institution, Washington, DC.

Sweet, D. 1992. Native resistance in eighteenth century Amazonia: The "abominable Muras" in war and peace. Radical History Review 53: 49–80.

Wagley, C. 1953. Amazon town: A study of man in the tropics. Macmillan, New York.

Wallace, A. R. 1905. My life: A record of events and opinions. 2 vols. Dodd, Mead & Co., New York.

———. 1912 [1853]. Travels on the Amazon and Rio Negro. Ward, Lock & Co., London.

Weinstein, B. 1985. Persistence of caboclo culture in the Amazon: The impact of the rubber trade, 1850–1920. Pages 89–114 *in* E. P. Parker, ed., The Amazon caboclo: Historical and contemporary perspectives. Studies in Third World Societies 32. College of William and Mary, Williamsburg, VA.

Whitehead, N. L. 1993a. Ethnic transformation and historical discontinuity in native Amazonia and Guyana, c. 1500–1900. L'Homme 33(2–4): 285–305.

———. 1993b. Recent research on the native history of Amazonia and Guyana. L'Homme 33(2–4): 495–506.

Williams, R. 1973. The country and the city. Oxford University Press, New York.

———. 1977. Marxism and literature. Oxford University Press, New York.

Zimmerer, K. S. 1994. Human geography and the "New Ecology": The prospect and promise of integration. Annals of the Association of American Geographers 84(1): 108–125.

Twelve Thousand Years of Human-Environment Interaction in the Amazon Floodplain

Anna C. Roosevelt

Introduction

Human prehistory provides basic data about long-term human–environment interaction in Amazonia. This paper reviews current knowledge of native land use and its relationships with the environment in the floodplains of Amazonia over the last 12,000 years, and outlines the scientific and policy implications for biodiversity, conservation, and development.

It is widely assumed that shifting cultivation, supplemented by foraging, represents the traditional native adaptation to the Amazon forest. In this system, small plots are slashed and burned, and crops are planted in the ashes for a few years, after which the forest is allowed to return. As such, it is seen as giving people access to the nutrients of the forest vegetation for their subsistence without having long-term deleterious effects on the habitat. Pure foraging is supposed to be impossible, due to the scarce and dispersed forest resources (Bailey et al., 1989), and permanent agriculture, the economic support of civilization, was impossible, due to infertile soil. Shifting cultivation is considered to have developed when Andean agriculturists migrated into Amazonia and could not maintain permanent cultivation systems (Meggers, 1971). In this view, the relations between native people and the forest have long been in an adaptive equilibrium.

Biologists and ecologists traditionally have seen the Amazon forest as a natural formation with characteristics produced through the interaction of climate history, ecological succession, and natural selection (Prance, 1982; Lovejoy, 1985; Whitmore & Prance, 1987; Wilson, 1988; Whitmore, 1990). To many natural scientists, the equilibrium of the forest was so fragile that even native foragers and horticulturists could destroy it. Consequently, researchers on Amazon ecology and biodiversity have sought study areas they believed were free of human impacts and have viewed forest people as threats to forest conservation. Consequently, biological reserves and sustainable development projects customarily take land-use decision-making away from local people and give it to government or nongovernmental organizations based outside the region (Davis & Wali, 1994). Paradoxically, the implications of their belief that indigenous people are dangerous to the survival of the forest is not incorporated in explanations of forest ecology and biodiversity patterns, which are attributed solely to natural factors.

Evidence of the nature and history of the Amazon and its use by humans, however, does not substantiate these approaches. The habitat turns out to be much more heterogeneous than most researchers realized, and it includes sizeable resource-rich areas as well as poor ones. Native land use has varied greatly over time and space, from foraging to permanent agriculture, and has adapted to factors of population, economy, and sociopolitical organization, as well as to environmental factors. Few or no regions can realistically be considered "virgin" of human impacts, and salient ecological and biodiversity patterns traditionally interpreted as natural are strongly associated with evidence of prehistoric human interventions. All human uses appear to have affected the environment, but the greatest lasting impacts come from use by large, hierarchical, centralized, supraregional organizations rather than from the activities of small, independent local groups.

The specific characteristics of the habitat, the different patterns of human organization and land use, and their effects on the habitat need to be integrated in scientific research and planning in Amazonia.

The Environment

Contrary to theoretical assessments of the Amazon tropical rain forest as uniformly poor in resources for human subsistence (Meggers, 1971; Goodland & Irwin, 1975), the empirical data reveal a highly heterogeneous habitat with varied climate, vegetation and fauna, geology, and soils. Evidence for habitat heterogeneity has come from both on-the-ground regional studies and from remote-sensing surveys. Emerging patterns in the empirical data point up the fallacy in long-held assumptions about the interaction of different environmental factors and their relationship to land-use capability.

Over and above the poor soils on the crystalline pre-Cambrian shields and Tertiary sediments of the Amazon high plain, several regions have large areas of high-quality soils formed from alkaline volcanic rocks, limestones, Pleistocene and Holocene alluvium, and prehistoric archaeological deposits (Sioli, 1984: 15–214, 537–580; Rasanen et al., 1991; Franzinelli & Latrubesse, 1993: 123–126). Although poor soils predominate, Amazonia's great size makes its higher-nutrient soils some of the most extensive in the world. The floodplains, or *várzea*, make up the largest area of good-quality, well-watered soils, a significant resource comprising about 25% of the area of Greater Amazonia.

Even Amazonia's poorer soils are suitable for cultivation. Nutrients are not washed out during cropping, as was originally hypothesized, but, rather, have become chelated and can be easily released by mulching (Jordan, 1985; National Research Council, 1994). Tropical soils in the Amazon and elsewhere are generally less weathered than expected (Tricart, 1972: 130–142, 167–219, 233–242).

Rather than the 3000–4000 mm of evenly distributed annual rain expected for tropical rain forest regions (Whitmore, 1990: 9–36), seasonally distributed annual rainfall of ca. 2000–2500 mm predominates in Amazonia, and edaphic factors of soil drainage and nutrient supply strongly mediate the effects of rainfall (Sioli, 1984: 85–126; Whitmore & Prance, 1987: fig. 2.4). Furthermore, the predominant vegetation types of Amazonia include semideciduous rain forest, woodland, and *várzea* forest in addition to evergreen rain forest (Schmink & Wood, 1984: 338; Whitmore & Prance, 1987: fig.

2.4). The large extent of Quaternary floodplain deposits means that várzea forests in particular are more extensive than they appear in older vegetation maps.

Individuals of plant species are not as dispersed as had been assumed but, rather, form distinctive clusters in both terra firme (dry land) and várzea microenvironments (Junk, 1970; Beckerman, 1979; Balée, 1989; Roosevelt, 1991: 7–26). On the areas of higher-quality soils, vegetation is more luxuriant, drought resistant, diverse, and productive of renewable biomass than is vegetation on the poor soils and consequently supports a higher faunal biomass.

Human Influence

Extensive human influence on landscape, forest diversity, and land-use capability is now documented for Amazonia. Ethnobotanists and geographers have identified widespread secondary forests in formerly cleared areas, the most diverse and luxuriant of which lie on the enriched soil of ancient human middens (Smith, 1980; Anderson, 1983; Balée, 1989; Balée & Campbell, 1990; Roosevelt, 1991: 170–179). Highly important economic species cluster at these abandoned habitation sites.

The human influence began at the end of the Pleistocene as the present forest was forming and intensified as population grew (Bush & Colinvaux, 1988; Van der Hammen, 1991; Roosevelt, 1993; Roosevelt et al., 1996). During the prehistoric human occupation, forest species were exploited intensively for fuel and food and consequently came to cluster at settlements, and native people increased forest diversity by spreading endemic species into other regions (Roosevelt et al., 1996). The increased clearing near larger settlements opened up the forest canopy and reduced rainfall, a process that is currently intensifying (Hagmann, n.d.; unpubl. longitudinal records, Monte Alegre, Pará). Patterns of vegetation and its diversity observed in the forest at present, therefore, cannot be assumed to be either ancient or purely naturally caused. Since the late Pleistocene, forest ecology has been part of human ecology, and forest history, part of human history.

Environmental Archaeology

Data on long-term land use and environment in Amazonia has come primarily from the fields of paleoecology and environmental archaeology, only recently applied widely in the tropics. Because of researchers' confidence in theories about prehistory and because of expectations of poor preservation, it has not until recently been the practice to collect biological remains systematically from archaeological sites in tropical lowlands to test the theories.

Environmental archaeology treats environmental change and biological diversity as having human components rather than only natural causes and relates human behavior to the environment in a nonlinear, nonreductionist fashion through analysis of comprehensive, definitive, cultural and environmental data. Causality in environmental patterning is sought not only in factors of present and past climate, soils, biology, and geology but also in past and present human impacts. Human development is related not only to environmental influences but also to changing human contexts and interactions.

This is an interdisciplinary approach that uses both quantitative and qualitative methods from both natural and human sciences and involves sustained collaborative research by international teams of scientists and students from different disciplines. The field emphasizes the compatibility of basic and applied research and relevance of research results to policy in conservation and sustainable development.

In terms of research strategy, environmental archaeology takes a diachronic and regional approach to gathering comprehensive data relevant to theory. Reconstruction of present and past biomes and land use relies on data from multiple segments of the biomes: plants, animals, soils, and rocks, from both natural and human cultural deposits, and from multiple periods in the sequence. In fieldwork and analysis, phenomena are studied with multiple methods of measurement and observation, to lessen experimental and taphonomic bias. The research collects a wider range of material from archaeological and paleontological sites than usual, employing exhaustive, total sieving. Research is highly controlled spatially. Sampling locations are instrument-mapped, satellite-located, and stratigraphically excavated. The data are therefore more precise and more definitive than those previously available.

Prehistoric biota are collected both from offsite sediment cores and from archaeological deposits. Archaeological sites provide numerous datable, identifiable biological macrospecimens not recovered in off-site cores. Research on archaeological macrospecimens gives evidence of environmental change, the resources available, and the uses to which they were put. Species representation in offsite palynological cores is biased by depositional variation and dating problems, but the assemblages are important comparative evidence for prehistoric environment and land use (Hather, 1994: 172–201). In environmental archaeology, all soil excavated is fine-screened, and all materials from screens are curated, not just a selection.

To solve the dating problems that have plagued Quaternary geology, palynology, and bioarchaeology (Rasanen et al., 1991; Roosevelt et al., 1991, 1996; Fritz, 1994), multiple dating methods are carried out on multiple specimens in a sample: botanical and faunal specimens, soil organic matter, and artifacts, to assess experimental error, contamination, bioturbation, and stratigraphic disturbance. Interpretively important specimens are directly dated, not simply dated by association or extrapolation. Microfossil distributions are assessed for statistical significance to avoid overinterpretation of rare, intrusive specimens (as happened in Bush & Colinvaux, 1988; Bush et al., 1989). The identification of archaeological biological assemblages and comparison with present-day ones are achieved through intensive ecological surveys and collections by field biologists and comparisons with existing systematic collections.

Cultural and human biological data, as well as prehistoric biological remains and sediments, are used to assess ecological patterns and land-use systems through time. If integrated with other evidence, artifact types can be associated with certain land-use activities (e.g., manioc griddles and graters, maize grinders, spear points, axes). Human skeletons also yield information about adaptations to environment and to social structures affecting quality of life. Comparisons through time inform on systematic changes in adaptation. Skeletal analysis provides preliminary information about human physiology, diet, and genetics in different periods (Roosevelt, 1991: 384–395).

Stable isotope analysis is performed on the materials collected, for information on ecological and cultural change. Research on a wide range of material has revealed subfossil tropical material to be suitable for analysis (Van der Merwe et al., 1981; Roosevelt, 1989: 54–55; Roosevelt et al., 1991). Material varies isotopically through time and space according to ecological conditions and organisms' nutrient sources. Accordingly, radiocarbon dates are corrected for stable isotope variation. In addition to routine analysis of stable carbon isotopes in soils, plants, and animal and human tissues for photosynthetic pathways, aquatic faunas are analyzed for oxygen, and terrestrial plants, animals, and human bones are analyzed for nitrogen, to track water quality and temperature and prehistoric diet patterns of both humans and animals as ecological relationships change (Lajtha & Michener, 1994).

Through environmental archaeology, many thousands of biological remains have come out of archaeological sites, and some of the earliest sites have the greatest abundance. With these methods, specimens of hundreds of wild and domestic herbs, legumes, shrubs, trees, palms, mammals, birds, reptiles, fish, shellfish, and more have been recovered (Roosevelt, 1989, 1991: 170–180, 373–384; Roosevelt et al., 1991, 1996). In most archaeological sites, the most common finds have been carbonized tree seeds and wood fragments; the vertebrae, teeth, and spines of small fishes; and pieces of the shells of small turtles.

The archaeological biological collections give a comprehensive, securely dated diachronic record of premodern Amazon forest composition, river ecology, and land use. Among the collections are the first finds ever of prehistoric specimens of important economic species in the Amazon and some important indicator species. The research has produced the first direct dates on indicator plant species previously found only in the form of undatable pollen. The faunal finds frequently constitute the only prehistoric examples of the taxa from the Amazon.

The Sequence of Human–Environment Interaction in Amazonia

The human ecological record emerging from archaeology does not show the simple human ecological sequence envisioned earlier but, rather, a dynamic, long-term occupation of habitats by people of diverse cultures that manifest marked political and economic changes, increasingly intensive land use, and increasing human impacts on the habitat. It documents a much wider range of different land-use systems and anthropogenic effects in Amazonia than recognized by earlier researchers relying on projections of present-day indigenous land-use patterns. Even regions considered poor in resources supported people from the earliest occupation, and even resource-rich habitats were changed by the human occupation. The sequence disproves long-accepted conclusions about the ecology and sequence of cultural development in the Americas, showing that, contrary to expectations, the resources of Amazonia supported early foragers, the earliest American pottery cultures, and complex agricultural societies in late prehistoric times.

Paleo-Indian Foragers

Neither primitive nor unable to adapt to forest, foragers came in soon after 12,000, as early as in temperate regions, and developed fine lithic tools, rock art, and a long-term subsistence adaptation focused on forest fruits and river fauna.

Paleo-Indians hunted, fished, and gathered tree fruits along the rich alluvium of major rivers and intensively harvested the fruits and game of adjacent forests and woodlands. Their artifacts are also found in the forests, savannas, and river floodplains of the resource-poor Tertiary high plain and geological shield areas of the Amazon. Their cutting and burning of vegetation probably had the greatest impact in the drier, poorer-soil forests, encouraging the proliferation of taxa that adapted well to disturbance, especially palms and legumes.

The only Paleo-Indian occupation site from which environmental and subsistence information has been collected systematically is Caverna da Pedra Pintada, in the Paleozoic sandstone mountains of Monte Alegre, overlooking the extensive floodplains of the Amazon and its tributaries in the Tapajos mouth area near Santarém. In this area of secondary tropical rain forest, secondary savanna woodland, and swamp vegetation, 11 contiguous excavations uncovered a four-part occupation dated between 11,200 and 9800 years ago by 56 radiocarbon dates on tree fruits and wood and 13 luminescence dates on lithics and soil (Roosevelt et al., 1996).

Many thousands of carbonized tree seeds and charcoal were recovered from the layers and features of the ca. 20 cm–thick Paleo-Indian deposit at the site, documenting intensive harvesting of fruits of trees that still occur in the relict forests outside the cave today. Among the most common carbonized tree seeds were beans of the pan-neotropical forest legume *Hymenea* cf. *parvifolia* and *H.* cf. *oblongifolia*, and the starchy fruit of *Sacoglottis guianensis*, both upper-stratum trees about 10–20 m tall. The pits of the fruits of *Talisia esculenta* and *Mouriri apiranga* were present. Also present, but rare, were the tiny seeds of *Byrsoniuma crispa*, a forest species of muruci, and testa fragments of Brazil nut, *Bertholletia excelsa* (Cavalcante, 1991; identified by Scott Mori, New York Botanical Garden). The oily, durable seeds of several palms adapted to the sandy soils of high-rainfall areas and to conditions of human disturbance were very common. *Attalea microcarpa* was the most common, present in all levels and numbering in the thousands of specimens (identified by Andrew Henderson, New York Botanical Garden). *Astrocaryum vulgare*, *Attalea spectabilis*, and *Acrocomia* spp. were also present. Over-represented due to their durability and indebility, these would nonetheless have been a rich source of fat and vitamins. Many of the trees and palms were used for fuel, judging from the abundant charcoal in the site. The preponderance of highly negative stable carbon isotope ratios between $-25o/oo$ and $-31o/oo$ are typical of closed canopy tropical rain forests but not of open forests and grass savannas, which in Monte Alegre today are associated with widespread clearing for subsidized cattle ranching and agriculture.

The importance of tree fruits in the assemblage contradicts optimal foraging hypotheses that vegetal collectibles would not be of importance in the diet of early foragers, who were assumed to have been big-game hunters. It shows that forest products were an important part of early subsistence, contradicting hypotheses that the forests were too poor in resources to support people without cultivation. The similarity of the ancient and modern forest shows that Monte Alegre was already well forested by the late Pleistocene and that the main species harvested in prehistory were not extinguished. So far no extinct species have been identified in the assemblages and no species particularly adapted to arid, savanna habitats or cold habitats.

It is of interest for the history of the Amazon forests that cultivated species and certain exotic species, such as *Anacardium occidentale*, are completely missing from the Paleo-Indian seeds. Since these are present in the late-prehistoric levels of the cave and are commonly found in the forests today, their absence at this early time suggests that they were part of the system of forestry and orchard cultivation that spread with horticulture and agriculture during later prehistoric times. Cashew in particular spreads rapidly at the expense of native species when forests are opened up by clearing. Such findings support the hypothesis that present forest diversity is less "natural" than is usually assumed, and more influenced by human impacts both intentional and unintentional (Balée, 1989).

Faunal remains were rarer than carbonized plant remains in the sandy soil, but most levels had small, burned fishbones (including both catfishes and characins, and fishes such as *Arapaima gigas*), small water-turtle and tortoise shells and bones, bones of snakes, toads, birds, and rodents, crushed mollusc shell, and a few unidentifiable splinters from larger, probably ungulate, species. The large and medium-sized projectile points, often with stems and sharp barbs, were probably for hunting large aquatic species: mammals such as manatee and dolphin, larger fish such as *Arapaima gigas* and the large catfishes, and perhaps terrestrial ungulates attracted by fruiting trees. To investigate the proportion of harvesting of different types of fauna, riverside Paleo-Indian sites, now lying under water, need to be sampled. Sea levels rose more than 60 m just at the end of the Paleo-Indian occupation and would be expected to have acted to preserve perishable remains in the inundated sites.

Since many the identified plant species from the Paleo-Indian layers are trees or palms that tend to fruit in the rainy season in the cave area, we can envision the foragers as camping seasonally. Just as the local forests have abundant fruits and game during the wet time of year, the fish and shellfish in the region are more accessible during the season of low water. Thus, Paleo-Indians could use their wide peregrinations (some of the lithic materials used for points come from hundreds of kilometers away) to scoop up the bounty of different habitats, an adaptation that would have increased the overall amount of food available and would have minimized the impact of their settlements on the habitat. It is certainly possible, however, that riverside sites, now under water, could have had larger groups of people in them than terra firme camps, due to the abundance of food from aquatic fauna and várzea forest, especially in the dry season. Thus, although the early communities on the Amazon mainstream may have broken up into small bands during forest camping in the rainy season, they could possibly have regrouped into villages in the dry season. Quite a different pattern of trekking would be expected in distant interfluvial areas, where dry-season changes tend to disperse and fragment resources.

There is no evidence that the early foragers were excluded from any particular Amazonian habitats. The triangular-stemmed points of the Monte Alegre culture have been found throughout the Lower Amazon from the Guiana coast to the Bay of Marajó and upstream as far as Santarém. Other styles of large, finely chipped projectile points associated with the Paleo-Indians occur in many areas, including some quite poor in resources. For example, throughout the Rio Negro drainage from Manaus to the Guyana Shield, numerous examples of large-stemmed Paleo-Indian points have been found.

To summarize, floodplain and nearby interfluvial resources were certainly adequate to support nomadic Paleo-Indian foragers for several thousand years, and the cutting and burning carried out by these early migrants is the earliest known human disturbance of the forest. The numerous identified archaeological tree species and their highly negative stable carbon isotope ratios document the presence of closed-canopy tropical rain forest in the area by this time. The abundant carbonized remains of common river fauna document the importance of riverine resources in Paleo-Indian subsistence, as a supplement to the forest fruits.

Early Holocene Foragers

The long Paleo-Indian occupation had petered out by about 8,000, and by this time, the cultural patterns of terra firme and river floodplains had diverged in certain ways, to specialized collecting of aquatic fauna along the mainstream and intensive broad-spectrum hunting and collecting in the upland interfluves away from large streams.

Soon after 8,000, people had settled down along the shores and estuaries of the mainstream Amazon and intensified fishing, leaving huge shell middens nowadays mined for lime. Far from being retarded culturally by their humid tropical ecological adaptation, these were the first people in the New World to make pottery (Simoes, 1981; Williams, 1981; Roosevelt et al., 1991; Roosevelt, 1995a). The age of these cultures has been determined by 40 radiocarbon dates on charcoal, bone, shells, and pottery and by thermoluminescence dates on pottery. The dates come from Taperinha, Pedra Pintada, and seven other sites along the Amazon mainstream and estuary.

The quantitative analysis of the biological remains from the cave and riverine sites indicate that these foragers had become very specialized fishers of aquatic shellfish and fish. They even carried these foods into upland sites like Caverna da Pedra Pintada, which lies an hour or two from the shore. In comparison to the Paleo-Indians and to later prehistoric people, these mainstream Archaic people were not as interested in tree fruits. Both in the cave and the beach sites, tree seeds are comparatively rare.

Both fish and shellfish were primary food sources for these floodplain people. The strong dietary focus on fish and shellfish, which resulted in the creation of the large garbage heaps, did not extinguish them, for all the species identified in the middens are still around today, although they are no longer used for food. The freshwater pearly mussels most common both then and today are *Triplodon corrugatus*, *Castalia ambigua*, and *Paxyodon ponderosus*. The shellfish appear to have been targeted because they are locally available in large quantities year-round.

This specialized subsistence system appears to be part of the process of settling down to greater permanence of habitation. Just when the large shell middens are building up along the rivers, the occupation in the Caverna da Pedra Pintada lessens, as if less time was spent roaming the interfluves and more time spent in the home base. This increase in settlement permanence may, in turn, be related to the slow filling in of human occupation throughout Amazonian regions, making peaceful, wide-ranging treks difficult. The Paleo-Indian adaptation of hit-and-run harvesting may have no longer been possible by this time, because traditional foraging territories may have been taken over by other migrants. Whatever the reason, people appear to maintain good overall health, for their bones and teeth show few or no signs of disease or dietary pathologies.

In contrast to the mainstreamers, early Holocene foragers in the upland forests south of the Amazon developed a very broad-spectrum subsistence system (F. Lopes et al., n.d.; Schmitz, 1987; Magalhaes, 1994; Imazio da Silveira, 1994). Even without systematic fine-processing of soil, archaeologists have been able to recover a large number of identifiable plant and animal specimens. The assemblages include several shellfish species, many fishes, many species of small animals and birds, rare larger game such as monkey, deer, peccary, and jaguar, and fruits from an extremely diverse group of plant species of the local evergreen and semi-evergreen forests and savannas. Along with the comparative rarity of large game, there were few or no projectile points, suggesting that subsistence was more focused on small-package, gathered foods than on game. A more nomadic settlement pattern in the terra firme sites away from large rivers compared to the floodplain sites is suggested by the lack of large sites and absence of pottery in the interior until the Formative period, with its evidence of horticulture.

To summarize, riverine faunal resources supported the earliest known Amazonian villages and the development of the craft of pottery. The adoption of pottery in the interior, however, did not come until the inception of regular food plant cultivation during the subsequent Formative period.

Early Horticulturalists

Perhaps the settled residence patterns of the river foragers led them in time to cultivate food plants, for they could stay in one area for more time each year. Their longer-lasting settlements would have had more of an impact in the environment, creating nutrient-enriched areas where weedy plants would flourish, and where people would be around to select, care for, and harvest crops. In any case, it is in the next period of occupation in Amazonia that we begin to get evidence for economies based on horticulture, supplemented by fishing and hunting.

Manioc cultivation may have begun at least 4000 years ago and cropping of a number of crops probably was widespread by 2000 years ago (Lathrap, 1970; Roosevelt, 1980, 1993; Bush et al., 1989). In the cave at Monte Alegre, people whose bones, hearth charcoal, and food remains are dated between 3600 and 3000 years ago start making a new type of pottery, a sandy, red, incised ware with shapes that include what is considered the traditional manioc griddle. At the same time, their foraging pattern shifts somewhat from reliance on fish and shellfish toward a broader spectrum of species, as the protein supplement for the starchy crops (Roosevelt et al., 1996). This subsistence of mixed foraging and horticulture, the most common among Amazonian Indians today, is a more-than-adequate subsistence base, and, as before, people's bodies show few or no pathologies associated with nutritional or infectious disease.

Elsewhere in Amazonia, cultures of this and related cultural horizons have not been sampled for paleodietary evidence, except in the Orinoco and Ucayali. In the Venezuelan Orinoco, people with a distantly related pottery style with manioc griddles, living about the time of Christ, have a bone chemistry consonant with a manioc-fish-game diet (Roosevelt, 1980; Van der Merwe et al., 1981). In the Peruvian Amazon tributary, the Ucayali, skeletons collected from sites of roughly the same period have a similar stable carbon isotope pattern (Lathrap, 1970; Roosevelt, 1989). In addition, their nitrogen isotope patterns fit the hypothesis that fish and game, rather than cultivated plants, still furnished most of the protein in the diet. There is no isotopic support for the hypothesis that the early horticultural cultures were intensive maize cultivators (contra Bush et al., 1989; and Lathrap, 1975).

Paleoecological evidence of this period in the Ecuadorian Amazon suggests that these early horticulturists altered forest composition, encouraging weeds and tree species, such as *Cecropia*, that are adapted to forest clearings (Bush & Colinvaux, 1988; Bush et al., 1989), and there is palynological evidence from several areas for reduced rainfall (Absy, 1979), perhaps as a consequence of the reduction of forest cover. The plant remains in archaeological sites of the period include economic tree fruits, but in quantities not as large as those in the Paleo-Indian site or in later prehistoric sites. It seems, therefore, that people were still collecting fruits from the forest and possibly planting or encouraging valuable trees, but they may not have been using the products as intensively as earlier and later, perhaps because of the increase in horticulture.

Early Chiefdoms

Horticulture supplemented by fishing and hunting continued to be important in subsistence for many millennia, until about A.D. 1000 in the areas that we know about. This was the economy, in fact, that supported the first known mound-building complex societies in the Amazon. Descendants of the pottery styles developed by the Formative horticulturist-foragers are present in the basal layers of the great archaeological mounds of both the mouth of the Amazon and the Bolivian Amazon (Dougherty & Calandra, 1981–1982; Roosevelt, 1991: 63–64).

By A.D. 1000, the complex cultures had transformed landscapes with such earthworks over tens of thousands of square kilometers (Denevan, 1966; Erickson, 1980; Porras, 1987; Athens, 1989; Roosevelt, 1991). Earlier researchers thought that cultures able to build large mounds would be expected to be complex chiefdom societies with intensive agricultural systems. However, there is as yet no convincing evidence for centralized political organization or intensive maize cultivation until very late in the first millennium A.D., after most of the mounds had already been built and occupied for more than 1000 years. Instead, people during the early mound-building period use a range of crops, including primitive popcorn maize, tropical beans, and some native seed-bearing herbs that we know little about (Brochado, 1980; Roosevelt, 1980, 1991). For instance, in the floodplain savanna of Marajó Island, the mound-building societies of the period A.D. 400–1100 had a mixed subsistence system of fishing, plant collecting, and cultivation (Roosevelt, 1991). The majority of Marajó faunal remains are of small fish species. Such a focus on what many people today consider trash fish may indicate an intensification of harvesting in the floodplain waters near the large mound sites. As later, we find only a few examples of larger fish species, such as *Arapaima gigas*, and these tend to be in ceremonial contexts. In the black soil middens where everyday food remains were discarded on the mounds, the majority of bones are tiny, 1–3 mm vertebrae, spines, and teeth of characins, such as *Hoplias malabaricus*, catfish, and shell fragments from small turtles.

In this period tree fruits are a common item in the archaeobotany, as later. Among common finds are seeds from legumes, such as *Inga* spp. and species of Sterculiaceae, and palm fruits of many species, including *Astrocaryum vulgare*, *Euterpe oleracea*, and *Acrocomia sclerocarpa*. Of great interest is the fact that some of the edible tree seeds excavated are larger than those of present-day naturalized species, suggesting that the archaeological seeds are from cultivated or at least protected stands. For example, some seeds of the valuable species *Euterpe oleracea* in Marajó mound sites are larger than the seeds commonly harvested on Marajó today, and the ancient seeds are very common in areas where today examples of this palm rarely occur due to seasonal lack of water, except where ranch owners plant and water them. Thus, although early archaeobotanical research focused more on the history of agricultural field crops (Roosevelt, 1980; Smith & Roosevelt, n.d.), it seems that tree harvesting and planting also were an important part of subsistence during the period.

In fact, still today, the mounds built during this period are inhabited by an anthropogenic forest much richer in economic tree species than the non–mound forests (Roosevelt, 1991: 170–172). The construction of the mounds appears to have the overall increased area of várzea forest by elevating the terrain. Today, the mounds support forest in seasonally flooded areas where only herbs and shrubs grow off of the mounds.

The Agricultural Chiefdoms of Late Prehistoric Times

Only in late prehistoric times do we have archaeobotanical and isotopic evidence that people established permanent field agriculture in some alluvial plains and uplands. The evidence suggests that in several regions subsistence turned predominantly agricultural, in contrast to the subsistence of earlier and later Amazonian Indians.

In areas with large alluvial floodplains or rich upland soils, settlements and mounds proliferate in late prehistoric times, and maize cultivation is implicated (Bush et al., 1989; Roosevelt, 1989; Roosevelt et al., 1996), although other crops also appear to have been important and sometimes more important than maize in certain areas. Carbonized maize cobs and kernels turn up in late prehistoric floodplain sites and dried cobs appear in caves. Although the races vary, sometimes relatively primitive (Coroico type at Caverna da Pedra Pintada at Monte Alegre) and sometimes quite modern (Chandelle type in Parmana), human bone chemistries in many areas become heavily influenced by C-4 plants such as maize. By this time, crops appear to have become so important in diets that the isotopes of plant protein predominate in human bone chemistry, indicating that the food economy was moving down the trophic pyramid. (The validity of stable isotope results relies on analyzing the ratios both of humans and of all their food sources for several different elements. For example, since shellfish often have stable carbon isotopes similar to maize, nitrogen isotopes must also be run to distinguish between their effects. Nitrogen isotopes of shellfish are much higher than those of maize, allowing consumption of the two different foods to be distinguished in archaeological bones.)

It seems that as populations increased, so did intensity of cultivation and exploitation of river and forest resources, and considerable deforestation occurred around large settlements (Bush & Colinvaux, 1988; Bush et al., 1989; Roosevelt, 1989; Roosevelt et al., 1996). Late prehistoric earthworks and anthropogenic enrichment of soils by human refuse become widespread in the period, showing that the extensive human occupation of the time had a significant lasting influence on landforms and soil quality. Concurrent with these demographic and human ecological processes, warlike political complexes and supraregional trade expanded in the region.

Paleoecologists' pollen profiles in the Ecuadorian Amazon reveal a measurable reduction of tall, mature forest in the Montana and a commensurate increase in the representation of pollen of maize, weed species, and secondary forest trees, such as cecropia. People living in the Peruvian montana during this time have maize-influenced bone chemistry, in contrast to earlier people there. The evidence suggesting permanent agriculture in the uplands is a surprise for those who assumed that the terra firme forests were too poor to support agriculture or that Indians would conserve the forest. The environmental context of terra firme deforestation, however, includes the rich volcanic-ash soils of the Ecuadorian Oriente and the limestone-based soils of the Peruvian Amazon, both soils that are not limited to shifting cultivation. The cultural context for the pollen evidence of intensive agriculture in the uplands is the building of increasingly larger mound groups and dwelling sites.

As mentioned above, we now know that complex societies were in existence in Amazonia long before maize became important in the food economy. It seems that chiefdom evolution and the demographic expansion that accompanied it may have led to maize intensification, not the reverse. No doubt the productive storable crop was an economic boon to the growing, sedentary populations in the floodplains and in richer-soil terra firme areas, but it also seems to have important ceremonial functions. Conquest-period records chronicle the association of maize with tithes to furnish beer for community ceremonies in paramount chiefdoms such as the Tapajo (Nimuendaju, 1949).

Fish continue to be an important protein supplement at floodplain sites, and the comparative abundance of fish in Amazonian waters seems to have led to a pattern in which maize amino acids were supplemented with fish protein rather than with cultivated beans, which serve as the important supplement today in Mexico and Andean Peru. Interestingly, in the areas where there are many large garbage dumps built up near human population concentrations, the majority of fish remains are those of the small species that make up the bulk of the fish biomass. Larger species, such as river turtles (*Podocnemis* spp.) and pirarucu (*Arapaima gigas*), and large land species, such as deer (*Mazama americana*), peccary (*Tayassu* sp.), and tapir (*Tapirus* sp.), appear in ceremonial deposits, apparently as feast food. This pattern of limiting use of certain less abundant species to special occasions could be argued to be a form of resource conservation of species important in the ritual and sociopolitical systems. In any case, the larger species that have been identified all survived the exploitation and are found in the region still today.

Quite a broad spectrum of hunting and collecting was carried out both in the floodplains and interfluves, with the main difference being that fish and turtles predominate quantitatively in the floodplain sites but not in the interfluves. But both regions' sites include piranhas (*Hoplias malabaricus*), numerous catfishes, numerous turtles, rays, numerous shellfish, land tortoises, crocodilians, numerous lizards (including *Iguana iguana*), deer, peccary, tapir, numerous rodents, monkeys, ducks, doves, gallinaceous birds, bats, and so forth.

The research also shows that numerous tree species exploited during the first known occupation, more than 11,000 years ago, continued to be exploited in late prehistory. These include *Bertholletia excelsa, Hymenea* spp., *Attalea microcarpa, Attalea spectabilis, Astrocaryum vulgare, Sacoglottis guianensis, Mouriri apiranga, Talisia esculenta, Vitex* spp., and *Byrsonima crispa*. Other species exploited in late prehistory are *Inga* spp., *Byrsonima crassifolia, Acrocomia sclerocarpa,* and *Mauritia flexuosa,* among others.

Other species, such as *Anacardium occidentale,* seem to come into the sequence for the first time in later prehistoric times. Today, these tree species are cultivated at current occupation sites, persist at ancient occupation sites in the forest, and flourish in large numbers in areas of disturbed or degraded forest. Thought by botanists to have been introduced into Amazonia from other New World areas, the distribution of such species through time suggests that some economic species that are very common today were not very common for most of prehistory. They appear to have spread during the late prehistoric expansion of human settlement, whether by purposeful introduction or by invasions of weedy species into increasingly disturbed areas.

The impact of human populations in late prehistoric times was not limited to deforestation, introduction of new tree species, encouragement of weedy species, and increased growing of cultivated species. The large garbage dumps of late prehistoric settlements spread almost continuously along the branches of major rivers and in the interior on the soils developed on volcanic rock and limestones. These middens enriched the soil with nutritive elements (Smith, 1980) and today support a type of diverse, tall anthropogenic forest that is rich in useful and domesticated species (Balée, 1989).

The building of more extensive earthworks at this time expanded the extent of forests in seasonally flooded areas, where alternation of flooding and drought had limited vegetation to herbs and shrubs. However, increased sedimentation from clearing for agriculture in late prehistoric times choked some streams and may have favored the development of vegetation and fish taxa, such as armored catfish, which are especially adapted to seasonally dry streams, over taxa requiring more permanent moisture sources.

Since important landscape features and the distributions of some very important and common economic species are anthropogenically influenced, rather than determined by purely "natural" factors, both biodiversity research and planning for economic development need to be more cognizant of the human factors involved in current biogeography and ecology.

Land Use During Colonial Expansion and National Development

The European conquest of the Amazon led to great changes in human–environment interaction.

During the disruptions caused by European conquest during the period 1600–1850, many native polities fell, and the area with a majority of Indians shifted away from the contact zone (Porro, 1994; Rodrigues de Oliveira, 1994). Subsistence reverted to horticulture and foraging, and the more permanent maize farms returned to forest. Settlers set up plantations, ranches, timbering operations, and extraction concerns with forced labor, but economic downturns late in the nineteenth century led to abandonment of many establishments, and cattle and manioc flour have become the main products since then, with sporadic booms in rubber and gold. The sites abandoned by the early settlers can often be recognized by the clusters of citrus and mango trees that they planted. The extent of colonial deforestation is unclear but could be elucidated through research on historical photographs and timber-shipping records. In any case, huge areas by transportation routes and around cities are being actively deforested now.

The most enduring post-Conquest economic process has been the establishment of cattle in floodplain areas, such as the Beni and Marajó. This development limited agriculture and even horticulture in some areas, due to the lack of fences, and the cattle tended to push out populations of large game. Predatory game has been kept at low densities through rancher extermination efforts. Sedimentation in watercourses, because of the trampling and erosion of land by cattle herds, has led to reduction in permanent habitats for trees. Moisture-loving palms such as *Euterpe oleracea* will not grow when streams turn seasonal due to sediment choking, and some upland forests have been reduced to open, secondary growth through clearing for pasture. Nevertheless, várzea cattle-raising has been economically sustainable, if not ecologically benign, for several hundred years because of the constant nutrient supply of the alluvium.

At the same time that settlers and cattle moved into the Amazon in increasing numbers, increased infectious disease, nutritional stress, and war deaths diminished the populations of Amazonian Indians. In response to the destruction of agricultural fields by ranching and to harassment for forced labor, Indian settlement mobility and size also decreased, and concurrently the intensity of cultivation decreased, and in some cases cultivation was suspended.

Maize, so important in some areas immediately before the conquest, is cultivated intensively by native groups in very few areas of Amazonia today (Roosevelt, 1994); instead, manioc or introduced starchy crops are staples. For example, in Jivaroan-language areas, where paleoecological evidence suggests long-term intensive maize cultivation in late-prehistoric times, manioc is the staple food. In the case of the Shipibo, manioc and bananas are staple plant foods, but the bone chemistry of late prehistoric humans buried in the ground under present-day Shipibo settlements shows that maize was the prehistoric staple (Roosevelt, 1989). As human ecologists have shown, Shipibo tend to produce annual seed crops for sale rather than to eat, for they live within the national economy, not isolated from civilization in the forest. In some areas on the Amazon mainstream where indigenous societies were substantially destroyed in the nineteenth century, rural peasants and small agribusinesses produce large quantities of maize for cattle feed and seed and manioc flour for human consumption (Swales, 1993).

In some areas, such as the Manu basin of the Peruvian Amazon, where internecine conflict has led to considerable instability of settlement, some people do not cultivate at all, presumably because they do not stay in one place long enough. Nevertheless, 70% of their food comes from crops that they harvest from others' abandoned fields

(Roosevelt, 1998). Some have called this hunting and gathering, but it is nonetheless dependent on the horticultural economy regardless.

In some areas where agriculture was important before the Conquest, such as in the Bolivian and Venezuelan floodplains, since the introduction of cattle has limited cultivation, some surviving native groups have given up the cultivation systems recorded for them in Conquest-period records (Roosevelt, 1993), and some of these groups have gone over to cattle rustling as a major subsistence strategy (Hurtado & Hill, 1989). In such areas, native harvesting of foreign cultivated fruit trees, such as mangos, clustered at abandoned Euro-American settlements is often an important subsistence supplement.

The patterns of human impact on forests has also changed after the Conquest. Although deforestation has sped up in the last decades, forest regeneration has occurred in some areas where formerly dense indigenous populations became sparser or more mobile, as in parts of the Ecuadorian Oriente. How the new and old forests in those areas compare biotically has not yet been studied, because it has not been widely recognized by forest specialists that the forest had been cut for farming previously.

The "primitive" horticulturist and forager lifeways of today are therefore as much a product of post-Conquest political economy as they are adaptations to forest ecology. Where Indians roam "virgin" forests today, prehistoric people farmed and built mounds, leaving charcoal from repeated forest burns in the soil, below. The causality in human–environment relations, then, is not linear and unidirectional but a complex pattern of mutual causality. Humans adapted not only to environment but also to their population history and relationships with others, and the forest both affected human behavior and was affected by it.

Land Use, Environmental Degradation, and Human Cultural Diversity

The long-term record of environment and land use in Amazonia has some lessons to teach about resources, land use, and social organization today.

Land use and settlement practices past and present have had a pervasive impact on the soils, drainage, and diversity and patterning of biota in the Amazon. Although research and conservation have treated the Amazon forest as primarily a natural entity to be studied by natural science methods, few Amazonian habitats have not been affected by the human occupation. The fact that many trees in the "virgin" forest are domesticates, escapees, or pre-adapted to human disturbance needs to be taken into account in forest ecology, biodiversity research, and conservation planning. Some of the biodiversity in Amazon forests exists today because of anthropogenic soil enrichment and plant introductions by humans. If the causes of such phenomena are misconstrued as natural, how can research and conservation be effective?

What is known of the forest and forest peoples' roughly 12,000-year history really needs to be applied to conservation and development planning. The record of the prehistoric and historic human occupations can inform conservation policy by providing data on the causes and history of the biological communities and patterns of biological diversity to be conserved. It provides concrete information on the quality of soils of the region and their ability to support both natural and cultivated vegetation and fauna. It also relates to sustainable development policy by showing the range of uses made of the environment over the last 12,000 years and their results. Past land use was conditioned by human culture, demography, and organization as well as by the environment, so sustainable development needs to consider the human factor. Environmental archaeology and historical ecology provide information on the way different types of human societies exploited Amazonian resources in the past and how their exploitation impacted those resources.

Native Amazonian people's land use varied through the millennia much more than anthropologists and natural scientists had thought possible, furnishing some practical alternatives for the management of tropical lands in the future. Given that the considerable native impact on Amazonian habitats escaped the notice of most natural scientists and conservationists, the impact was less than expected. It is likely, therefore, that native management methods even at their most intensive could be compatible with reasonable biodiversity-conservation goals. All economic species documented in intensive use in prehistoric times are still around today, so native use of them did not extinguish them, and most surviving forests are in the lands occupied by Indians and Mestizo rural people (Clay, 1988). New species and valuable economic plants are discovered because scientists explore with local guides. Because their futures are inseparable, conservation and development need to work for both.

Native populations and cultures in Amazonia are actually in more danger of extinction than nonhuman species. In contrast to the economic plant and animal species, most human cultures and populations that once occupied the Amazon floodplains are now extinct, showing that cultural diversity has been considerably more fragile than biological diversity. Native Amazonians were much more greatly impacted by the Conquest than had earlier been recognized, and the forced acculturation and demographic decimation are accelerating (Clay, 1988). The disappearance of indigenous peoples is dangerous for natural resources because cultural and biological diversity are interdependent. Neither Indians nor rural peasants are conservationists, but they are more resourceful and knowledgeable resource managers than most outsiders, and their input is greatly needed in conservation and development (Anderson, 1990; Nations, 1990; Redford & Padoch, 1992).

Contrary to expectation, local peoples and cultures were fully capable of managing resources in their regions without outside direction. Archaeological research shows that large-scale cultivation, fishing, earthwork construction, fine art manufacture, and trade were carried out successfully for long periods by regional societies without the centralized control and bureaucratic hierarchies characteristic of state societies. Since the Conquest, local smallholders have achieved large-scale, sustained production of manioc, maize, fish, and forest fruits for local consumption and regional export (Swales, 1993). Their agriculture is productive, and rural families with little technology and cash produce most of the food sold in the markets of the Brazilian Amazon. (Land tenure in Amazonia varies greatly. Largeholders dominate some areas like Marajó, but smallholders dominate in the Santarém–Monte Alegre region. Agricultural subsidies, unfortunately, have for the most part benefitted the largeholders rather than smallholders.)

Research on current human occupations shows that the largest environmental impacts came when comparatively more centralized and hierarchical societies occupied the land. These findings about the differential impact of different types of political and economic organization can inform attempts to create conservation programs and organize sustainable land use in the future.

The earthworks, garbage dumps, and species introductions of the late-prehistoric chiefdoms are the most conspicuous prehistoric human alterations to the landscape. Since the conquest, extensive deforestation and erosion is associated with government-subsidized large-scale ranching, agriculture, urbanization, hydroelectric projects, mines, and highway projects. Although the economic species survived 12,000 years with native people, research shows that current deforestation, agriculture, ranching, and urbanization carried out under the aegis of nation-states have eliminated some species locally. One example is *Astrocaryum vulgare*, which is no longer available as a raw material for basket-makers around the town of Monte Alegre. Other examples include *Arapaima gigas* and *Podocnemis expansa*, which also have been extinguished in some localities.

Since human factors of settlement pattern, culture, and political organization have had a powerful influence on variation in land use through time, development and conservation plans that focus on natural factors to the exclusion of human factors are doomed to failure and may even have effects that are detrimental to their goals. For in the human community lie both the sources of both problems and their solutions. Scientifically detectable extinctions and irreversible environmental degradation primarily correlate with the imposition of Conquest state rule, stratified national cultures, and specialized and hierarchical socioeconomic organizations with overseas connections. Planning that tries to solve biodiversity conservation problems with similar institutional organizations is likely only to intensify the problems. As the environmental conservation movement has spread and intensified, resource damage has also spread and intensified, in part because conservation projects have supplanted or disempowered local peoples and reduced cultural diversity. Purist environmentalism has not worked. Forests and forest peoples have disappeared ever more rapidly during the post–World War II movement.

Conservation and development programs have not taken into account the evidence of the long human occupation of the forest, and rural subsistence farmers and Indians are often presented as enemies of biodiversity (Lovejoy, 1985; Wilson, 1988; CNIE, 1993). Top-down management by large, centralized organizations is promoted, and local groups are left out or pushed out (Moran, 1993; Nugent, 1993; Davis & Wali, 1994). Reserves, often only on paper, are rapidly invaded by agencies, companies, and migrants, when long-term residents leave, and tenure for migrant smallholders is too insecure for sustained land use (Schmink & Wood, 1992: 37–57; Schneider, 1993). Large organizations' technology and large territories protect them from the consequences of local overuse, and conspicuous damage stems from government-aided dams, mines, ranches, forestry, migration, road-building, and urbanization, more than from local smallholders (Uhl, 1982; Roosevelt, 1990; Whitmore & Sayer, 1992: 129–140; Dove, 1993).

Options for the Future

Popular treatments sometimes present a picture of Amazonia as a virgin forest under siege by a wide range of attackers, from Indians and peasants to large companies and governments, and imply that this forest needs to be declared a nature reserve to keep all people out. The prehistory, history, and present data, however, show that Amazonia is fully occupied and has been occupied for a very long time. There is no way that the people of Amazonia are going to go away and leave it all for the naturalists. Therefore, the real problem for conservation and development in Amazonia is to facilitate an optimal mix of uses. Concentrating on the creation of purely natural reserves is not going to be a useful direction, because reserves need local people's presence and expertise and because that strategy would leave the majority of Amazonia, which lies outside reserves, unprotected from destructive land use.

The archaeological record shows that there are many more land-use options than development experts have traditionally recognized. The most common rural land use in Amazonia today is open-range cattle. Because this business is neither reliably profitable economically nor sustainable ecologically, other systems are going to be needed to support the population and regional economies. The most obvious options include natural and cultural history reserves and facilities, agroforestry projects, shifting cultivation, rotation field agriculture, improved cattle ranching, sustainable fishing, and ecological and cultural tourism.

Reserves can include a range of use zones (Robinson & Redford, 1991: xv–xvii): mature-forest parks solely for research and guided hiking, secondary forest and lakes where smallholder farms, orchards, and fishing would continue, and an urban zone for services, museums, and public parks.

Ecotourism is highly regarded for sustainable production because well-managed projects cause less ecological damage than cattle or mining and provide income by showing off the very resources that need to be preserved. Resources for ecotourism include the forests, savannas, rivers, hot springs, semiprecious rock lodes, rock paintings, native artifacts, local artists, and historic buildings. The lodging and transportation infrastructure for ecotourism is present in the area, as is local expertise for guarding and guiding. Local artists both indigenous and Mestizo create salable paintings, sculpture, and crafts.

The long history of sustained forest use, from the archaeological record, is an incentive to pursue agroforestry in the future. There is a wide variety of possible directions in this area of conservation and development. Management of valuable species such as Brazil nut can be improved by using the new knowledge of its adaptation to the soils of human habitation sites. Certain várzea forests, such as the morichales, have a high potential for sustainable production (Heinen, 1991). In addition to improving the management of existing forests, local Amazonian agronomists are interested in speeding reforestation by planting useful trees in highly degraded landscapes. At present, growers often plant fruit trees that are expensive or difficult to grow, such as cupuaçú (*Theobroma cupuacu*) (Smith et al., 1992: 68–75). However, the archaeological remains show that many lesser-known fruit trees and palms flourished and even expanded their range under human exploitation for many thousands of years. They are vigorous in the climate, resistant to pests, and highly productive without fertilizer. Their products are valued locally and also could be exported in forms appropriate for external market conditions. By encouraging such native species, forest regeneration and sustainable production can be fostered and rainfall increased, a benefit for farmers. Concurrent with reforestation, fauna extinguished by ranchers in the last decades can be reintroduced in wildlife reserves.

One land-use method that is not going to go away is cattle ranching, mentioned above. Floodplain cattle ranching, seldom treated in discussions of sustainable development, has been sustainable economically for hundreds of years. Improved management can make it ecologically more sustainable. The enormous ranches on the open range of Marajó are poorly managed. Palatable native grasses are being extinguished locally by overgrazing, and inedible or poisonous species (such as *algodao bravo*, *Ipomoea* spp.) are being encouraged. If the range were enclosed, and ranch size and stocking rates reduced, productivity per hectare would improve and the impact of ranching on the environment would be lessened—both desirable outcomes, given the fact that subsidies for cattle production are diminished and the industry needs to cut costs. In addition, with the land divided into more sensible-sized parcels, it could be better cared for and at the same time unemployment could be reduced.

Farming and fishing for subsistence and regional export is a long tradition in the region and an obvious direction for development (Smith, 1981; Wood, 1993). The viability of crops in Amazonia is demonstrated by archaeological evidence for several thousand years of cultivation. Intensive crop production in selected areas frees other land to be utilized less intensively (Smith et al., 1992). Both maize and manioc are very productive in the area and yield products valued in a range of local and international markets, but other root and seed crops also have potential for the future. Native wet-loving taros (such as *Xanthosoma sagittifolium*) could be grown on floodplain land and provide industrial starch or alternative fodder for cattle. There are also numerous native edible floodplain grasses and molluscs, intensively exploited during times in prehistory (Brochado, 1980; Roosevelt et al., 1991), that have not yet been looked at as possibilities for either local subsistence or industrial uses.

To help forest people, international banks and nongovernmental organizations have started to give small grants for sustainable production (Rainforest Alliance, 1994; World Bank, 1994). This is a good direction, but uninformed searches for sustainable uses can be dangerous, insofar as short-term economic and political goals contrary to survival of species and cultures become paramount. The archaeological record of the forest shows thousands of years of use under varying demographic and sociopolitical conditions, primary evidence of management that was sustainable over the long term. Without long-term knowledge of environment and land use and involvement of local residents, sustainable development is impossible.

Conclusion

Knowledge of the history of human–environment interactions in Amazonia under various conditions from the arrival of humans to today can illuminate present processes in ways no synchronic, monocausal studies can. Research has shown that ways of using the habitat changed over time, depending on density of human population, sociopolitical organization, settlement patterns, and local resource patterning. Seeing the result of different land uses over time can show better than any short-term study how they worked and what effect they had on the habitat. Human ecological approaches can also illuminate social and political factors in the interaction, which are left out of purely natural science approaches. Already, history and prehistory show more land-use options than most conservation and development experts recognize. Application of information about the long-term relationship of people with environments is crucial for understanding the causes and solutions for critical environmental problems in Amazonia.

Literature Cited

Absy, M. L. 1979. A palynological study of Holocene sediments in the Amazon basin. Ph.D. dissertation, University of Amsterdam.

Anderson, A. B. 1983. The biology of *Orbignya martiana* (Palmae): A tropical dry forest dominant in Brazil. Ph.D. dissertation, University of Florida, Gainesville.

————, ed. 1990. Alternatives to deforestation. Columbia University Press, New York.

Athens, S. 1989. Pumpuentsa and the pastaza phase of southeastern lowland Ecuador. Nawpa Pacha 24: 1–29.

Bailey, R., G. Head, M. Jenike, M. Owen, R. Rechtman & E. Zechenter. 1989. Hunting and gathering in tropical rain forest: Is it possible? American Anthropologist 91: 59–82.

Balée, W. 1989. Culture of Amazonian forests. Pages 1–21 *in* D. A. Posey & W. Balée, eds., Resource management in Amazonia: Indigenous and folk strategies. Advances in Economic Botany 7. New York Botanical Garden, Bronx.

———— **& D. G. Campbell.** 1990. Evidence for the successional status of liana forest (Xingu Basin, Amazonian Brazil). Biotropica 22(1): 36–47.

Beckerman, S. 1979. The abundance of protein in Amazonia: A reply to Gross. American Anthropologist 81: 533–560.

Brochado, J. P. 1980. The social ecology of the Marajoara culture. M.S. thesis, University of Illinois, Urbana.

Bush, M. & P. Colinvaux. 1988. A 7000-year pollen record from the Amazon lowlands, Ecuador. Vegetatio 76: 141–154.

————, **D. Piperno & P. Colinvaux.** 1989. A 6,000 year history of Amazonian maize cultivation. Nature 340: 303–305.

Cavalcante, P. B. 1991. Frutas comestiveis da Amazonia. Museu Paraense Emilio Goeldi/Editora CEJUP, Belém.

Clay, J. 1988. Indigenous peoples and tropical forests: Models of land use and management from Latin America. Cultural Survival, Cambridge, MA.

Committee for the National Institute for the Environment. 1993. A proposal for a National Institute for the Environment: Need, rationale, and structure. Committee for the National Institute for the Environment, Washington, DC.

Davis, S. H. & A. Wali. 1994. Indigenous land tenure and tropical forest management in Latin America. Ambio 23(8): 485–490.

Denevan, W. 1966. An aboriginal cultural geography of the Llanos de Mojos de Bolivia. Ibero-Americana 48. University of California Press, Berkeley.

Dougherty, B. & H. Calandra. 1981–1982. Excavaciones arqueologicas en La Loma Alta de Casarabe, Llanos de Moxos, Departamento del Beni, Bolivia. Relaciones de La Sociedad Argentina de Antropologia, n.s. 14(2): 9–48.

Dove, M. 1993. Smallholder rubber and swidden agriculture in Borneo: A sustainable adaptation to the ecology and economy of the tropical forest. Economic Botany 47(2): 136–147.

Erickson, C. 1980. Sistemas agricolas prehispanicas en los Llanos de Mojos. America Indigena 40(4): 731–755.

Franzinelli, E. & E. Latrubesse, eds. 1993. Geologia Quaternario de Amazonia. Conferencia Internacional. Resumos. Universidade Federal de Amazonas, Manaus.

Fritz, G. 1994. Are the first American farmers getting younger? Current Anthropology 35(3): 305–310.

Frois Lopes, M. Magalhaes, M. Imazio da Silveira. N.d. Caverna do Gaviao. Manuscript.

Goodland, R. & H. Irwin. 1975. Amazon jungle: Green hell to red desert? Elsevier, New York.

Hagman Family. N.d. Rainfall statistics from Taperinha. Taperinha, Santarém.

Hather, J. G., ed. 1994. Tropical archaeobotany: Applications and new developments. Routledge, London.

Heinen, D. 1991. El ecosistema morichalero del delta Orinoco y su utilizacion humana. Pages 481–526 *in* J. San Jose & J. Celecia, eds., Enfoques ecologia humana aplicados a los sistemas agricolas tradicionales del tropico americano. Centro Internacional de Ecologia Tropical Programma sobre el Hombre y la

Biosfera, Instituto Venezolano de Investigaciones Cientificas, Inparques, Universidade de los Andres, Universidad Simon Bolivar, Caracas.

Hurtado, A. M. & K. Hill. 1989. Seasonality in a foraging society: Variation in diet, work effort, fertility, and sexual division of labor among the Hiwi of Venezuela. Journal of Anthropological Research 46(3): 293–346.

Imazio da Silveira, M. 1994. Estudo sobre estrategias de subsistencia de cacadores-coletores pre-historicos do Sitio Gruta do Gaviao, Carajas (Pará). M.S. thesis, University of São Paulo.

Jordan, C. F. 1985. Nutrient cycling in tropical forest ecosystems. Wiley Interscience, New York.

Junk, W. J. 1970. Investigations on the ecology and production biology of the "floating meadows" (Paspalum–Echinochloetum) on the middle Amazon. Amazoniana 4: 9–102.

Lajtha, K. & R. H. Michener, eds. 1994. Stable isotopes in ecology and environmental science. Blackwell Scientific, Oxford.

Lathrap, D. 1970. The upper Amazon. Praeger, New York.

———. 1975 Ancient Ecuador: Culture, clay, and creativity. Field Museum, Chicago.

Lovejoy, T. 1985. Amazonia, people and today. Pages 328–338 in G. Prance & T. Lovejoy, eds., Amazonia. Pergamon Press, Oxford.

Magalhaes, M. P. 1994. Archaeology of Carajas: The prehistoric presence of man in Amazonia. Companhia Vale do Rio Doce, Rio de Janeiro.

Meggers, B. J. 1971. Amazonia: Man and nature in a counterfeit paradise. Aldine, Chicago.

Moran, E. 1993. Through Amazonian eyes: The human ecology of Amazonian populations. University of Iowa Press, Iowa City.

Nations, J. D. 1990. Protected areas in tropical rainforests. Pages 208–216 in S. Head & R. Heinzman, eds., Lessons of the rainforest. Sierra Club Books, San Francisco.

Nimuendaju, C. 1949. Os Tapajo. Boletim do Museu Paraense Emilio Goeldi 10: 93–106.

Nugent, S. 1993. Amazonian caboclo society: An essay on invisibility and peasant economy. Berg Press, Oxford.

Porras, P. 1987. Investigaciones archaeologicas a las faldas de Sangay, Provincia Morona Santiago. Artes Graficas Senal, Impresenal Cia, Quito.

Porro, A. 1994. Social organization and political power in the Amazon floodplain: The ethnohistorical sources. Pages 79–94 in A. C. Roosevelt, ed., Amazonian Indians from prehistory to the present. University of Arizona Press, Tucson.

Prance, G. T., ed. 1982. Biological diversification in the tropics. Columbia University Press, New York.

Rainforest Alliance. 1994. Annual report. Rainforest Alliance, New York.

Rasanen, M., J. S. Salo & H. Jungner. 1991. Holocene floodplain lake sediments in the Amazon: ^{14}C dating and paleoecological use. Quaternary Science Reviews 10: 363–372.

Redford, K. H. & C. Padoch, eds. 1992. Conservation of neotropical forests: Working from traditional resource use. Columbia University Press, New York.

Robinson, J. H. & K. H. Redford, eds. 1991. Neotropical wildlife use and conservation. University of Chicago Press, Chicago.

Rodrigues de Oliveira, A. 1994. Evidence of the nature of the process of indigenous deculturation and destabilization in the Brazilian Amazon in the last three hundred years: Preliminary data. Pages 95–122 in A. C. Roosevelt, ed., Amazonian Indians from prehistory to the present. University of Arizona Press, Tucson.

Roosevelt, A. C. 1980. Parmana: Prehistoric maize and manioc subsistence along the Amazon and Orinoco. Academic Press, New York.

———. 1989. Resource management in Amazonia before the Conquest: Beyond ethnographic projection. Pages 30–62 in D. A. Posey & W. Balée, eds., Resource management in Amazonia: Indigenous and folk strategies. Advances in Economic Botany 7. New York Botanical Garden, Bronx.

———. 1990. The historical perspective on resource use in Latin America. Economic catalysts to ecological change. Working Papers of the Tropical Conservation and Development Program. University of Florida Latin American Studies Center, Gainesville.

———. 1991. Moundbuilders of the Amazon: Geophysical archaeology on Marajó Island, Brazil. Academic Press, San Diego.

———. 1993. The rise and fall of the Amazon chiefdoms. Pages 255–384 in A. C. Taylor & P. Descola, eds., Le remantée de l'Amazone: Anthropologie et histoire des socières Amazoniennes. L'Homme 33(special issue): 126–128.

———, ed. 1994. Amazonian Indians from prehistory to the present. University of Arizona Press, Tucson.

———. 1995a. Early pottery in Amazonia: 20 years of obscurity. *In* W. Barnett & J. Hoopes, eds., The emergence of pottery: Technology and innovation in ancient societies. Smithsonian Institution, Washington, DC.

———. 1995b. The excavations at Corozal, Venezuela: Stratigraphy and ceramic seriation. Yale University Publications in Anthropology no. 82. Yale University Press, New Haven.

———. 1998. Ancient and modern hunter-gatherers in the Amazon. *In* W. Balée (ed.). Advances in Historical Ecology. Columbia University Press, New York.

———, **R. A. Housley, M. Imazio da Silveira, S. Maranca & R. Johnson.** 1991. Eighth millennium pottery from a prehistoric shell midden in the Brazilian Amazon. Science 254: 1621–1624.

———, **M. Lima da Costa, W. Barnett, J. Feathers, M. Michab, H. Valladas, C. Machado, M. Imazio da Silveira, A. Henderson, N. Toth & K. Schick.** 1996. Cave dwellers in the Amazon: The peopling of the Americas. Science 272: 373–384.

Schmink, M. & C. H. Wood. 1992. Contested frontiers in Amazonia. Columbia University Press, New York.

——— & ———, eds. 1984. Frontier expansion in Amazonia. University of Florida Press, Gainesville.

Schmitz, P. 1987. Cacadores antigos no sudoeste de Goias, Brasil. Pages 16–36 *in* L. Nunez & B. J. Meggers, eds., Investigaciones paleoindias al sur de la linea ecuatorial. Estudios Atacameños no. 8. Universidad del Norte, San Pedro de Atacama, Chile.

Schneider, R. 1993. Land abandonment, property rights, and agricultural sustainability in the Amazon. Latin American Dissemination Note #3. The World Bank Latin American Technical Department, Environmental Division, Washington, DC.

Simoes, M. 1981. Coleltores-pescadores ceramistas do litoral do Salgado (Pará). Nota preliminar. Boletim do Museu Paraense Emilio Goeldi, n.s. 78: 1–26.

Sioli, H., ed. 1984. The Amazon: Limnology and landscape ecology of a mighty tropical river. W. Junk, Dordrecht.

Smith, C. E. & A. C. Roosevelt. N.d. Prehistoric plant use in the middle Orinoco basin: Tree products. Manuscript.

Smith, N. 1980. Anthrosols and human carrying capacity in Amazonia. Annals of the American Association of Geographers 70: 553–566.

———. 1981. Man, fishes, and the Amazon. Columbia University Press, New York.

———, **J. T. Williams, D. L. Plucknett & J. P. Talbot.** 1992. Tropical forests and their crops. Cornell University Press, Ithaca, NY.

Swales, S. 1993. Agricultural practices in the Monte Alegre District, Brazil. M.A. thesis, University of Illinois, Chicago.

Tricart, J. 1972. The landforms of the humid tropics, forests and savannas. C. J. Kiewiet de Jonge, trans. Geographies for Advanced Study. St. Martin's Press, New York.

Uhl, C. 1982. Recovery following disturbances of different intensities in Amazonian rain forest. Interciencia 7: 19–24.

Van Der Hammen, T. 1991. Palaeoecology of the Neotropics: An overview of the state of affairs. Boletin IG-USP, Publicacion Especial 8: 35–55.

Van der Merwe, N., A. C. Roosevelt & J. Vogel. 1981. Isotopic evidence for prehistoric subsistence change at Parmana, Venezuela. Nature 292: 536–538.

Whitmore, T. C. 1990. An introduction to tropical rain forests. Clarendon Press, Oxford.

——— & **G. T. Prance, eds.** 1987. Biogeography and Quaternary history in tropical America. Clarendon Press, Oxford.

——— & **J. A. Sayer, eds.** 1992. Tropical deforestation and species extinction. Chapman & Hall and International Union for Conservation of Nature and Natural Resources, London.

Williams, D. 1981. Excavations of the Barambina shell mound northwest district: An interim report. Journal of the Walter Roth Museum of Archaeology and Anthropology 41(1–2): 13–36.

Wilson, E. O., ed. 1988. Biodiversity. National Academy Press, Washington, DC.

Wood, D. 1993. Restoring tropical lands to agriculture. Land Use Policy (April): 91–107.

World Bank. 1994. Update. Newsletter of the Pilot Program to Conserve the Brazilian Rainforest 2(3): 1–3.

Section 6:
The Case of the Vanishing Stingless Bee

The Probable Consequences of the Destruction of Brazilian Stingless Bees

Warwick Estevam Kerr, Gislene Almeida Carvalho, and Vania Alves Nascimento

Species Extinction

As we begin the third millennium, the extinction of species is unrelenting, and bees are being subjected to particularly high rates. The problem is not confined to the loss of Meliponid bees; it affects all Hymenoptera of crossed fecundity. In a study of all the Aculeata in England, where 250 of the 520 total species of fauna are bees, population decline has been such that almost 50% of all described species are in danger of extinction (Falk, 1991).

There are five principal causes of destruction of Meliponid colonies:

1. Agricultural clearing using fire: Meliponid bees are particularly vulnerable to fires, because although the wings of the queens are similar in size to those of the workers, their highly developed abdomens weigh much more, making flight (and escape) impossible. Consequently, entire nests are lost in fires. In addition, the 2–3% of Meliponid species that make nests near the soil surface are killed after just one plowing. Only *Melipona quinquefasciata* and some *Triginoid* species have nests deep enough to escape fatal damage.

2. Honey collection: Collectors of honey from wild hives tend to harvest destructively. In Uberlândia, honey collectors have caused local extinction of at least four species: *Melipona rufiventris, M. bicolor, M. marginata*, and *Cephalotrigona femorata*. One collector informed us that in 20 years he destructively harvested honey from an estimated 200 colonies within 90 hectares of forest, including areas within the Reserva Universitária do Panga.

3. Inadequate reserve areas: IBAMA requires that a certain forest area be protected from destructive activities. However, work by Marcio Oliveira (pers. comm.) indicates that most reserves in Amazonia are too small to support bee reproduction and maintenance of an adequate number of xo heteroalleles, which are the primary genetic determinants of sex in Meliponid bees.

4. Habitat destruction: Large sawmills and timber extractors favor cutting old-growth trees, which, with their large branches and trunks, often serve as important locations of adequate-size cavities for bee nests.

5. Deforestation: A general loss of native forests limits bee-inhabitable areas. For example, in the state of São Paulo only 6% of the original forest remains. We can assert that only 5% of the Meliponid species remain.

The Role of the XO Heteroalleles, Primary Sex Determinants

The great majority of Hymenoptera are parthenogenetic and arrhenotokous; that is, males arise from unfertilized eggs and females arise from fertilized eggs. In terms of breeding systems, Hymenopterans can be divided into endogametics (obligatory mating between male and female siblings) and panmictics (mating between random individuals). The endogametic species have male and female genes that determine the sex both of male haploids (with n chromosomes) and of female diploids (with $2n$ chromosomes). Male genes (M) are totally or partially nonadditive and female genes (F) are totally or partially additive. The result is this: males = M>F and females = 2F>M. This hypothesis (Cunha & Kerr, 1957) was borne out in the work of Oliveira (1992) and Oliveira et al. (1992), who found that 17.4% of the proteins examined had the same number of molecules in males as in females (i.e., they are nonadditive). In other words, compensation occurs in the females.

In panmictic Hymenoptera, one of the two determinant genes of femaleness xo^0 mutated to xo^1, then gradually mutated to xo^2, xo^3, and so on, and then came to control the formation of testicles versus ovaries (Table I; Gonçalves & Kerr, 1970).

When the xo heteroalleles are homozygous (e.g., $xo^7 xo^7$), diploid males, rather than females, will form. Each species has evolved a method to inhibit diploid males from breeding and thereby avoid the creation of triploid individuals ($3n$), which would undermine the entire breeding system. Diploid males are totally or partially sterile, are nearly blind, have a greatly reduced number of sperm, are eliminated during the larval stage, or are killed as adults by workers.

Carvalho et al. (1995) observed a gradual decrease in the number of xo heterosexuals in a population comprising an aggregation of 18 colonies of *Melipona scutellaris* and 50 of their descendants. Having at least six different alleles of the xo gene is very important for the maintenance of our bee populations. Kerr and Vencovsky (1982) reported that a minimum of 44 colonies is needed to maintain six alleles in a population; with fewer than 44 colonies, a population can be totally eliminated in 15 generations (Yokoyam & Nei, 1979). With this in mind, in the last 14 years we have warned Meliponid beekeepers to raise at least 44 colonies of the same Meliponid species. Such advice does not apply to *Apis mellifera*. (It is also possible that it does not apply to many

Table I. Mutation of one of the two determinant genes in panmictic Hymenoptera

Haploid male (n)	Diploid female ($2n$)	Diploid male ($2n$)
xo^1	$xo^1 xo^2$	$xo^1 xo^1$
xo^2	$xo^2 xo^3$	$xo^2 xo^2$
xo^3	$xo^3 xo^4$	$xo^3 xo^3$
...	...	...
xo^{20}	$xo^{19} xo^{20}$	$xo^{20} xo^{20}$

Meliponid species.) The reasoning behind this advice is simple. Kerr and Vencovsky (1982) confirmed, using formulas from Wright (1969) and Cornuet (1980), that an active genetic population varied from 115 to 150 colonies in a reproductive pool. Woyke (1980) proved that the minimum of 6 xo alleles is needed for a population of *Apis mellifera* to maintain itself. Our experience indicates that 6 xo alleles is also acceptable for Meliponids. Therefore, 44 colonies is the minimum number of colonies needed in a given reproductive area. In the case of jataí (*Tetragonisca angustula*) in Ribeirão Preto, the number of colonies is great so that an amateur Meliponid beekeeper could have just one colony at home because it would be within a reproductive area with many more than 44 colonies. In contrast, an individual colony of mandaçaia (*Melipona quadrifasciata*) would be the only one in its reproductive area and therefore would have approximately a 50% probability of breeding with the same xo allele. With various additional colonies, we might have just 3 alleles and 33% of the crosses would fail (i.e., they would produce diploid males).

The reproductive area of Africanized queens and males has a radius of approximately 40 km (i.e., 4800 km²; the mean of 20 km for queens and 60 km for males). Each queen is fertilized by 15 to 17 males (see literature in Lobo & Kerr, 1993). In Barra do Garça, Goiás, Kerr (1974) estimated the occurrence of 42,000 *Apis mellifera* colonies in a 3800 km² area of campo cerrado. This density is sufficient to maintain 20 or more xo heteroalleles in the population. Consequently, a novice beekeeper could have just one *Apis* colony without running the risk of losing it from producing diploid males.

However, the risk is much greater for the Meliponids because their colonies raise diploid males until the end of their development. Hours after they are born (or days after, depending on the species) they are killed by the workers, which also kill the queen. Only strong colonies can resist such a trauma.

Why Save Meliponids?

Why is it so important to save these bees? Bees are an integral part of the ecosystems in which they live. Their primary function in nature is pollinating flowers, and consequently, they facilitate fruit and seed production. Work by Absy et al. (1984) showed that 8% of the trees (from a sample of 363 plants in the margins of the Trombetas and the Tapajós Rivers, Pará) are pollinated by at the very least five species of Meliponids. Furthermore, 50 plant species (14%) from the same sample were pollinated by only one species of bee. The elimination of that bee would impair the pollination of those species' flowers.

In an excellent study by Pedro and Camargo (1991), 4086 bees were collected from flowers of species in the cerrado vegetation of Fazenda Santa Carlota (Cajuru, São Paulo) over the course of one year, of which 1855 (45.4%) were Meliponids. It is evident from these data that the constant preferential selection of the *Melipona* genus by honey collectors has probably caused the reduction of the importance of Meliponids, especially of the *Melipona* genus. From 1934 to 1945 in the forests of Parnaiba, Pirapora, and Cabreuva (São Paulo), it was found that (in descending order, and with asterisks denoting species found by Pedro and Camargo [1991] in Cajuru) *Melipona quadrifasciata*,

Scaptotrigona xanthotricha, Melipona marginata, Melipona rufiventris, Tetragonisca angustula★, *Trigona spinipes*★, and *Melipona bicolor* commonly frequented the flowers of cambará (*Moquinia polymorpha*), assapeche (*Veronia polyanthes*), and vassourinha (*Baccharis dracunculifolia*).

In the forests of Brazil, stingless bees are the primary pollinators of 40–90% of the trees, depending on the ecosystem. Solitary bees, butterflies, beetles and other insects, bats, and birds (not to mention wind, water, and some mammals) that pollinate the remaining 10–60% can fly to escape from forest fire. The Meliponids, however, cannot escape, for the queens do not fly and are unable even to leave the tree interiors because of their highly developed abdomens. Although the nearly 2% of the Meliponids that have subterranean nests escape damage from fire, they can be destroyed by just one plowing.

The consequences of the reduction and/or extinction of these bee species are evident. The various types of vegetation (Amazonian forest, Atlantic forest, pantanal, agreste, caatinga, cerrado, tropical dry forest, pampas, mangroves, palm forests, auracarias) harbor actual groups of species that depend on the intra- and interspecific competition that has existed for thousands of years. The presence of each plant species depends on the soil, the climate, and the genetic composition; each species' presence in the tropics or subtropics is a direct result of its ability to produce fertile seeds (to be naturally dispersed by the wind, birds, fish, mammals, and so forth)—an ability that in turn hinges on pollination. Depending on the forest type, anywhere from 40% to 90% of plant species in this region rely on one or more species of bee from the tribe Meliponini for pollination.

For thousands of generations, plants and bees adapted to one another to the point that many Meliponid species require long periods of time to adapt after being transferred to new habitats. In 1984, Kerr took 30 colonies of *Melipona subnitida* to the island Fernando de Noronha, where a species of Convolvulaceae was at the peak of flowering. After a few days, honey overflowed from the hives because the bees were not adapted and thus collected more nectar than they were actually able to store.

Reforestation: A Difficult Task

Environmental preservation projects commonly place an emphasis on large mammals, which are called "charismatic megafauna" by Mark J. Plotkin. Plotkin (1988) showed that any attempt to save the panda bear will be invalid if measures are not first taken to preserve the bamboo that they feed on.

The conservation and restoration of a tropical or subtropical forest is a much more complicated task than it might first appear. Nearly 90% of tropical plants need to outcross; 40–90% of the tropical and subtropical phanerogams are pollinated by bees. The rest are pollinated by Lepidoptera, Diptera, ants, beetles, other insects, molluscs, birds, bats, and other animals, in addition to wind and water.

Most Meliponid species make their nests in tree cavities. Reporting on a forest sample from Reserva Ducke, near Manaus, Rodrigues and Valle (1964) noted that 145 trees (32% of the sample) with dbh ≥10 cm have cavities. We found a similar situation in April 1993, with logs that had been extracted in cerrado near Corumbaiba (Goiás). We counted 28 (28.9%) trunks with cavities and 69 without cavities. Many of the cavities were made by insect larvae (mostly Coleoptera, Lepidoptera, and Diptera) when the trees are very young.

For many tree species, long-term preservation—second, third, fourth generations and beyond—is guaranteed by outcross pollination of their flowers for the production of fertile seeds, which also maintains genetic diversity. One *Gliricidium sepium* (madre-de-cacau) produces almost 600 seeds with bee pollination. Without bee pollination, wind and butterfly pollination only yields 10 seeds. This is a decrease in adaptive value from 1 to 0.017, which is similar to the effect of a quasi-lethal gene.

Therefore, studies of bee biology and management, especially those related to controlled reproduction and division of colonies, are extremely important not only for the conservation of bee species but also for the conservation of the remaining forests and the fauna that depend on them.

What would happen if the bees were destroyed? Without bees, the reproductive capacity of many plant species would be reduced in a similar way as though there were a detrimental, quasi-lethal, or even lethal gene. Within four or five generations, natural selection would favor plants whose flowers are self-fertilized or pollinated by butterflies, flies, wind, and even beetles, which plants stopped selecting for in the past due to the high cost of those pollination systems. As a result, the industrial integrity of today might have time for four to five generations.

Bee colonies are dying out from deforestation and human population increase. In nearly 30 years, 9% of Amazonian forest has been destroyed; 94% of Serra do Mar has been cut in the last 50 years, and 70% of the cerrado in the past 80 years. Such destruction of the flora is already causing climate changes and species losses. Walker (1991) alerted us to the danger of similar great changes that have occurred recently in Somalia and in the past 2000 years in Europe and Africa. She suggests that the reforestation of southern Europe, Asia Minor, and North Africa is crucial to improving conditions in those areas. Similarly in Brazil, the southern margin of Amazonia and the dry areas of the northeast should be reforested to finally return water to many areas, thereby avoiding desertification (Walker, 1991).

Forest destruction impacts fish, birds, mammals, bees, and more. Goulding (1980) brilliantly showed that fish species depend, directly or indirectly, on flooded forests (and many várzea forest areas are flooded more than 100 km from the river bank). Goulding (1980: 253) collected more than 3000 fish specimens of 50 species along the margin of the Río Machado (Rondônia), practically all of which had leaves, flowers, and fruits of local plants in their stomachs. To identify the plant species eaten by the fish he used many methods, including the obvious and intelligent method of planting seeds found in the fish digestive tracts. All of the plants were identified and deposited in the herbarium at INPA. Goulding's finding that 75% of the plants collected were from várzea and igapó lent greater emphasis to the enormous importance of these forest types. The total or partial destruction of flooded forests and nutrient-poor river systems will at the same time destroy the fish fauna as we know it today. These data are used to show that bees are important even to Amazonian fish populations.

Attempts at Salvation

On a recent trip to Estação Ecológica Mamirauá near Tefé, Amazonas, we found that the principle enemies in várzea and igapó are the honey collectors. We estimated that 60–90 people, from the southern one-third of Reserva Mamirauá, in the past 10 years removed and destroyed 600–900 nests of just three species: *Melipona crinita*, *Melipona seminigra*, and *Melipona* cf. *rufiventris*. It so happens that these are the three largest bee species in the reserve, and are therefore those that produce the largest quantities of honey. The destruction of large bees will in the medium-term modify the forest structure in the area: first, by favoring species with flowers pollinated by small bees and reducing species with large flowers like the Fabaceae, Caesalpinioideae, Bixaceae, Chrysobalanaceae, Ebenaceae, and some palms; and second, by consequently increasing some species of Mimosoideae, Myrtaceae, and Polygonaceae. In synthesis, the destruction of large bees is causing disequilibrium in nature with unpredictable consequences for the preservation of the fauna.

We believe that if there were 90 Meliponid beekeepers at the Estação Ecológica Mamirauá, each one with 40 hives of large Meliponids (for a total of 3600 hives), then the forest would return to the same equilibrium that existed before 1960 while also creating a mini-industry of honey production (the honey collectors have an obvious talent for working with bees). We also suggest that environmental education books include special sections about bees.

In addition to our suggestions for bee conservation at Mamirauá, we are conducting three conservation programs: one in Espírito Santo with *Melipona capixaba* (Moure & Camargo, 1994), one in Maranhão with *M. compressipes*, and one in Uberlândia (through the collaboration with dozens of Meliponid beekeepers of the northeast) with *M. scutellaris*.

In an area of 90 km radius centered in Domingo Martins, Espírito Santo, there is a black bee (uruçu-preto, *Melipona capixaba*) larger than *Apis mellifera* that pollinates *Cassia* species with medium to large flowers. There, the vegetable farmers have 1–8 colonies each, which alone cannot save the species from extinction (only 10–20% of the original forest flora remains). We convinced these farmers to have 50 colonies each, and we are supplying some of the colonies and teaching methods of division and management.

For the uruçu (*Melipona scutellaris*) in the northeast, we obtained an initial population of 22 nests from Lencois, Bahia. These were multiplied using different methods of division. In 1991, 13 fisogastric queens from Piata, Bahia (Chapada, Diamantina) were introduced, and in 1992, three more queens from Batu, Bahia, were introduced. From that population, we attained 84 colonies by 1994 that are used in studies at the Genetics Laboratory of the Universidade Federal de Uberlândia, at the Department of Biology of the Universidade Federal de Vicosa, and at the Departamento de Zootecnia of the Universidade Estadual do São Paulo – Jaboticabal.

Murilo Drummond and José Ribamar Silva Barros worked with *Melipona compressipes* in the state of Maranhão. *Melipona compressipes* and *M. scutellaris* are the only two bees that were domesticated in Brazil before 1830.

In the last 15 years, and especially since 1988, we have succeeded at developing an effective method of dividing colonies and ensuring good nest structures, and it has been easy to convince dozens of Meliponid beekeepers in Bahia, Pernambuco, Maranhão, and Espírito Santo to raise 50 or more colonies, placing their orphan colonies in the beehives of friends and exchanging queens with transport cages similar to those used for *Apis mellifera*. Unfortunately, in many places economic resources are lacking to buy hives. Funds from CNPq and from the Fundação Banco do Brasil have helped.

Bierregaard et al. (1994) suggested that forest fragments in Amazonia should be larger than 100 hectares because the distribution of many species is not homogeneous and they need large areas to be able to survive in small fragments. Kerr (1987) found 20 xo alleles in one population of *Melipona compressipes*, and Carvalho et al. (1994) found 14 in one restricted population of *M. scutellaris*, so it is clear that bees, too, need larger forest fragments. To maintain so many xo alleles, approximately 315 existing colonies of *M. compressipes* and approximately 200 existing colonies of *M. scutellaris* are needed in these areas (Kerr & Vencovsky, 1982).

Economic Examples

Until now there has been no economic analysis relative to these bees. We present here two activities for economic gain:

1. Pollination: In Uberlândia and Maranhão, we use *Scaptotrigona postica* for efficient pollination for seed production of carrots. In Uberlândia, we use *Trigona spinipes* for pollination of cabbage and pineapple.
2. Honey production: Recently, Barros and Krogh (1990) and Barros (1994) have conducted studies on shifting Meliponid apiculture and found that *Melipona compressipes* produces more than 1 kilogram during just a few days of flowering and that workers of *Melipona scutellaris* visit and collect an extraordinarily delicious honey from the flowers of orange trees, pollinating those that need seeds to ensure fruiting.

In the Bonfim neighborhood of Arari, Maranhão, José de Souza has almost 200 colonies of *Melipona compressipes*, and in good years he collects nearly 600 liters of honey, making his income the highest one in his neighborhood.

In Jardim do Seridó, Río Grande do Norte, Ezequiel Madeiros de Macedo owns 120 *Melipona subnitida* colonies, and he earns nearly 600 reais per month by selling honey from active hives. He, too, earns the highest income in his town.

Acknowledgments

We thank CNPq, FAPEMIG, and Fundação Banco do Brasil for funding; and Projeto Mamirauá for logistic support in Amazonia. José Márcio Ayres hosted us in Tefé and recommended a very efficient guide, Tito Jonas Martins Cavalcante. Alvino Pianzolli provided support in Espírito Santo; his son, Alvino Pianzolli Filho, was our efficient guide in that region. Débora Lima Ayres from the Projeto Mamirauá, Ana Maria Lopes Menezes from Piata, Bahia, Rogério Marcos de Oliveira Alves from Catu,

Bahia, and João Luis Aleixo de Caruaru all helped us with local infrastructure and supplied complete nests, queens, and much data about uruçu.

Literature Cited

Absy, M. L., J. M. F. Camargo, W. E. Kerr & I. P. A. Miranda. 1984. Espécies de plantas visitadas por Meliponinae (Hymenoptera; Apoidea), para coleta de pólen na região do médio Amazonas. Revista Brasileira de Biologia 44(2): 227–237.

Barros, J. R. S. 1994. Genética da capacidade de produção de mel com abelhas *Melipona scutellaris* com meliponicultura migratória e sua adaptabilidade no sudeste do Brasil. M.Sc. dissertation, UNESP (JA-BOTICABAL).

———— **& H. Krogh.** 1990. Apicultura migratória com a abelha tiúba (*Melipona compressipes fasciculata*). Ciência e Cultura 42(10–12): 846–847.

Bierregaard, R. O., Jr., A. A. dos Santos & R. W. Hutchings. 1994. Biological Dynamics of Forest Fragments Project. Biodiversity Program, NMNH no. 180. Smithsonian Institution, Washington, DC; INPA, Manaus.

Carvalho, G. A., W. E. Kerr & V. A. Nascimento. 1995. Sex determination in bees. XXXIII. Decrease of xo heteroalleles in a finite population of *Melipona scutellaris* (Apidae, Meliponini). Revista Brasileiro de Genética 18(1): 13–16.

Cornuet, J. M. 1980. Rapid estimation of the number of sex alleles in panmictic honeybee populations. Journal of Apicultural Research 19: 3–5.

Cunha, A. B. de & W. E. Kerr. 1957. A theory to explain sex determination by arrhenotokous parthenogenesis. Forma et Functio 1(4): 33–36.

Falk, S. 1991. A review of the scarce and threatened wasps and ants of Great Britain. *In* Research and Survey in Nature Conservation, no. 35. Nature Conservancy Council for England, Peterborough, UK. [Apicultural Abstracts 1992, vol. 43(3): 196–197.]

Gonçalves, L. S. & W. E. Kerr. 1970. Noções sobre genética e melhoramento de abelhas. Pages 8–36 *in* Anais 1° Congresso Brasileiro Apicultura. Florianópolis, SC.

Goulding, M. 1980. The fishes and the forest. University of California Press, Berkeley.

Kerr, W. E. 1974. Genetik des Polymorphismus bei Bienen. Pages 94–109 *in* G. Schmidt, ed., Sozialpolymorphismus bei Insekten. Wissenschaftliche Vergagsgesellschaft, Stuttgart.

———— **& R. Vencovsky.** 1982. Melhoramento genético em abelhas. I. Efeito do número de colônias sobre o melhoramento. Revista Brasileira de Genética 5(2): 279–285.

Lobo, J. A. & W. E. Kerr. 1993. Estimation of the number of matings in *Apis mellifera*: Extension of the model and comparison of different estimates. Ethology, Ecology & Evolution 5: 337–345.

Moure, J. S. & J. M. F. Camargo. 1994. *Melipona (Michnelia) capixaba*, uma nova espécie de Meliponinae do sudeste do Brasil. Revista Brasileira de Zoologica 11(2): 289–296.

Oliveira, D. A. G. de. 1992. M.Sc. thesis, Universidade Estadual do São Paulo, Rio Claro.

————, **W. E. Kerr & M. S. Palma.** 1992. Compensação de dose em *Scaptotrigona postica* (Hym., Apidae). Revista Brasileira de Genética 15(1): 256–260.

Pedro, S. R. M. & J. M. F. Camargo. 1991. Interactions on floral resources between the Africanized honey bee *Apis mellifera* L. and the wild bee community (Hymenoptera, Apoidea) in natural "cerrado" ecosystem in southeast Brazil. Apidologie 22(4): 397–415.

Plotkin, M. J. 1988. The outlook for new agricultural and industrial products from the tropics. Pages 106–116 *in* E. O. Wilson, ed., Biodiversity. National Academy Press, Washington DC.

Rodrigues, W. & R. C. Valle. 1964. Ocorrência de ocos em mata de baixo da região de Manaus, Amazonas. Estudo preliminar. Publ. INPA, Série Botânica, 16: 1–8.

Vencovsky, R. & W. E. Kerr. 1982. Melhoramento genético em abelhas. II. Teoria e avaliação de alguns métodos de seleção. Revista Brasileira de Genética 5(3): 493–502.

Walker, I. 1991. Algumas considerações sobre um programa de zoneamento da Amazonia. Pages 37–46 *in* A. L. Val et al., eds., 2° capítulo de "Bases científicas para estratégias de preservação e desenvolvimento da Amazonia: fatos e perspectivas," vol. 1. INPA, Manaus.

Woyke, J. 1980. Effect of sex allele homo- and heterozygosity on honeybee colony populations and on their honey production. 1. Favourable development conditions and unrestricted queens. Journal of Apicultural Research 19: 51–63.

Wright, S. 1969. The theory of gene frequencies. Evolution and the genetics of populations, vol. 2. University of Chicago Press, Chicago.

Yokoyama, S. & M. Nei. 1979. Population dynamics of sex-determining alleles in honey bees and self-incompatibility alleles in plants. Genetics 91: 609–626.

List of Contributors

Ana Luisa K. M. Albernaz, Sociedade Civil Mamirauá, Caixa Postal 38, Tefé, AM 69470-000, Brazil.

Ana Rita Alves, Sociedade Civil Mamirauá, Caixa Postal 38, Tefé, AM 69470-000, Brazil.

Anthony B. Anderson, Ford Foundation, Praia do Flamengo 100, Rio de Janeiro 22.210-030, Brazil.

Scott Douglas Anderson, Museu Paraense Emílio Goeldi, Departamento de Ciências Humanas, Caixa Postal 893, Belém 66.017, PA, Brazil.

J. Márcio Ayres, Sociedade Civil Mamirauá, Caixa Postal 38, Tefé, AM 69470-000, Brazil.

Aline Azevedo, Sociedade Civil Mamirauá, Caixa Postal 38, Tefé, AM 69470-000, Brazil.

Ana Cristina Barros, Instituto do Homem e Meio Ambiente da Amazônia (IMA-ZON), Caixa Postal 1015, Belém 66017-000, PA, Brazil.

Ronaldo B. Barthem, Sociedade Civil Mamirauá, Caixa Postal 38, Tefé, AM $69470-000, Brazil.

Richard E. Bodmer, Tropical Conservation and Development Program, Center for Latin American Studies, University of Florida, 319 Grinter Hall, Gainesville, FL 32611, U.S.A.

Evandro Câmara, Graduate Program in Biology and Environmental Sciences, Federal University of Pará, Belém, PA, Brazil.

Gislene Almeida Carvalho, Departamento de Genética e Bioquímica, Universidade Federal de Uberlândia, Uberlândia 38400-900, MG, Brazil.

Fábio de Castro, School of Public and Environmental Affairs, University of Indiana, Bloomington, IN 47405, U.S.A.

William G. R. Crampton, Sociedade Civil Mamirauá, Caixa Postal 38, Tefé, AM 69470-000, Brazil.

Doris R. Dias, Proyecto Pacaya-Samiria, c/o Fundación Peruana para la Conservación de la Naturaleza, Napo 449, Iquitos, Peru.

Eduardo Durand, The Nature Conservancy, Latin American and Caribbean Division, 1815 North Lynn St., Arlington, VA 22209, U.S.A.

Célia Futemma, School of Public and Environmental Affairs, University of Indiana, Bloomington, IN 47405, U.S.A.

Juan E. Garcia, Proyecto Pacaya-Samiria, c/o Fundación Peruana para la Conservación de la Naturaleza, Napo 449, Iquitos, Peru.

Michael Goulding, Museu Paraense Emílio Goeldi, Departamento de Ciências Humanas, Caixa Postal 893, Belém 66.017, PA, Brazil.

cf3Mario Hiraoka, Department of Geography, Millersville University, Millersville, PA 17551, U.S.A.

Andrew Henderson, Insitute of Systematic Botany, The New York Botanical Garden, Bronx, NY 10458, U.S.A.

Peter A. Henderson, Animal Behaviour Research Group, Department of Zoology, University of Oxford, South Parks Rd., Oxford, OX1 3PS, U.K.

Päivi Jokinen, Department of Biology, University of Turku, FIN-20500 Turku, Finland.

Wolfgang J. Junk, Max Planck Institute for Limnology, Working Group on Tropical Ecology, P.O. Box 165, 24302 Plön, Germany.

Francis Kahn, ORSTOM, Caixa Postal 09747, Brasília 70001-970, DF, Brazil.

Risto Kalliola, Department of Geography, University of Turku, FIN-20500 Turku, Finland.

Warwick Estevam Kerr, Departamento de Genética e Bioquímica, Universidade Federal de Uberlândia, Uberlândia 38400-900, MG, Brazil.

Deborah de Magalhães Lima, Departamento de Antropologia, Universidade Federal do Pará, Belém, Pará, Brazil.

Domingos S. Macedo, Departamento de Ciências Florestais, ESALQ/USP, Caixa Postal 9, São Paulo 13400, Brazil.

Míriam Marmontel, Sociedade Civil Mamirauá, Caixa Postal 38, Tefé, AM 69470-000, Brazil.

Fernando Luiz Tavares Marques, Museu Paraense Emílio Goeldi, Departamento de Ciências Humanas, Caixa Postal 893, Belém 66.017, PA, Brazil.

Donald Masterson, Sociedade Civil Mamirauá, Caixa Postal 38, Tefé, AM 69470-000, Brazil.

Dennis McCaffrey, The Nature Conservancy, Latin American and Caribbean Division, 1815 North Lynn St., Arlington, VA 22209, U.S.A.

David McGrath, Núcleo de Altos Estudos Amazônicos, Federal University of Pará, Belém, PA, Brazil.

Edila Arnaud Moura, Sociedade Civil Mamirauá, Caixa Postal 38, Tefé, AM 69470-000, Brazil.

Igor Mousasticoshvily Jr., Instituto Socioambiental, Av. Higienópolis 901, São Paulo 01238-001, Brazil.

Vania Alves Nascimento, Deartamento de Genética e Bioquímica, Universidade Federal de Uberlândia, Uberlândia 38400-900, MG, Brazil.

Daniel Nepstad, Woods Hole Research Center, P.O. Box 296, Woods Hole, MA 02543, U.S.A.

Manoel Fernandes Martins Nogueira, Universidade Federal do Pará, Departamento de Engenharia Mecânica, Caixa Postal 459, Belém 66.017, PA, Brazil.

Jörg J. Ohly, Max Planck Institute for Limnology, Working Group on Tropical Ecology, P.O. Box 165, 24302 Plön, Germany.

Christine Padoch, Institute of Economic Botany, The New York Botanical Garden, Bronx, NY 10458, U.S.A.

Charles M. Peters, Institute of Economic Botany, The New York Botanical Garden, Bronx, NY 10458, U.S.A.

Miguel Pinedo-Vasquez, Center for Environmental Research and Conservation,

Columbia University, 1200 Amsterdam Ave., MC 5557, New York, NY 10027, U.S.A.

Pablo E. Puertas, Instituto Veterinario de Investigaciones Tropicales y de Altura, P.O. Box 621, Iquitos, Peru.

Helder Lima de Queiroz, Sociedade Civil Mamirauá, Caixa Postal 38, Tefé, AM 69470-000, Brazil.

Hugh Raffles, Yale School of Forestry and Environmental Studies, Yale University, New Haven, CT 06511, U.S.A.

Marise Reis, Sociedade Civil Mamirauá, Caixa Postal 38, Tefé, AM 69470-000, Brazil.

Cesar Reyes, Facultad de Ciencias Biológicas, Universidad Nacional de la Amazonía Peruana, Plaza Serafín Filomeno, Iquitos, Peru.

Barbara A. Robertson, Instituto Nacional de Pesquisa da Amazonia, Caixa Postal 478, Manaus 69.083-000, AM, Brazil.

John G. Robinson, Wildlife Conservation Society, 2300 Southern Blvd., Bronx, NY 10460, U.S.A.

Anna C. Roosevelt, Field Museum of Natural History, Roosevelt Rd. at Lake Shore Dr., Chicago, IL 60605, U.S.A.

Mauro Luis Ruffino, IARA/IBAMA Project, Av. Tapajós, 2267, Santarém 68.040-000, PA, Brazil.

Pedro Simões Santos, Sociedade Civil Mamirauá, Caixa Postal 38, Tefé, AM 69470-000, Brazil.

Ronis da Silveira, Sociedade Civil Mamirauá, Caixa Postal 38, Tefé, AM 69470-000, Brazil.

Nigel J. H. Smith, Department of Geography, P.O. Box 117315, 3141 Turlington Hall, Gainesville, FL 32611, U.S.A.

Eeva Tuukki, Department of Biology, University of Turku, FIN-20500 Turku, Finland.

Christopher Uhl, Department of Biology, Pennsylvania State University, University Park, PA 16802, U.S.A.

Daniel J. Zarin, Department of Natural Resources, University of New Hampshire, 56 College Rd., Durham, NH 03842, U.S.A.